networks™

A Social Studies Learning System

INDIANA

DISCOVERING

WORLD

GEOGRAPHY

Western Hemisphere

Richard G. Boehm, Ph. D.

Mc
Graw
Hill
Education

MHEonline.com

Send all inquiries to:
McGraw-Hill Education
8787 Orion Place
Columbus, OH 43240

ISBN: 978-0-02-136073-4
MHID: 0-02-136073-1

Printed in the United States of America.

2 3 4 5 6 7 8 9 QVS 20 19 18 17 16 15

SENIOR AUTHOR

Richard G. Boehm, Ph.D., was one of the original authors of *Geography for Life: National Geography Standards,* which outlined what students should know and be able to do in geography. He was also one of the authors of the *Guidelines for Geographic Education*, in which the Five Themes of Geography were first articulated. Dr. Boehm has received many honors, including "Distinguished Geography Educator" by the National Geographic Society (1990), the "George J. Miller Award" from the National Council for Geographic Education (NCGE) for distinguished service to geographic education (1991), "Gilbert Grosvenor Honors" in geographic education from the Association of American Geographers (2002), and the NCGE's "Distinguished Mentor Award" (2010). He served as president of the NCGE, has twice won the *Journal of Geography* award for best article, and also received the NCGE's "Distinguished Teaching Achievement." Presently, Dr. Boehm holds the Jesse H. Jones Distinguished Chair in Geographic Education at Texas State University in San Marcos, Texas, where he serves as director of The Gilbert M. Grosvenor Center for Geographic Education. His most current project includes the production of the video-based professional development series, Geography: Teaching With the Stars. Available programs may be viewed at www.geoteach.org.

CONTRIBUTING AUTHORS

Jay McTighe has published articles in a number of leading educational journals and has coauthored 10 books, including the best-selling *Understanding by Design* series with Grant Wiggins. McTighe also has an extensive background in professional development and is a featured speaker at national, state, and district conferences and workshops. He received his undergraduate degree from the College of William and Mary, earned a master's degree from the University of Maryland, and completed postgraduate studies at the Johns Hopkins University.

Dinah Zike, M.Ed., is an award-winning author, educator, and inventor recognized for designing three-dimensional, hands-on manipulatives and graphic organizers known as Foldables®. Foldables are used nationally and internationally by parents, teachers, and other professionals in the education field. Zike has developed more than 150 supplemental educational books and materials. Her two latest books, *Notebook Foldables®* and *Foldables®, Notebook Foldables®, & VKV®s for Spelling and Vocabulary 4th–12th,* were each awarded *Learning Magazine's* Teachers' Choice Award for 2011. In 2004 Zike was honored with the CESI Science Advocacy Award. She received her M.Ed. from Texas A&M, College Station, Texas.

ACADEMIC CONSULTANTS

William H. Berentsen, Ph.D.
Professor of Geography and
    European Studies
University of Connecticut
Storrs, Connecticut

David Berger, Ph.D.
Ruth and I. Lewis Gordon
Professor of Jewish History
Dean, Bernard Revel Graduate
    School
Yeshiva University
New York, New York

R. Denise Blanchard, Ph.D.
Professor of Geography
Texas State University–San Marcos
San Marcos, Texas

Brian W. Blouet, Ph.D.
Huby Professor of Geography and
    International Education
The College of William and Mary
Williamsburg, Virginia

Olwyn M. Blouet, Ph.D.
Professor of History
Virginia State University
Petersburg, Virginia

Maria A. Caffrey, Ph.D.
Lecturer, Department of
    Geography
University of Tennessee, Knoxville
Knoxville, Tennessee

So-Min Cheong, Ph.D.
Associate Professor of Geography
University of Kansas
Lawrence, Kansas

Alasdair Drysdale, Ph.D.
Professor of Geography
University of New Hampshire
Durham, New Hampshire

**Rosana Nieto
    Ferreira, Ph.D.**
Assistant Professor of Geography
    and Atmospheric Science
East Carolina University
Greenville, North Carolina

Eric J. Fournier, Ph.D.
Associate Professor of Geography
Samford University,
Birmingham, Alabama

Matthew Fry, Ph.D.
Assistant Professor of Geography
University of North Texas
Denton, Texas

Douglas W. Gamble, Ph.D.
Professor of Geography
University of North Carolina,
    Wilmington
Wilmington, North Carolina

Gregory Gaston, Ph.D.
Professor of Geography
University of North Alabama
Florence, Alabama

Jeffrey J. Gordon, Ph.D.
Associate Professor of Geography
Bowling Green State University
Bowling Green, Ohio

Alyson L. Greiner, Ph.D.
Associate Professor of Geography
Oklahoma State University
Stillwater, Oklahoma

William J. Gribb, Ph.D.
Associate Professor of Geography
University of Wyoming
Laramie, Wyoming

Joseph J. Hobbs, Ph.D.
Professor of Geography
University of Missouri
Columbia, Missouri

Ezekiel Kalipeni, Ph.D.
Professor of Geography and
    Geography Information Science
University of Illinois at
    Urbana–Champaign
Urbana, Illinois

Pradyumna P. Karan, Ph.D.
University Research Professor of
    Geography
University of Kentucky
Lexington, Kentucky

Christopher Laingen, Ph.D.
Assistant Professor of Geography
Eastern Illinois University
Charleston, Illinois

Jeffrey Lash, Ph.D.
Associate Professor of Geography
University of Houston–Clear Lake
Houston, Texas

Jerry T. Mitchell, Ph.D.
Research Professor of Geography
University of South Carolina
Columbia, South Carolina

Thomas R. Paradise, Ph.D.
Professor, Department of
    Geosciences and the King Fahd
    Center for Middle East Studies
University of Arkansas
Fayetteville, Arkansas

David Rutherford, Ph.D.
Assistant Professor of Public Policy
    and Geography
Executive Director, Mississippi
    Geographic Alliance
University of Mississippi
University, Mississippi

Dmitrii Sidorov, Ph.D.
Professor of Geography
California State University, Long
    Beach
Long Beach, California

Amanda G. Smith, Ph.D.
Professor of Education
University of North Alabama
Florence, Alabama

Jeffrey S. Ueland, Ph.D.
Associate Professor of Geography
Bemidji State University
Bemidji, Minnesota

Fahui Wang, Ph.D.
Professor of Geography
Louisiana State University
Baton Rouge, Louisiana

TEACHER REVIEWERS

**Precious Steele Boyle,
    Ph.D.**
Cypress Middle School
Memphis, TN

Jason E. Albrecht
Moscow Middle School
Moscow, ID

Jim Hauf
Berkeley Middle School
Berkeley, MO

Elaine M. Schuttinger
Trinity Catholic School
Columbus, OH

Mark Stahl
Longfellow Middle School
Norman, OK

**Mollie Shanahan
    MacAdams**
Southern Middle School
Lothian, MD

Sara Burkemper
Parkway West Middle Schools
Chesterfield, MO

Alicia Lewis
Mountain Brook Junior High
    School
Birmingham, AL

Steven E. Douglas
Northwest Jackson Middle School
Ridgeland, MS

LaShonda Grier
Richmond County Public Schools
Martinez, GA

Samuel Doughty
Spirit of Knowledge Charter
    School
Worcester, MA

CONTENTS

Reference Atlas Maps **RA1**

World: Political. RA2
World: Physical RA4
North America: Political RA6
North America: Physical RA7
United States: Political. .RA8
United States: Physical. RA10
Canada: Physical/Political RA12
Middle America: Physical/Political RA14
South America: Political RA16
South America: Physical RA17
Europe: Political. RA18

Europe: Physical. RA20
Africa: Political. RA22
Africa: Physical. RA23
Middle East: Physical/Political RA24
Asia: Political RA26
Asia: Physical RA28
Oceania: Physical/Political RA30
World Time Zones RA32
Polar Regions RA34
A World of Extremes RA35
Geographic Dictionary. RA36

Scavenger Hunt **RA38**

UNIT ONE

Our World: The Western Hemisphere 1

What Is Geography? 15

ESSENTIAL QUESTION

How does geography influence the way people live?

LESSON 1 Thinking Like a Geographer 18

LESSON 2 Tools Used by Geographers 26
Thinking Like a Geographer Relief 29
Think Again The Height of Mount Everest. 31
What Do You Think? Are Street-Mapping Technologies an Invasion of People's Privacy? 34

Earth's Physical Geography 39

ESSENTIAL QUESTION

How does geography influence the way people live?

LESSON 1 Earth and the Sun. 42

LESSON 2 Forces of Change 52

LESSON 3 Landforms and Bodies of Water 58

CONTENTS

Earth's People **69**

CHAPTER 3

ESSENTIAL QUESTIONS

How do people adapt to their environment? • What makes a culture unique? • Why do people make economic choices?

LESSON 1 A Changing Population **72**
Think Again Immigration 78

LESSON 2 Global Cultures **82**
Global Connections Social Media in a
  Changing World 90

LESSON 3 Economic Systems **94**
Think Again Command Economy 96

UNIT TWO

North America 105

**The United States East of the
Mississippi River** **113**

CHAPTER 4

ESSENTIAL QUESTIONS

How does geography influence the way people live? • Why is history important? • What makes a culture unique?

LESSON 1 Physical Features **116**

LESSON 2 History of the Region **124**

LESSON 3 Life in the Region **132**
Thinking Like a Geographer Population
  Centers 134
Thinking Like a Geographer The Nation's
  Capital 136
What Do You Think? Is Fracking a Safe Method
  for Acquiring Energy Resources? 140

**The United States West of the
Mississippi River** **145**

CHAPTER 5

ESSENTIAL QUESTIONS

How does geography influence the way people live? • How do people make economic choices? • How does technology change the way people live?

LESSON 1 Physical Features **148**
Think Again Bison and Buffalo 149
Think Again Deserts 155

LESSON 2 History of the Region **156**
Thinking Like a Geographer Exploring the
  Louisiana Territory 158

**LESSON 3 Life in the United States
  West of the Mississippi** **164**
Thinking Like a Geographer Making the
  Desert Bloom 167

Canada 175

ESSENTIAL QUESTIONS

How do people adapt to their environment? • What makes a culture unique?

LESSON 1 The Physical Geography of Canada **178**
Think Again   Northern Canada 181

LESSON 2 The History of Canada **184**
Thinking Like a Geographer   Canadian Mounties 187

LESSON 3 Life in Canada **190**

Mexico, Central America, and the Caribbean Islands 199

ESSENTIAL QUESTIONS

How does geography influence the way people live? • Why does conflict develop? • Why do people trade?

LESSON 1 Physical Geography **202**
Think Again   The Panama Canal 205
Thinking Like a Geographer   The Caribbean Islands 208

LESSON 2 History of the Region **210**

LESSON 3 Life in the Region **216**
Think Again   Mexican Food 217
Global Connections   NAFTA and Its Effects 222

UNIT THREE   South America 229

Brazil 237

ESSENTIAL QUESTIONS

How does geography influence the way people live? • How do governments change? • What makes a culture unique?

LESSON 1 Physical Geography of Brazil **240**
Think Again   Seasons 245

LESSON 2 History of Brazil **248**

LESSON 3 Life in Brazil **256**
Global Connections   Rain Forest Resources.... 264

(t) Michelle Gilders Canada West/Alamy; (c) ©Sergio Pitamitz/Robert Harding World Imagery/Corbis; (b) Silvia Izquierdo/AP Images

CONTENTS

The Tropical North 271

ESSENTIAL QUESTIONS

How does geography influence the way people live? • Why does conflict develop? • What makes a culture unique?

LESSON 1 Physical Geography of the Region 274
Think Again Angel Falls 277

LESSON 2 History of the Countries ... 280

LESSON 3 Life in the Tropical North . 286

Andes and Midlatitude Countries 295

ESSENTIAL QUESTIONS

How does geography influence the way people live? • Why do civilizations rise and fall? • What makes a culture unique?

LESSON 1 Physical Geography of the Region 298
Think Again The Atacama Desert 301

LESSON 2 History of the Region 306
Thinking Like a Geographer Bolivia's Two Capitals 309

LESSON 3 Life in the Region 312
What Do You Think? Is Globalization Destroying Indigenous Cultures? 318

UNIT FOUR Europe 323

Western Europe 331

ESSENTIAL QUESTIONS

How does geography influence the way people live? • Why do civilizations rise and fall? • How do governments change?

LESSON 1 Physical Geography of Western Europe 334
Think Again London Fog 337

LESSON 2 History of Western Europe 342
Thinking Like a Geographer The English Channel 343
Think Again Johannes Gutenberg 345

LESSON 3 Life in Western Europe 350
What Do You Think? Is the European Union an Effective Economic Union? 358

CHAPTER 12

Northern and Southern Europe · · · · · · · · 363

ESSENTIAL QUESTIONS

How do people adapt to their environment? • Why do civilizations rise and fall? • How do new ideas change the way people live?

LESSON 1 Physical Geography of the Regions 366
Thinking Like a Geographer   The Sautso Dam . . 371

LESSON 2 History of the Regions 372
Thinking Like a Geographer   Eratosthenes 374
Think Again   Vatican City 376

LESSON 3 Life in Northern and Southern Europe 378
Thinking Like a Geographer   The Euskera Language 380
Global Connections   Aging of Europe's Population 386

CHAPTER 13

Eastern Europe and Western Russia · · · · · · 393

ESSENTIAL QUESTIONS

How does geography influence the way people live? • How do governments change?

LESSON 1 Physical Geography 396

LESSON 2 History of the Regions 402

LESSON 3 Life in Eastern Europe and Western Russia 410

FOLDABLES® LIBRARY 421

GAZETTEER 428

ENGLISH-SPANISH GLOSSARY 436

INDEX 453

(t) Christophe Boisvieux/age fotostock; (b) Dmitry Kostyukov/AFP/Getty Images

FEATURES

Think Again?

The Height of Mount Everest **31**
Immigration **78**
Command Economy **96**
Bison and Buffalo **149**
Deserts **155**
Northern Canada **181**
The Panama Canal **205**
Mexican Food **217**
Seasons **245**
Angel Falls **277**
The Atacama Desert **301**
London Fog **337**
Johannes Gutenberg **345**
Vatican City **376**

EXPLORE the CONTINENT

Explore the World **2**
North America **106**
South America **230**
Europe **324**

Indiana CONNECTION

Locating Capital Cities **21**
Tools of Discovery **25**
Saving and Investing **97**
Nation-States **344**
England's Political Heritage **345**
Democracy **373**

Thinking Like a Geographer

Relief **29**
Population Centers **134**
The Nation's Capital **136**
Exploring the Louisiana Territory **158**
Making the Desert Bloom **167**
Canadian Mounties **187**
The Caribbean Islands **208**
Bolivia's Two Capitals **309**
The English Channel **343**
The Sautso Dam **371**
Eratosthenes **374**
The Euskera Language **380**

GLOBAL CONNECTIONS

Social Media in a Changing World **90**
NAFTA and Its Effects **222**
Rain Forest Resources **264**
Aging of Europe's Population **386**

What Do You Think?

Are Street-Mapping Technologies an Invasion of
    People's Privacy? **34**
Is Fracking a Safe Method for Acquiring Energy Resources? **140**
Is Globalization Destroying Indigenous Cultures? **318**
Is the European Union an Effective Economic Union? **358**

MAPS

REFERENCE ATLAS MAPS

World: Political...**RA2**
World: Physical...**RA4**
North America: Political...**RA6**
North America: Physical...**RA7**
United States: Political...**RA8**
United States: Physical...**RA10**
Canada: Physical/Political......................................**RA12**
Middle America: Physical/Political..............................**RA14**
South America: Political..**RA16**
South America: Physical...**RA17**
Europe: Political...**RA18**
Europe: Physical..**RA20**
Africa: Political...**RA22**
Africa: Physical..**RA23**
Middle East: Physical/Political.................................**RA24**
Asia: Political...**RA26**
Asia: Physical..**RA28**
Oceania: Physical/Political.....................................**RA30**
World Time Zones..**RA32**
Polar Regions: Physical...**RA34**
A World of Extremes...**RA35**
Geographic Dictionary...**RA36**

UNIT 1: OUR WORLD: THE WESTERN HEMISPHERE

The World: Physical...**4**
The World: Political..**6**
The World: Population Density.......................................**8**
The World: Economic..**10**
The World: Climate...**12**
The World..**16**
Tectonic Plate Boundaries..**40**
Earth's Wind Patterns..**47**
Human Geography..**70**
Top 15 Countries by Social Networking Users........................**93**

UNIT 2: NORTH AMERICA

North America: Physical...**108**
North America: Political..**109**
North America: Population Density...................................**110**
North America: Economic Resources..................................**111**
North America: Climate..**112**
Cities and States of the Region....................................**115**
Subregions...**117**

Mississippi River..**129**
Cities and Towns of the Region.....................................**147**
Westward Expansion...**159**
Canada...**177**
Regions of Canada..**179**
Map of the Region..**201**
Columbian Exchange...**214**

UNIT 3: SOUTH AMERICA

South America: Physical...**232**
South America: Political..**233**
South America: Population Density...................................**234**
South America: Economic Resources..................................**235**
South America: Climate..**236**
Brazil...**239**
Colonies in Brazil...**252**
Global Impact: The World's Rain Forests............................**267**
The Tropical North...**273**
Winning Independence...**284**
Andes and Midlatitude Countries....................................**297**
Native American Civilizations......................................**308**
The Pan-American Highway..**316**

UNIT 4: EUROPE

Europe: Physical...**326**
Europe: Political..**327**
Europe: Population Density...**328**
Europe: Economic Resources...**329**
Europe: Climate..**330**
Western Europe...**333**
British Isles..**336**
Axis Control in World War II.......................................**348**
European Union...**353**
Western European Empires, 1914.....................................**356**
Northern and Southern Europe.......................................**365**
Greek and Roman Empires..**373**
Eastern Europe and Western Russia..................................**395**
Climate Regions..**399**
Expansion of Russia..**403**
Eastern Bloc...**408**
Slavic Settlement..**413**

CHARTS, GRAPHS, DIAGRAMS, AND INFOGRAPHICS

UNIT 1: OUR WORLD: THE WESTERN HEMISPHERE

Diagram: Latitude and Longitude **21**

Chart: The Six Essential Elements **24**

Diagram: Earth's Hemispheres **27**

Diagram: Earth's Layers **43**

Diagram: Seasons **45**

Diagram: Rain Shadow. **48**

Infographic: Saltwater vs. Freshwater **60**

Graph: Population Pyramid. **73**

Chart: Major World Religions **84**

Graph: Facebook Users Worldwide **93**

Infographic: Understanding Economic Choices **95**

Diagram: Factors of Production **98**

Graph: GDP Comparison. **99**

UNIT 2: NORTH AMERICA

Graph: Ethnic Origin **138**

Graph: Climate Zones **206**

Infographic: Top 10 Countries in Exports **225**

UNIT 3: SOUTH AMERICA

Graph: The Slave Trade **253**

Infographic: Emerald Mining **278**

Infographic: Effects of Altitude **302**

UNIT 4: EUROPE

Infographic: Black Death **344**

Chart: English Words From Other Languages **354**

Infographic: An Aging Population **389**

Graph: Europe, 2050 . . . . **389**

Infographic: Feudalism in Europe and Russia **404**

Graph: Population of Russia **412**

networks ONLINE RESOURCES

Videos

Every lesson has a video to help you learn more about your world!

Infographics

Chapter 2
Lesson 3 Fresh and Salt Water in the World

Chapter 9
Lesson 2 How a Hacienda Works

Interactive Charts/Graphs

Chapter 2
Lesson 1 Rain Shadow; Climate Zones

Chapter 3
Lesson 1 Understanding Population
Lesson 3 Economic Questions

Chapter 5
Lesson 3 Hispanic Population Growth

Animations

Chapter 1
Lesson 1 The Earth; Regions of Earth
Lesson 2 Elements of a Globe

Chapter 2
Lesson 1 Earth's Daily Rotation; Earth's Layers; Seasons on Earth
Lesson 3 How the Water Cycle Works

Chapter 3
Global Connections: Social Media

Chapter 4
Lesson 2 Down the Mississippi

Chapter 5
Lesson 1 Chinooks

Chapter 7
Lesson 1 Waterways as Political Boundaries
Global Connections: NAFTA

Chapter 8
Global Connections: Rain Forest Resources

Chapter 12
Lesson 1 How the Alps Formed
Global Connections: Aging of Europe's Population

Slide Shows

Chapter 1
Lesson 1 Spatial Effect; Places Change Over Time
Lesson 2 Special Purpose Maps; History of Mapmaking

Chapter 2
Lesson 1 Effects of Climate Change
Lesson 2 Human Impact on Earth

Chapter 3
Lesson 1 Different Places in the World

Chapter 4
Lesson 1 Agriculture East of the Mississippi
Lesson 2 Early Settlements; Immigrate/Emigrate

Chapter 5
Lesson 1 Bodies of Water: West of the Mississippi; How the Hoover Dam Works
Lesson 3 NAFTA; Economic Terms: Private and Public Sector

Chapter 6
Lesson 2 The First Peoples in Canada

Chapter 8
Lesson 1 Landforms of Brazil
Lesson 2 Brazil's Natural Products
Lesson 3 Brazilian Culture

Chapter 9
Lesson 1 Mining Emeralds
Lesson 3 What Is Carnival?

Chapter 10
Lesson 1 Landforms: Andean Region
Lesson 3 Traditional and Modern Lifestyles

Chapter 11
Lesson 1 Agriculture in Western Europe
Lesson 2 WWII

Chapter 12
Lesson 2 The Renaissance
Lesson 3 Recreation in Northern and Southern Europe

Jochen Schlenker/Photographer's Choice/Getty Images

netw○rks ONLINE RESOURCES

 Interactive Maps

Chapter 3
Lesson 1 Landforms, Waterways, and Population

Chapter 4
Lesson 2 Cultures and Communities on the Mississippi River

Chapter 5
Lesson 1 Resources West of the Mississippi
Lesson 2 Agricultural and Industrial Expansion; Trail of Tears

Chapter 6
Lesson 1 How the St. Lawrence Seaway Works

Chapter 7
Lesson 3 Hispaniola

Chapter 9
Lesson 2 Colonialism and Independence: Tropical North

Chapter 10
Lesson 1 Comparing Mountain Ranges
Lesson 3 Comparing Highway Systems

Chapter 11
Lesson 3 Language: Western Europe; Empires: Western Europe

Chapter 12
Lesson 2 Exploration to the New World; Three Religions in
            Northern and Southern Europe

Chapter 13
Lesson 2 Russian Historical Changes
Lesson 3 Russia's Population

 Interactive Images

Chapter 1
Lesson 1 360° View: Times Square

Chapter 2
Lesson 3 Ocean Floor; Garbage

Chapter 3
Lesson 1 Refugee Camps
Lesson 2 Cultural Change

Chapter 4
Lesson 3 360° View of New York City; Festival

Chapter 5
Lesson 1 360° View: Crater Lake
Lesson 2 Ranching in the West; Chicago: Before and After
            Industrialization

Chapter 6
Lesson 3 Winter Festival in Ottawa; 360° View: Toronto

Chapter 7
Lesson 1 360° View: Mexico City
Lesson 2 Tenochtitlán; Chinampas

Chapter 8
Lesson 1 Brazilian Rain Forest
Lesson 3 360° View: São Paulo

Chapter 9
Lesson 2 Caracas Cityscape
Lesson 3 360° View: Bogotá

Chapter 10
Lesson 1 Uses of Wool

Chapter 11
Lesson 1 How Windmills Work; How the Chunnel Was Built
Lesson 3 Dublin; 360° View: The Louvre

Chapter 13
Lesson 1 The Ural Mountains; Using Resources

Games

Chapter 1
Lesson 1 True or False
Lesson 2 Concentration; Map Legends

Chapter 2
Lesson 1 Fill in the Blank
Lesson 2 Tic-Tac-Toe
Lesson 3 Columns

Chapter 3
Lesson 1 Crossword
Lesson 2 Identification
Lesson 3 Flashcard; Economic Systems; Bartering and Trade

Chapter 4
Lesson 1 Concentration
Lesson 2 Identification
Lesson 3 Columns

Chapter 5
Lesson 1 Climate Vocabulary; Flashcard
Lesson 2 Tic-Tac-Toe
Lesson 3 Fill in the Blank

Games

Chapter 6
Lesson 1 Columns
Lesson 2 True or False
Lesson 3 Crossword

Chapter 7
Lesson 1 Columns
Lesson 2 Flashcard
Lesson 3 Tic-Tac-Toe

Chapter 8
Lesson 1 Crossword
Lesson 2 Fill in the Blank
Lesson 3 True or False

Chapter 9
Lesson 1 Concentration
Lesson 2 Identification
Lesson 3 Tic-Tac-Toe

Chapter 10
Lesson 1 Crossword
Lesson 2 Columns

Chapter 11
Lesson 1 Concentration
Lesson 2 Flashcard
Lesson 3 Identification

Chapter 12
Lesson 1 Columns
Lesson 2 Fill in the Blank
Lesson 3 Crossword

Chapter 13
Lesson 1 Columns
Lesson 2 Tic-Tac-Toe
Lesson 3 Concentration

REFERENCE ATLAS

Reference Atlas Maps	RA1	Europe: Physical	RA20
World: Political	RA2	Africa: Political	RA22
World: Physical	RA4	Africa: Physical	RA23
North America: Political	RA6	Middle East: Physical/Political	RA24
North America: Physical	RA7	Asia: Political	RA26
United States: Political	RA8	Asia: Physical	RA28
United States: Physical	RA10	Oceania: Physical/Political	RA30
Canada: Physical/Political	RA12	World Time Zones	RA32
Middle America: Physical/Political	RA14	Polar Regions	RA34
South America: Political	RA16	A World of Extremes	RA35
South America: Physical	RA17	Geographic Dictionary	RA36
Europe: Political	RA18		

ATLAS KEY

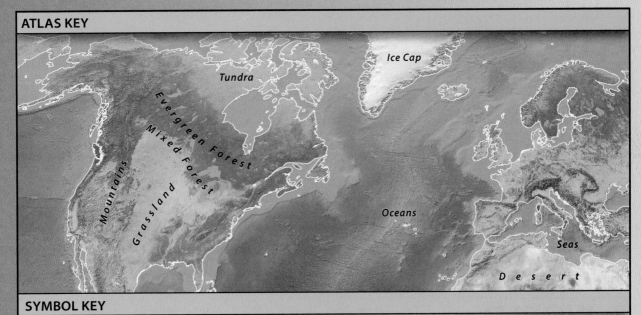

SYMBOL KEY

·········· Claimed boundary	✪ National capital	Dry salt lake
——— International boundary (political map)	○ State/Provincial capital	Lake
——— International boundary (physical map)	• Towns	⤳ Rivers
	▼ Depression	⊢⊢⊢⊢⊣ Canal
	▲ Elevation	

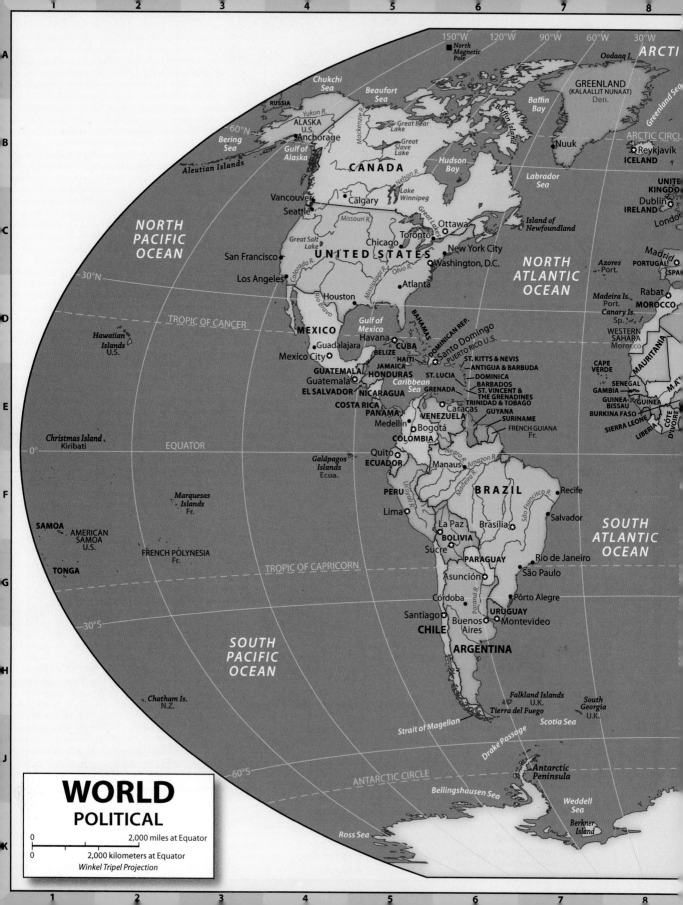

WORLD
POLITICAL

0 ———— 2,000 miles at Equator
0 ———— 2,000 kilometers at Equator
Winkel Tripel Projection

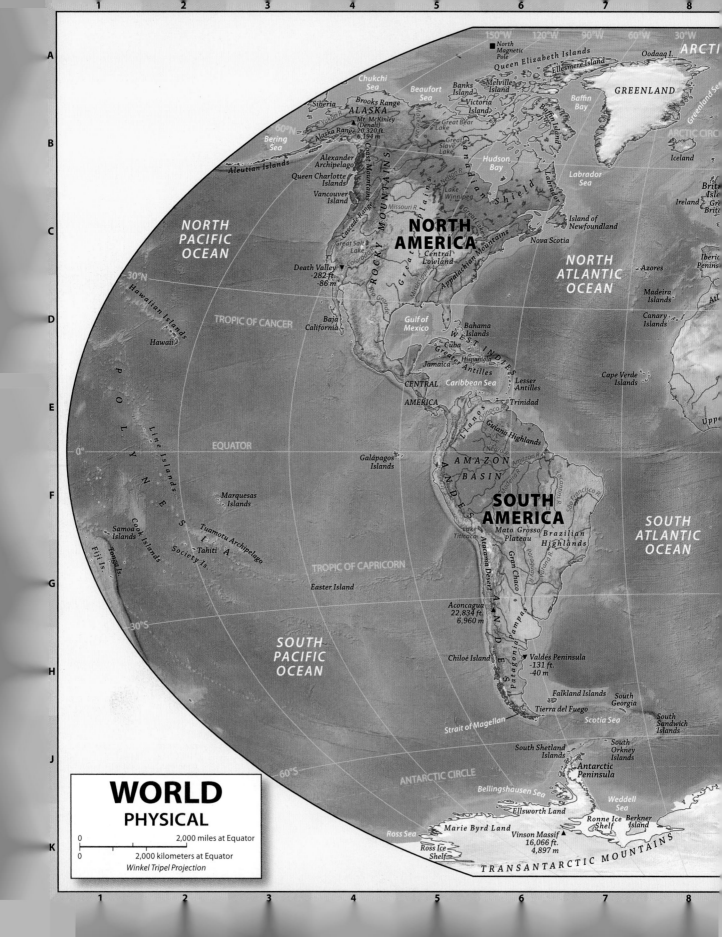

WORLD

PHYSICAL

0   2,000 miles at Equator

0   2,000 kilometers at Equator

Winkel Tripel Projection

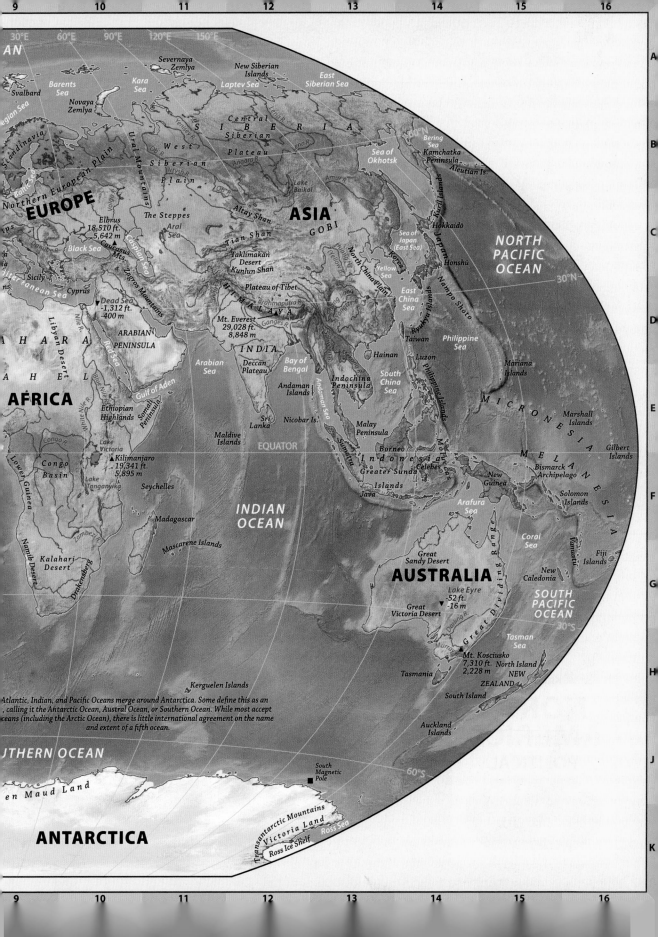

A

30°E 60°E 90°E 120°E 150°E

AN

Svalbard

Scandinavia

Barents
Sea

Novaya
Zemlya

Severnaya
Zemlya

Kara
Sea

New Siberian
Islands

Laptev Sea

East
Siberian
Sea

B

gian Sea

Baltic Sea

Northern European Plain

Ural Mountains

West
Siberian
Plain

Yenisey R.

Ob R.

S I B E R I A

Central
Siberian
Plateau

Lena R.

Amur R.

Bering
Sea

Sea of
Okhotsk

Kamchatka
Peninsula

Aleutian Is.

C

EUROPE

Volga

The Steppes

Aral
Sea

Altay Shan

Tian Shan

ASIA

GOBI

Lake
Baikal

Angara

Sea of
Japan
(East Sea)

Kuril Islands

Hokkaidō

NORTH
PACIFIC
OCEAN

Elbrus
18,510 ft.
5,642 m

Caucasus
Mts.

Black Sea

Caspian Sea

Zagros Mountains

Taklimakan
Desert
Kunlun Shan

Plateau of Tibet

North China Plain

Korea

Honshū

Nampo Shoto

30°N

D

Alps

ns

Medi terranean Sea

Sicily

Cyprus

Dead Sea
-1,312 ft.
-400 m

ARABIAN
PENINSULA

Red Sea

H I M A L A Y A

Brahmaputra R.

Ganges R.

Indus R.

Yellow R.

Yangtze R.

Yellow
Sea

East
China
Sea

Taiwan

Ryukyu Islands

Philippine Islands

Philippine
Sea

Mariana
Islands

Mt. Everest
29,028 ft.
8,848 m

I N D I A

Deccan
Plateau

Bay of
Bengal

Hainan

Luzon

E

SAHARA

Libyan Desert

Nile R.

SAHEL

AFRICA

White Nile

Blue Nile

Ethiopian
Highlands

Somali Peninsula

Gulf of Aden

Arabian
Sea

Arabian
Sea

Andaman
Islands

Sri
Lanka

Nicobar Is.

Indochina
Peninsula

South
China
Sea

M I C R O N E S I A

Marshall
Islands

Gilbert
Islands

Maldive
Islands

Andaman Sea

Malay
Peninsula

M E L A N E S I A

Bismarck
Archipelago

EQUATOR

Congo R.

Congo
Basin

Lake
Victoria

Kilimanjaro
19,341 ft.
5,895 m

Lake
Tanganyika

Seychelles

Sumatra

Borneo

Indonesia

Celebes

Greater Sunda

Islands

Java

New
Guinea

Solomon
Islands

F

Lower Guinea

Zambezi R.

Madagascar

Mascarene Islands

INDIAN
OCEAN

Arafura
Sea

Coral
Sea

Vanuatu

Fiji
Islands

Namib Desert

Kalahari
Desert

Drakensberg

Great
Sandy Desert

AUSTRALIA

New
Caledonia

G

Lake Eyre
-52 ft.
-16 m

Great
Victoria Desert

Great Dividing Range

SOUTH
PACIFIC
OCEAN

30°S

Darling R.

Murray R.

Tasman
Sea

Mt. Kosciusko
7,310 ft.
2,228 m

North Island

NEW
ZEALAND

H

Kerguelen Islands

Tasmania

South Island

Auckland
Islands

J

UTHERN OCEAN

60°S

South
Magnetic
Pole

en Maud Land

Transantarctic Mountains

Victoria Land

Ross Sea

ANTARCTICA

Ross Ice Shelf

K

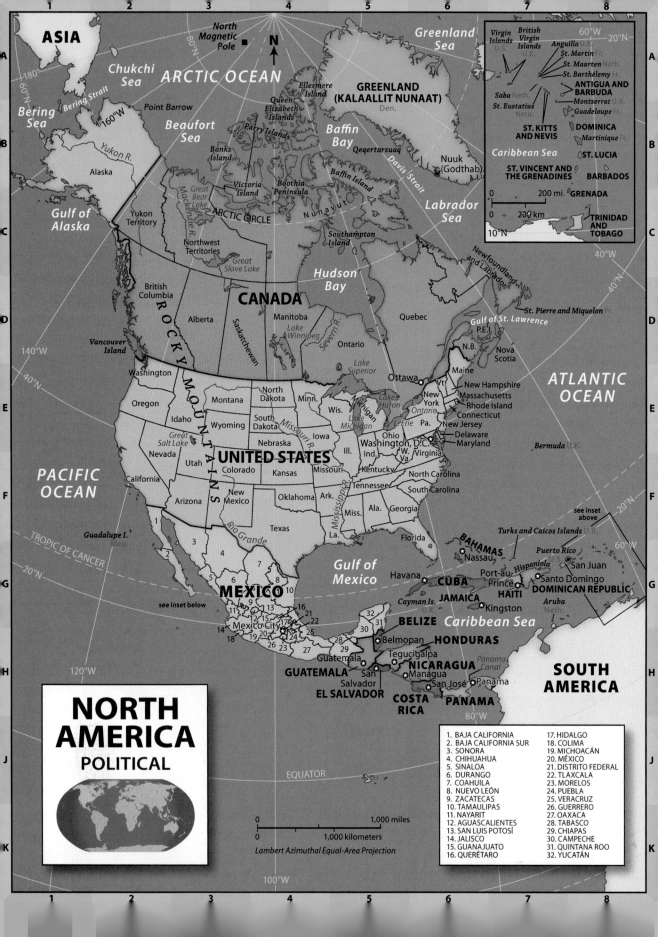

NORTH AMERICA

POLITICAL

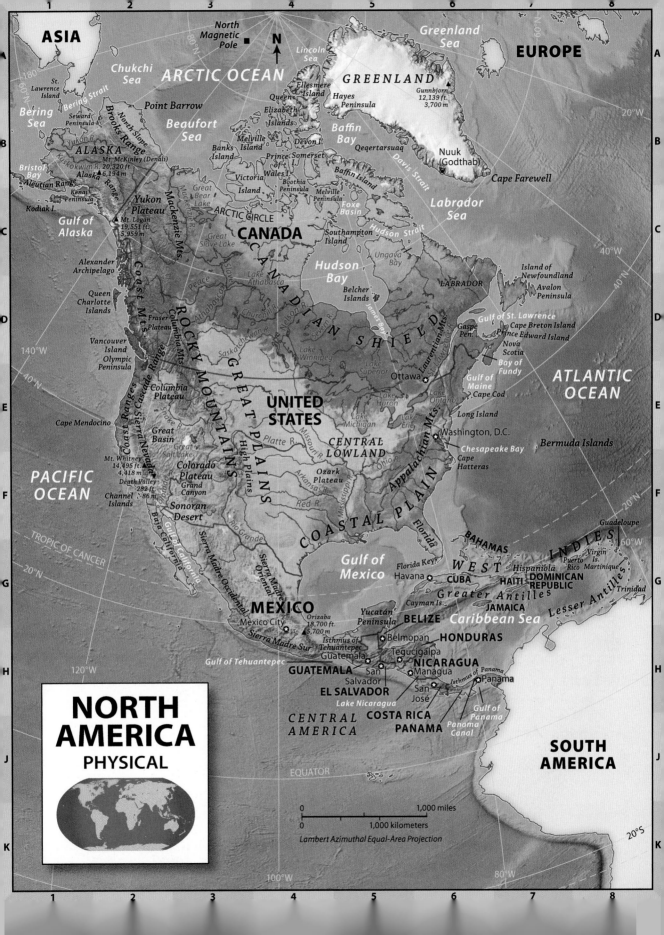

NORTH AMERICA

PHYSICAL

ASIA

EUROPE

ARCTIC OCEAN

North Magnetic Pole

North Slope

GREENLAND

Greenland Sea

Gunnbjorn 12,139 ft. 3,700 m

Chukchi Sea

Point Barrow

Beaufort Sea

Lincoln Sea

Queen Elizabeth Islands

Ellesmere Island

Hayes Peninsula

Nuuk (Godthab)

Bering Sea

St. Lawrence Island

Seward Peninsula

Bering Strait

Brooks Range

ALASKA

Mt. McKinley (Denali) 20,320 ft. 6,194 m

Alaska Range

Melville Island

Banks Island

Victoria Island

Prince of Wales I.

Boothia Peninsula

Devon I.

Baffin Bay

Qeqertarsuaq

Davis Strait

Cape Farewell

Bristol Bay

Aleutian Range

Kenai Peninsula

Kodiak I.

Kuskokwim R.

Yukon R.

Yukon Plateau

Great Bear Lake

Melville Peninsula

Baffin Island

Southampton Island

Foxe Basin

Hudson Strait

Labrador Sea

Gulf of Alaska

Mt. Logan 19,551 ft. 5,959 m

Mackenzie Mts.

ARCTIC CIRCLE

Great Slave Lake

Ungava Bay

Island of Newfoundland

Avalon Peninsula

Alexander Archipelago

Coast Mts.

Peace R.

Liard R.

Lake Athabasca

CANADA

CANADIAN SHIELD

Hudson Bay

Belcher Islands

James Bay

LABRADOR

Gulf of St. Lawrence

Cape Breton Island

Prince Edward Island

Nova Scotia

Queen Charlotte Islands

Fraser Plateau

Columbia Mts.

Churchill R.

Saskatchewan R.

Nelson R.

Severn R.

Lake Winnipeg

Laurentian Mts.

Gaspé Pen.

Bay of Fundy

ATLANTIC OCEAN

Vancouver Island

Olympic Peninsula

Coast Ranges

Cascade Range

Columbia R.

ROCKY MOUNTAINS

GREAT PLAINS

Lake Superior

Ottawa

Lake Huron

Gulf of Maine

Cape Cod

Columbia Plateau

Snake R.

Platte R.

Missouri R.

Lake Michigan

Lake Ontario

Lake Erie

Long Island

Washington, D.C.

Cape Mendocino

Sierra Nevada

Great Basin

Great Salt Lake

UNITED STATES

CENTRAL LOWLAND

Ohio R.

Appalachian Mts.

Chesapeake Bay

Cape Hatteras

Bermuda Islands

PACIFIC OCEAN

Mt. Whitney 14,495 ft. 4,418 m

Death Valley 282 ft. 86 m

Colorado Plateau

Grand Canyon

Colorado R.

Arkansas R.

Ozark Plateau

Mississippi R.

Red R.

COASTAL PLAIN

Channel Islands

Sonoran Desert

Rio Grande

Sierra Madre Occidental

Florida

Florida Keys

Havana

BAHAMAS

WEST INDIES

Guadeloupe

Virgin Is.

TROPIC OF CANCER

Baja California

Gulf of California

Sierra Madre Oriental

Gulf of Mexico

CUBA

Cayman Is.

HAITI

Hispaniola

DOMINICAN REPUBLIC

Puerto Rico

Martinique

Greater Antilles

JAMAICA

Lesser Antilles

Trinidad

MEXICO

Mexico City

Orizaba 18,700 ft. 5,700 m

Yucatán Peninsula

BELIZE

Caribbean Sea

Sierra Madre Sur

Isthmus of Tehuantepec

Belmopan

HONDURAS

Gulf of Tehuantepec

Guatemala

Tegucigalpa

NICARAGUA

Managua

Isthmus of Panama

GUATEMALA

San Salvador

EL SALVADOR

San José

Gulf of Panama

Panama

CENTRAL AMERICA

Lake Nicaragua

COSTA RICA

PANAMA

Panama Canal

SOUTH AMERICA

EQUATOR

0   1,000 miles

0   1,000 kilometers

Lambert Azimuthal Equal-Area Projection

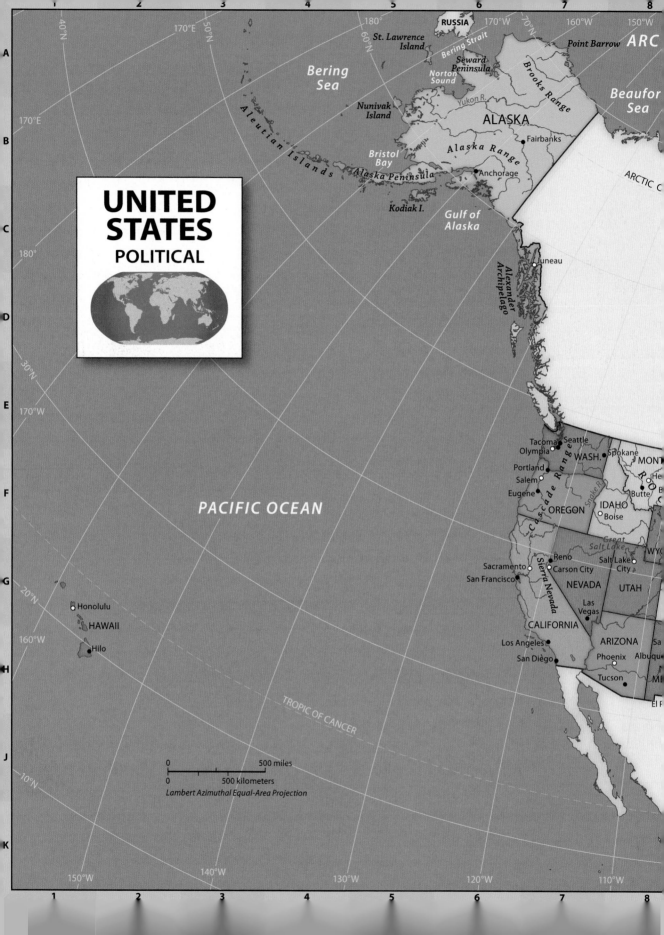

UNITED STATES
POLITICAL

RUSSIA

ARC

ARC

Point Barrow

Beaufor Sea

ARCTIC C

St. Lawrence Island

Bering Strait

Seward Peninsula

70°N

Bering Sea

Norton Sound

Yukon R.

ALASKA

Brooks Range

Nunivak Island

Alaska Range

Fairbanks

Bristol Bay

Alaska Peninsula

Anchorage

Aleutian Islands

Kodiak I.

Gulf of Alaska

Juneau

Alexander Archipelago

ARCTIC C

PACIFIC OCEAN

Tacoma • Seattle
Olympia
WASH. • Spokane
MONT
Portland
Salem
Eugene
OREGON
Cascade Range
Snake R.
He
Butte
IDAHO
Boise
R

Great Salt Lake

WYO

Sacramento
Reno
Carson City
Salt Lake City

San Francisco
Sierra Nevada
NEVADA
UTAH

Las Vegas

Honolulu

HAWAII

CALIFORNIA

ARIZONA

Sa

Hilo

Los Angeles
San Diego
Phoenix
Albuqu

Tucson

MI

El F

TROPIC OF CANCER

0   500 miles
0   500 kilometers
Lambert Azimuthal Equal-Area Projection

40°N
50°N
60°N
170°E
180°
30°N
170°W
160°W
150°W
170°E
20°N
160°W
10°N

150°W
140°W
130°W
120°W
110°W
180°
170°W
160°W
150°W

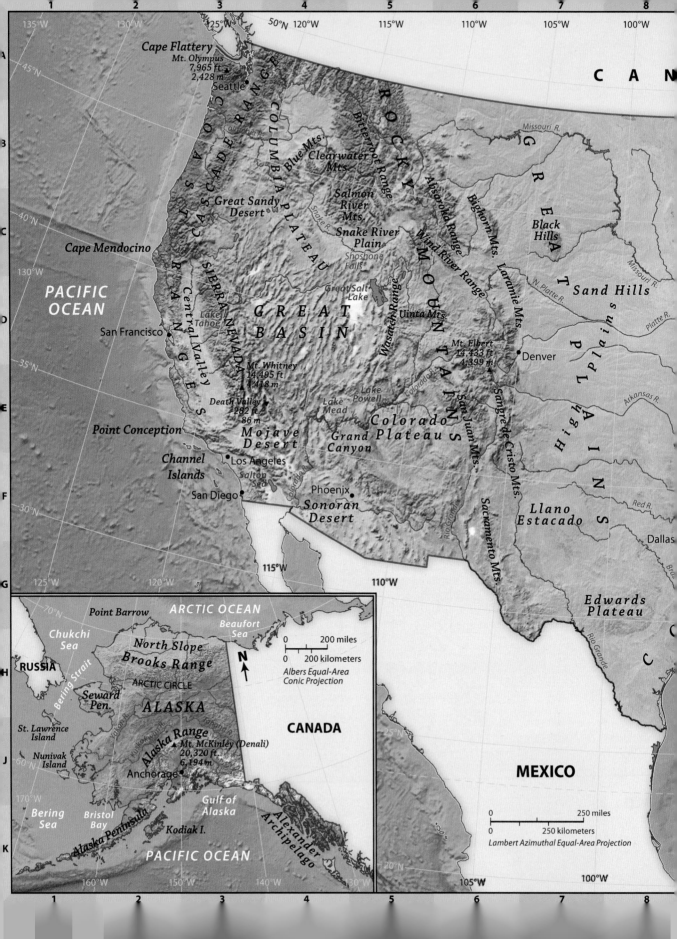

UNITED STATES
PHYSICAL

CANADA

Lake of the Woods

Isle Royale

Lake Superior

Upper Peninsula

Lake Michigan

Lower Peninsula

Lake Huron

Minneapolis

Milwaukee

Chicago

Mississippi R.

Detroit

Cleveland

Lake Erie

Niagara Falls

Lake Ontario

Lake Champlain

Adirondack Mts.

Green Mts.

White Mts.

APPALACHIAN MOUNTAINS

Gulf of Maine

Boston

Cape Cod

Hudson R.

Connecticut R.

CENTRAL LOWLAND

Pittsburgh

Appalachian Plateau

Allegheny Mts.

Long Island

New York City

Philadelphia

Baltimore

Delaware Bay

Washington, D.C.

Chesapeake Bay

Indianapolis

Wabash R.

St. Louis

Ohio R.

Cumberland R.

Blue Ridge

Piedmont

Flint Hills

Ozark Plateau

Boston Mts.

Memphis

Ouachita Mts.

Mt. Mitchell 6,684 ft 2,037 m

Tennessee R.

Cumberland Plateau

APPALACHIAN PLAINS

Cape Hatteras

ATLANTIC OCEAN

Black Belt

Atlanta

Savannah R.

Mississippi R.

Red R.

COASTAL

Houston

New Orleans

Mississippi River Delta

Gulf of Mexico

Jacksonville

Cape Canaveral

Lake Okeechobee

The Everglades

Miami

Florida Keys

Straits of Florida

TROPIC OF CANCER

CUBA

PRINCIPAL HAWAIIAN ISLANDS

N

Kauai

Niihau

Honolulu

Oahu

Molokai

Lanai

Kahoolawe

Maui

Hawaii

Mauna Kea 13,796 ft 4,205 m

PACIFIC OCEAN

0   100 miles
0   100 kilometers
Albers Equal-Area Conic Projection

CANADA
PHYSICAL / POLITICAL

Ellesmere Island

GREENLAND
(KALAALLIT NUNAAT)
Den.

ICELAND

Devon Island

Baffin Bay

Arctic Bay

Baffin Island

Davis Strait

Igloolik

Melville Peninsula

Foxe Basin

N U N A V U T

Repulse Bay

Iqaluit

Southampton Island

Chesterfield Inlet

Hudson Strait

Labrador Sea

Ungava Bay

Kuujjuaq

Nain

NEWFOUNDLAND AND LABRADOR

Cartwright

Hudson Bay

Belcher Islands

Schefferville

Happy Valley-
Goose Bay

Smallwood Reservoir

Churchill Falls

Island of
Newfoundland

Fort Severn

Kuujjuarapik

Labrador City

St. John's

Q U E B E C

Avalon Peninsula

James Bay

Manicouagan Reservoir

Anticosti I.

St.-Pierre & Miquelon
Fr.

Sept-Îles

Gulf of
St. Lawrence

O N T A R I O   S H I E L D

Gaspé Pen.

PRINCE EDWARD ISLAND

Sydney

Cape Breton I.

Lake Nipigon

Chicoutimi

Charlottetown

ATLANTIC
OCEAN

Thunder Bay

Timmins

Rouyn-Noranda

Quebec

NEW BRUNSWICK

Fredericton

NOVA SCOTIA

Lake Superior

Sudbury

North Bay

Montreal

St. Lawrence R.

Saint John

Halifax

Bay of Fundy

Ottawa

Lake Michigan

Lake Huron

Toronto

L. Ontario

Niagara Falls

London

Lake Erie

80°N 40°W

Arctic Circle

90°W

80°W

70°W

60°W

| | 1 | 2 | 3 | 4 | 5 | 6 | 7 | 8 |

120°W 110°W 100°W 90°W

A

Tijuana
Mexicali
*Sonoran
Desert*

UNITED STATES

30°N

B

Baja California
BAJA
CALIFORNIA
SONORA

Ciudad
Juárez

CHIHUAHUA

Chihuahua

Rio Bravo

C

Gulf of California
Sierra Madre Occidental

COAHUILA

Nuevo
Laredo

Gulf of Mexico

BAJA
CALIFORNIA
SUR

La Paz

SINALOA

DURANGO

M
E

Sierra Madre Oriental

Monterrey
NUEVO
LEÓN

Matamoros

D

Cape Falso

Mazatlán

ZACATECAS

SAN
LUIS
POTOSÍ

X

TAMAULIPAS

20°N

E

NAYARIT

AGUASCALIENTES

Guadalajara
JALISCO

San
Luis
Potosí

León

I

C

Ciudad Madero
Tampico

QUERÉTARO

VERACRUZ

HIDALGO

MORELOS

Mérida • YUCATÁN
*Yucatán
Peninsula*

Cozum
Island

QUINTAN
ROO

Revillagigedo Islands
Mex.

GUANAJUATO

COLIMA

MICHOACÁN

DISTRITO FEDERAL

Mexico City
MÉXICO

TLAXCALA

Veracruz

Bay of Campeche

CAMPECHE

F

*Popocatépetl
17,887 ft.
5,450 m*

PUEBLA

*Orizaba
18,700 ft.
5,700 m*

O

TABASCO

Beliz
City

GUERRERO

Acapulco

Sierra Madre del Sur

OAXACA

*Isthmus of
Tehuantepec*

Belmopan

BELIZE
*Gulf of
Hondur*

CHIAPAS

Sierra Madre

GUATEMALA

HON

G

*Gulf of
Tehuantepec*

Guatemala

Tegucigalpa

San Salvador

10°N

EL SALVADOR

Leó

H

CENTRAL

MIDDLE AMERICA
PHYSICAL / POLITICAL

AMERICA

J

PACIFIC OCEAN

Cocos Island
C.R.

K

110°W 100°W 90°W

| | 1 | 2 | 3 | 4 | 5 | 6 | 7 | 8 |

N

9 10 11 12 13 14 15 16

80°W 70°W 60°W

ATLANTIC OCEAN

30°N

Freeport

Nassau

TROPIC OF CANCER

Andros
Island

Straits of Florida

Turks &
Caicos Islands
U.K.

20°N

Havana

CUBA Camagüey

Isle of Youth

Holguín

Santiago de Cuba

Santiago

St. Kitts & Nevis

Cayman Islands
U.K.

Greater

Hispaniola

Santo Domingo

San Juan

*Virgin
Islands*
U.S. & U.K.

**ANTIGUA &
BARBUDA**

Montego Bay

HAITI

Port-au-Prince

**DOMINICAN
REPUBLIC**

Puerto
Rico
U.S.

JAMAICA

Kingston

Antilles

Guadeloupe
Fr.

Bird I.
Venez.

DOMINICA

Martinique
Fr.

Caribbean Sea

ST. LUCIA

BARBADOS

**ST. VINCENT &
THE GRENADINES**

GRENADA

RAS

Coco R.

Aruba
Neth.

Curaçao Neth.
Bonaire

Tobago

CARAGUA

Managua

*Lake
Nicaragua*

Lesser Antilles

TRINIDAD & TOBAGO

Port-of-Spain

Trinidad

10°N

Puerto
Limón

San José

Mosquito Gulf

**COSTA
RICA**

Isthmus of Panama

David

Panama

PANAMA

*Gulf of
Panama*

SOUTH AMERICA

0   500 miles

0   500 kilometers

Lambert Azimuthal Equal-Area Projection

80°W 70°W 60°W

9 10 11 12 13 14 15 16

A commonly accepted division between Asia and Europe—here marked by a gray line—is formed by the Ural Mountains, Ural River, Caspian Sea, Caucasus Mountains, and the Black Sea with its outlets, the Bosporus and the Dardanelles.

Europe/Asia boundary

ASIA

ASIA

Barents Sea

Tobseda

Pechora

URAL MOUNTAINS

LAPLAND

Murmansk

Ivalo · Kirovsk
Kiruna

Kola Peninsula

Umba

White Sea

Kem'

Arkhangel'sk

Severodvinsk

Syktyvkar

Kemi
Luleå · Oulu
Umeå

FINLAND

Northern Dvina R.

Perm

Vaasa · Kuopio

Pori · Tampere

Lake Onega

Kirov

Ufa

Turku · Helsinki

Lake Ladoga

RUSSIA

Kazan'

Sea

St. Petersburg

Tallinn

ESTONIA

Novgorod

Yaroslavl'

Orenburg

Nizhniy Novgorod

Samara

LATVIA

Riga

Tver' · Moscow

Ural R.

Daugavpils

Oral

LITHUANIA

Vilnius

Vitsyebsk

Smolensk

Ryazan'

Penza

Saratov

Kaunas

Kaliningrad

Minsk

Bryansk

KAZAKHSTAN

BELARUS

Homyel'

Kursk

Don R.

Volga R.

Warsaw

Chernihiv

Volgograd

Sumy

Vistula R.

Kyiv (Kiev)

Kharkiv

Poltava

Astrakhan

L'viv

Dnieper R.

UKRAINE

Donets'k

Vinnytsya

Dniester R.

Dnipropetrovs'k

Rostov

Caspian Sea

MOLDOVA

Carpathian Mts.

Chișinău

Sea of Azov

Stavropol'

Odessa

Kerch

Caucasus Mountains

AZERBAIJAN

Crimea

Simferopol'

Grozny

ROMANIA

Belgrade · Bucharest

Sevastopol' · Yalta

GEORGIA

Baku

Constanța

Balkan Mts.

Varna

Black Sea

SERBIA

KOSOVO

Priština

BULGARIA

Sofia

Bosporus

İstanbul

Skopje

MACEDONIA

Thessaloníki

T U R K E Y

GREECE

Dardanelles

Sea of Marmara

Aegean Sea

Athens

Peloponnese

ASIA

Rhodes

Nicosia

Iraklíon

Crete
Greece

CYPRUS

400 miles

400 kilometers

Lambert Azimuthal Equal-Area Projection

Sea

EUROPE
PHYSICAL

N

Reykjavík
ICELAND

Faeroe Islands

Shetland Islands

Norwegian Sea

SCANDIN

ARCTIC CIRCLE

PRIME MERIDIAN

NORWAY

Oslo

SWEDEN

Stockholm

Gulf of

Outer Hebrides

Orkney Islands

Highlands

British Isles

Edinburgh

Belfast

UNITED KINGDOM

IRELAND

Dublin

Irish Sea

Great Britain

North Sea

Skagerrak

Kattegat

Jutland

DENMARK
Copenhagen

Zealand

Gotland

Baltic

Celtic Sea

Cardiff

London

Thames R.

NETHERLANDS
Amsterdam

Elbe R.

Berlin

N R T H

POLAND

Land's End

ATLANTIC OCEAN

English Channel

Brussels

BELGIUM

Rhine R.

GERMANY

Oder R.

Brittany

Seine R.

Paris

Luxembourg

LUXEMBOURG

Prague

CZECH REPUBLIC

Danube R.

Bratislava

SLOVAKI

FRANCE

LIECHTENSTEIN

Vienna

Budapest

Loire R.

Bern

Vaduz

AUSTRIA

HUNGARY

Bay of Biscay

Mont Blanc
15,771 ft.
4,807 m

SWITZERLAND

A L P S

SLOVENIA

Ljubljana

Drava R.

Massif Central

Po R.

Zagreb

Cantabrian Mountains

Pyrenees

Andorra la Vella

MONACO

Riviera

CROATIA

BOSNIA & HERZEGOVINA

Douro R.

Ebro R.

ANDORRA

SAN MARINO

Adriatic Sea

Sarajevo

I B E R I A N

Madrid

A p e n n i n e s

MONTENEGRO

Podgorica

SPAIN

Corsica

ITALY

Lisbon

PORTUGAL

Tagus R.

Rome

Tiranë

P E N I N S U L A

VATICAN CITY
(within Rome)

ALBANIA

Cape St. Vincent

Baetic Mountains

Balearic Islands

Sardinia

Tyrrhenian Sea

Ionian Sea

Strait of Gibraltar

GIBRALTAR

M e d i t e r r a n e a n

Sicily

Etna
10,902 ft.
3,323 m

MALTA
Valletta

AFRICA

30°W   20°W   10°W   70°N   60°N   50°N   40°N   20°W   30°N   10°W   30°N   0°   10°E   20

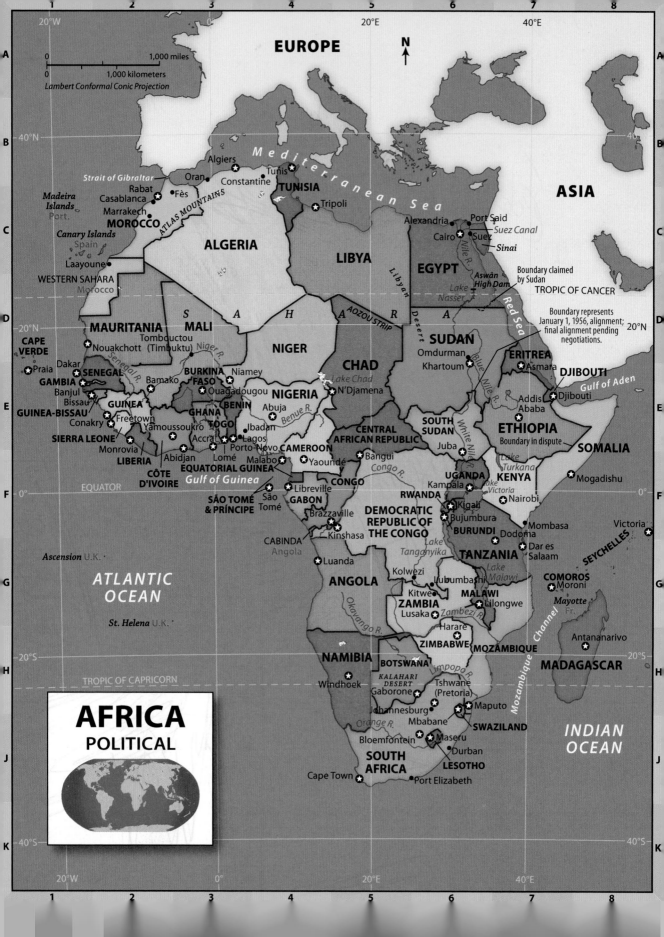

AFRICA
POLITICAL

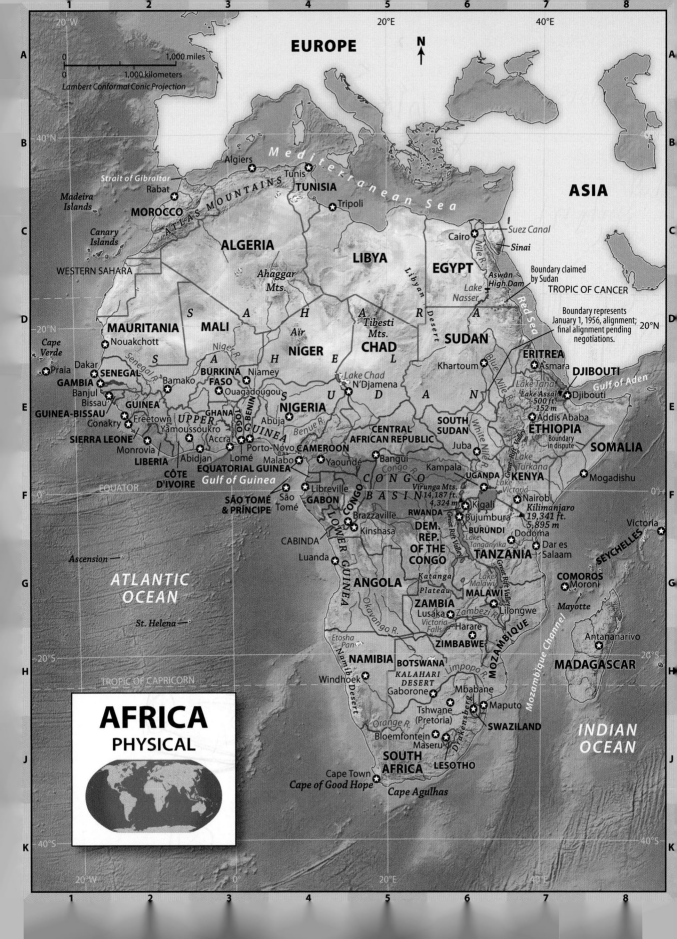

AFRICA
PHYSICAL

EUROPE

N

ASIA

Scale:
0 — 1,000 miles
0 — 1,000 kilometers
Lambert Conformal Conic Projection

20°W · 20°E · 40°E

Madeira Islands

Strait of Gibraltar
Algiers ✪
Tunis
TUNISIA
Rabat ✪
Tripoli ✪
MOROCCO
ATLAS MOUNTAINS
Mediterranean Sea

ALGERIA
LIBYA
EGYPT
Cairo ✪
Suez Canal
Sinai

Canary Islands

WESTERN SAHARA

Boundary claimed by Sudan
TROPIC OF CANCER

Ahaggar Mts.
S A H A R A
Tibesti Mts.
Libyan Desert
Aswān High Dam
Lake Nasser
Red Sea

Boundary represents January 1, 1956, alignment; final alignment pending negotiations.

20°N

MAURITANIA
MALI
Nouakchott ✪
Aïr
NIGER
CHAD
SUDAN
Khartoum ✪
Cape Verde
Praia ✪
Dakar ✪
Niger R.
ERITREA
Asmara ✪
DJIBOUTI
Gulf of Aden
SENEGAL
Senegal R.
BURKINA FASO
Bamako ✪
Niamey
Lake Chad
N'Djamena
Blue Nile R.
Lake Tana
Lake Assal -500 ft. -152 m
Djibouti ✪
GAMBIA
Banjul
Ouagadougou ✪
Addis Ababa ✪
GUINEA-BISSAU
Bissau
GUINEA
Freetown ✪
NIGERIA
Abuja ✪
White Nile R.
ETHIOPIA
GHANA
BENIN
TOGO
Yamoussoukro
CENTRAL AFRICAN REPUBLIC
SOUTH SUDAN
Juba ✪
Boundary in dispute
SIERRA LEONE
Conakry ✪
UPPER GUINEA
Accra ✪
Benue R.
SOMALIA
Monrovia ✪
Porto-Novo
CAMEROON
Bangui ✪
LIBERIA
Abidjan
Lomé
Malabo
Yaoundé ✪
UGANDA
Kampala ✪
Mogadishu ✪
CÔTE D'IVOIRE
EQUATORIAL GUINEA
Gulf of Guinea
Virunga Mts. 14,187 ft. 4,324 m
Lake Victoria
KENYA

EQUATOR
SÃO TOMÉ & PRÍNCIPE
São Tomé
GABON
Libreville ✪
C O N G O   B A S I N
Kigali
Nairobi ✪
Kilimanjaro 19,341 ft. 5,895 m
0°
SEYCHELLES
Victoria ✪
CONGO
LOWER GUINEA
Brazzaville ✪
RWANDA
Bujumbura
DEM. REP. OF THE CONGO
Kinshasa
BURUNDI
Dodoma ✪
Dar es Salaam
Great Rift Valley
CABINDA
Luanda ✪
Lake Tanganyika
TANZANIA
COMOROS
Moroni ✪
Mayotte

Ascension

ATLANTIC OCEAN

Katanga Plateau
ANGOLA
Okavango R.
Lake Malawi
MALAWI
Lilongwe ✪
Antananarivo ✪
ZAMBIA
Lusaka ✪
Zambezi R.
Victoria Falls
MADAGASCAR

St. Helena

Etosha Pan
Harare ✪
ZIMBABWE
MOZAMBIQUE
Mozambique Channel

20°S
TROPIC OF CAPRICORN
Namib Desert
NAMIBIA
BOTSWANA
Limpopo R.
Windhoek ✪
KALAHARI DESERT
Gaborone ✪
Orange R.
Mbabane
Maputo ✪
Tshwane (Pretoria)
SWAZILAND
Drakensberg
Bloemfontein
Maseru
INDIAN OCEAN

SOUTH AFRICA
LESOTHO

Cape Town ✪
Cape of Good Hope
Cape Agulhas

40°S

20°W · 0° · 20°E · 40°E

MIDDLE EAST
PHYSICAL / POLITICAL

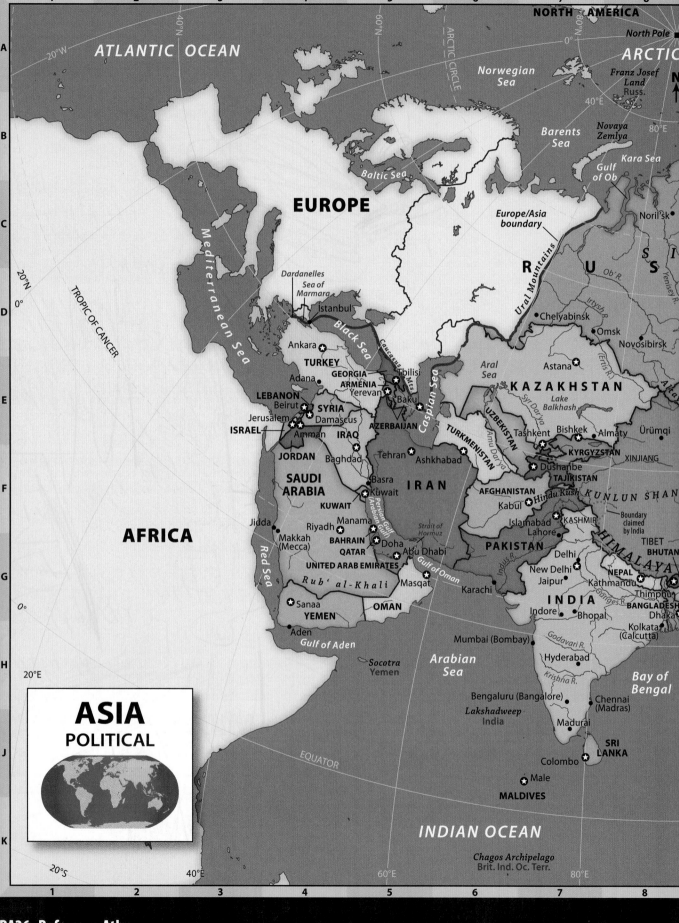

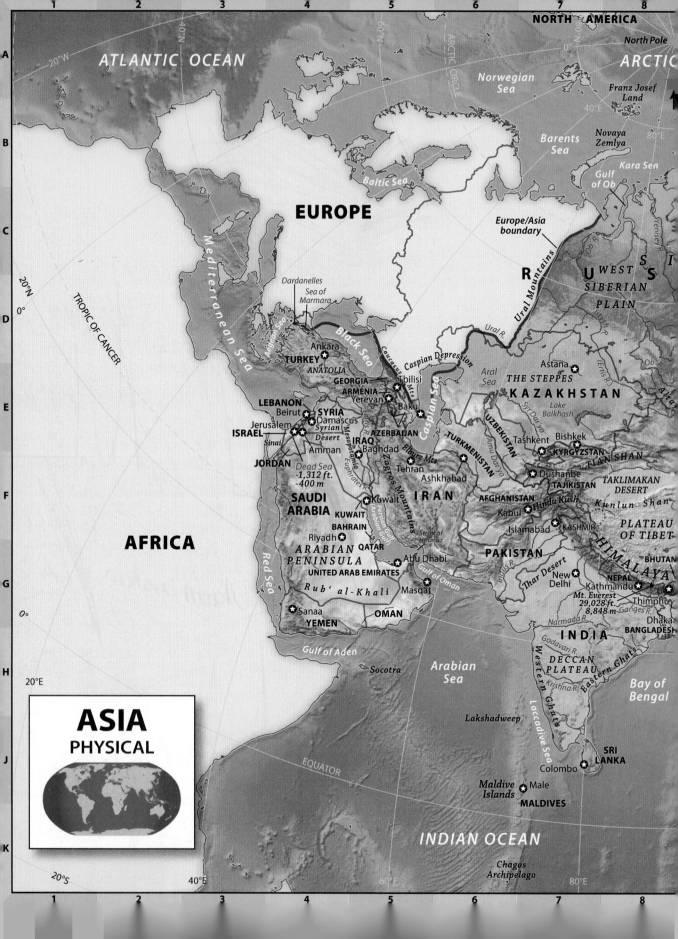

ASIA
PHYSICAL

NORTH AMERICA

OCEAN

Chukchi Sea
Chukchi Peninsula
Wrangel I.
Gulf of Anadyr
Bering Strait
Bering Sea

New Siberian Islands
East Siberian Sea

evernaya Zemlya
Laptev Sea
aymyr eninsula

Cherski Range
Kolyma Mts.
Kolyma R.

Commander Is.
Kamchatka Peninsula

CENTRAL SIBERIAN PLATEAU

Verkhoyanski Mts.
Lena R.
Aldan R.

Sea of Okhotsk

Aleutian Islands

S I B E R I A

Lena R.
Lake Baikal
Angara R.
Yablonovyy Range

Amur R.
Greater Khingan Range
Manchurian Plain
Sikhote-Alin' Range

Sakhalin

Kuril Islands

Hokkaidō

Honshū

TROPIC OF CANCER

Ulaanbaatar
MONGOLIA
GOBI

NORTH KOREA
P'yŏngyang
Sea of Japan (East Sea)

JAPAN
Tokyo

Nampo-shoto

Beijing
Huang He (Yellow R.)
North China Plain
SOUTH KOREA
Seoul
Yellow Sea

Shikoku
Kyūshū

Qilian Shan
Qaidam Basin

C H I N A
Chang Jiang (Yangtze)
Sichuan Basin
Gongga Shan 24,790 ft. 7,556 m
Xi R.

East China Sea

Okinawa
Ryukyu Islands

Taipei
TAIWAN

PACIFIC OCEAN

Mariana Islands

Mekong R.
Salween R.

Hanoi
MYANMAR (BURMA)
Nay Pyi Taw
LAOS
Vientiane
Hainan

Luzon

Philippine Sea

Caroline Islands

EQUATOR

Andaman Islands
Andaman Sea

THAILAND
Bangkok
VIETNAM
South China Sea

Manila
Mindoro
PHILIPPINES
Samar

P H I L I P P I N E   I S L A N D S

Nicobar Islands

CAMBODIA
Phnom Penh
Gulf of Thailand
Malay Peninsula

Palawan
Mindanao
Sulu Sea

Kuala Lumpur
M A L A Y S I A

Bandar Seri Begawan
BRUNEI
Celebes Sea

New Guinea

Mentawai Islands

SINGAPORE
Borneo
Sulawesi (Celebes)
Buru
Ceram
Aru Is.
Dolak

Moluccas

Sumatra
I N D O N E S I A
Greater Sunda Islands
Java Sea
Tanimbar Is.
Arafura Sea

Jakarta
Java
Dili
EAST TIMOR (TIMOR-LESTE)
Timor
Timor Sea
AUSTRALIA

0   1,000 miles
0   1,000 kilometers
Two-Point Equidistant Projection

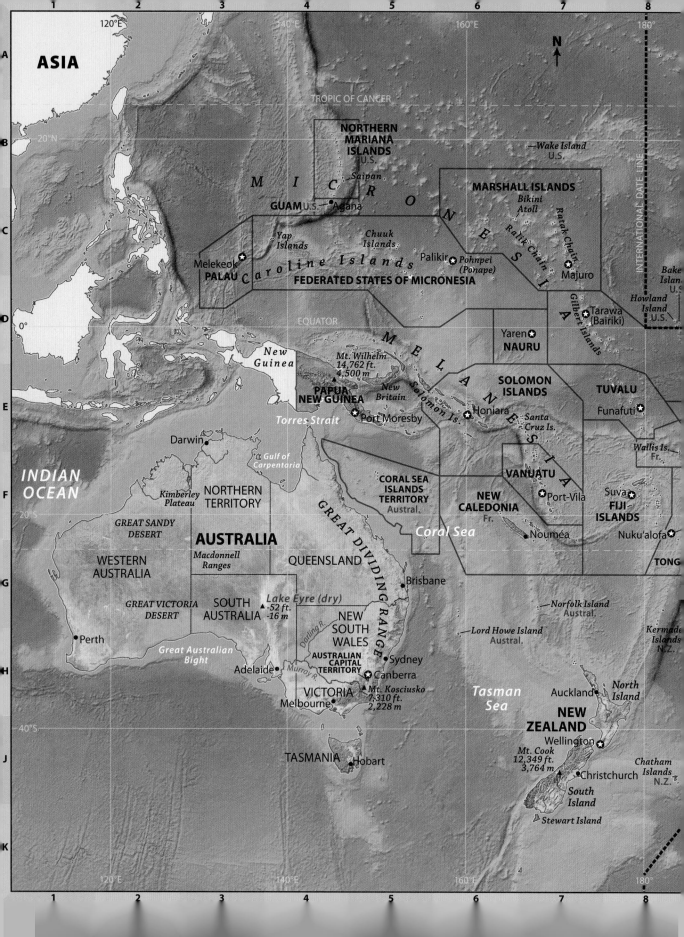

ASIA

NORTHERN
MARIANA
ISLANDS
U.S.

Saipan

M I C R O N E S I A

Wake Island
U.S.

GUAM U.S. • Agana

MARSHALL ISLANDS

Bikini
Atoll

Ratak Chain

Yap
Islands

Chuuk
Islands

Palikir

Pohnpei
(Ponape)

Raiik Chain

Majuro

Bake
Islan.
U.S

Melekeok ✪

PALAU

Caroline Islands

FEDERATED STATES OF MICRONESIA

Howland
Island
U.S.

INTERNATIONAL DATE LINE

Gilbert Islands

Tarawa
(Bairiki)

EQUATOR

New
Guinea

Mt. Wilhelm
14,762 ft.
4,500 m

M E L A N E S I A

Yaren ✪

NAURU

SOLOMON
ISLANDS

TUVALU

PAPUA
NEW GUINEA

New
Britain

Solomon Is.

Honiara ✪

Funafuti ✪

Torres Strait

Port Moresby

Santa
Cruz Is.

Wallis Is.
Fr.

Darwin •

Gulf of
Carpentaria

CORAL SEA
ISLANDS
TERRITORY
Austral.

VANUATU

NEW
CALEDONIA
Fr.

Port-Vila

Suva ✪

FIJI
ISLANDS

INDIAN
OCEAN

Kimberley
Plateau

NORTHERN
TERRITORY

GREAT DIVIDING RANGE

Coral Sea

Nouméa •

Nuku'alofa ✪

GREAT SANDY
DESERT

AUSTRALIA

Macdonnell
Ranges

QUEENSLAND

Norfolk Island
Austral.

TONG

WESTERN
AUSTRALIA

Brisbane •

Lord Howe Island
Austral.

Kermade
Islands
N.Z.

GREAT VICTORIA
DESERT

SOUTH
AUSTRALIA

Lake Eyre (dry)
▲ -52 ft.
-16 m

NEW
SOUTH
WALES

Darling R.

• Perth

Great Australian
Bight

Adelaide •

Murray R.

AUSTRALIAN
CAPITAL
TERRITORY

• Sydney

Canberra ✪

VICTORIA

Melbourne •

Mt. Kosciusko
▲ 7,310 ft.
2,228 m

Auckland •

North
Island

Tasman
Sea

NEW
ZEALAND

TASMANIA

Hobart •

Wellington • ✪

Mt. Cook
12,349 ft.
3,764 m

Christchurch •

Chatham
Islands
N.Z.

South
Island

Stewart Island

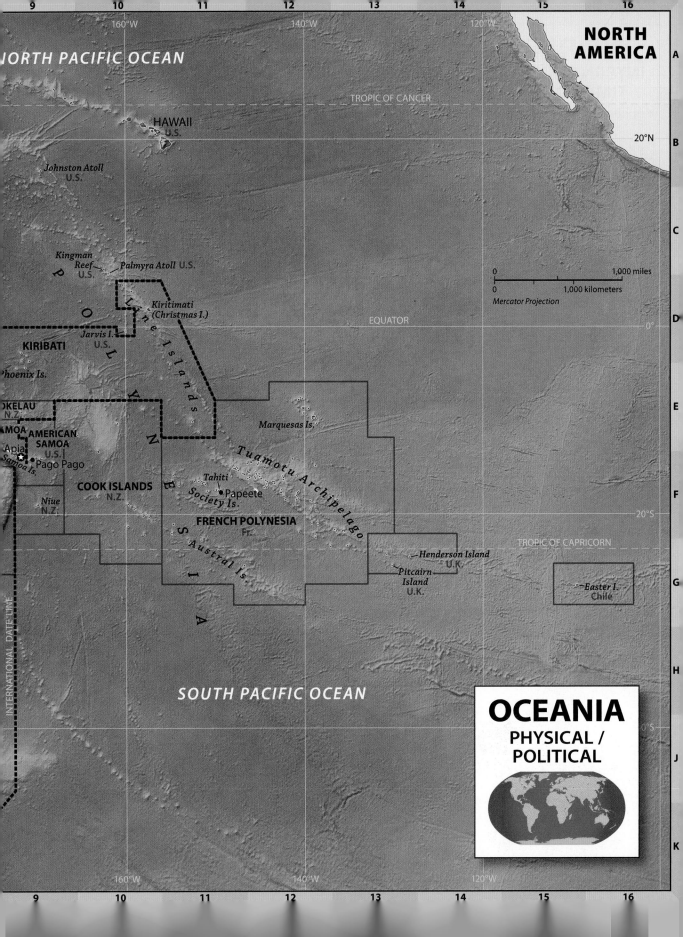

9 10 11 12 13 14 15 16

160°W   140°W   120°W

NORTH PACIFIC OCEAN

A

NORTH
AMERICA

TROPIC OF CANCER

HAWAII
U.S.

20°N   B

Johnston Atoll
U.S.

C

Kingman
Reef   Palmyra Atoll U.S.
U.S.

P   Kiritimati
            (Christmas I.)

EQUATOR   0°   D

O

Jarvis I.
U.S.

KIRIBATI

L

Phoenix Is.

Marquesas Is.

E

OKELAU
N.Z.

Y

MOA

AMERICAN
SAMOA
Apia   U.S.

Samoa Is. Pago Pago

N

Tuamotu Archipelago

Tahiti

COOK ISLANDS
N.Z.

Society Is.

Papeete

E

FRENCH POLYNESIA
Fr.

20°S   F

Niue
N.Z.

S

Austral Is.

Henderson Island
U.K.

TROPIC OF CAPRICORN

I

Pitcairn
Island
U.K.

Easter I.
Chile

G

A

H

INTERNATIONAL DATE LINE

SOUTH PACIFIC OCEAN

30°S

OCEANIA

PHYSICAL /
POLITICAL

J

K

9 10 11 12 13 14 15 16

160°W   140°W   120°W

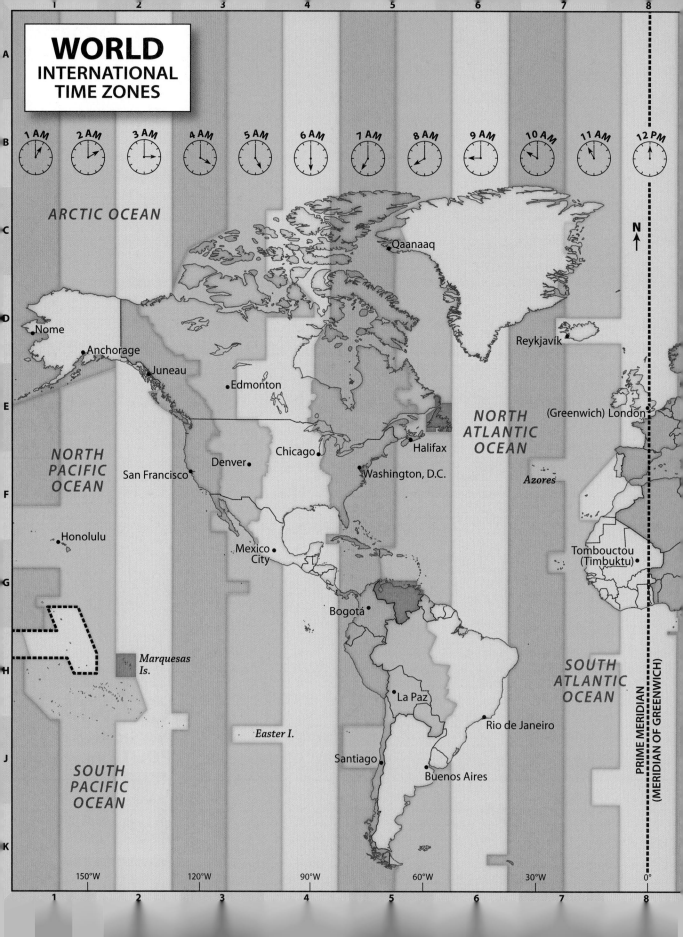

WORLD
INTERNATIONAL
TIME ZONES

ARCTIC OCEAN

1 AM 2 AM 3 AM 4 AM 5 AM 6 AM 7 AM 8 AM 9 AM 10 AM 11 AM 12 PM

N

Qaanaaq

Nome

Anchorage

Juneau

Edmonton

Reykjavík

(Greenwich) London

NORTH
PACIFIC
OCEAN

San Francisco

Denver

Chicago

Halifax

NORTH
ATLANTIC
OCEAN

Washington, D.C.

Azores

Honolulu

Mexico
City

Tombouctou
(Timbuktu)

Marquesas
Is.

Bogotá

SOUTH
ATLANTIC
OCEAN

La Paz

Easter I.

Rio de Janeiro

SOUTH
PACIFIC
OCEAN

Santiago

Buenos Aires

PRIME MERIDIAN
(MERIDIAN OF GREENWICH)

150°W   120°W   90°W   60°W   30°W   0°

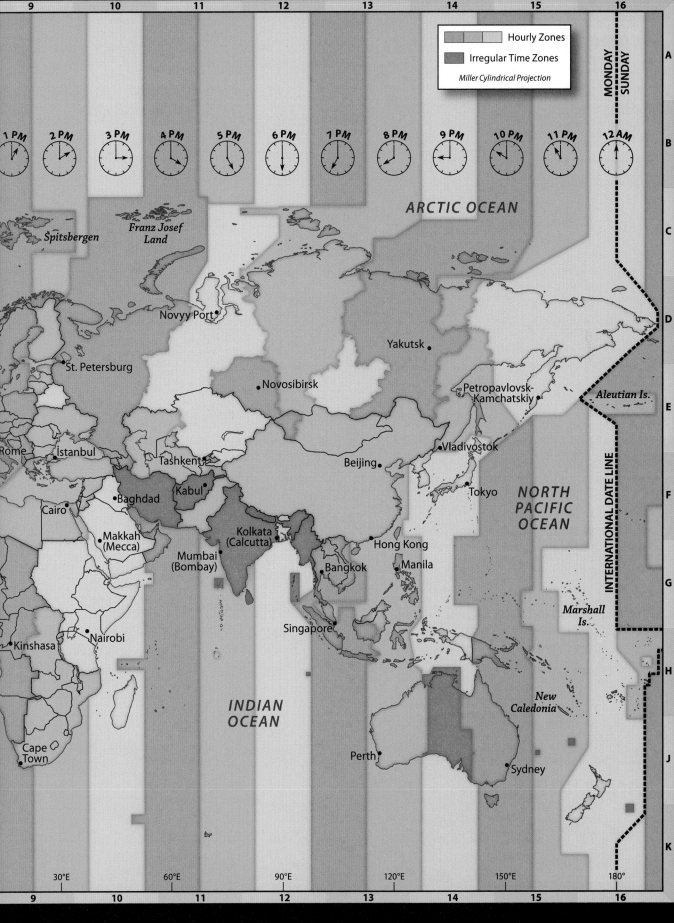

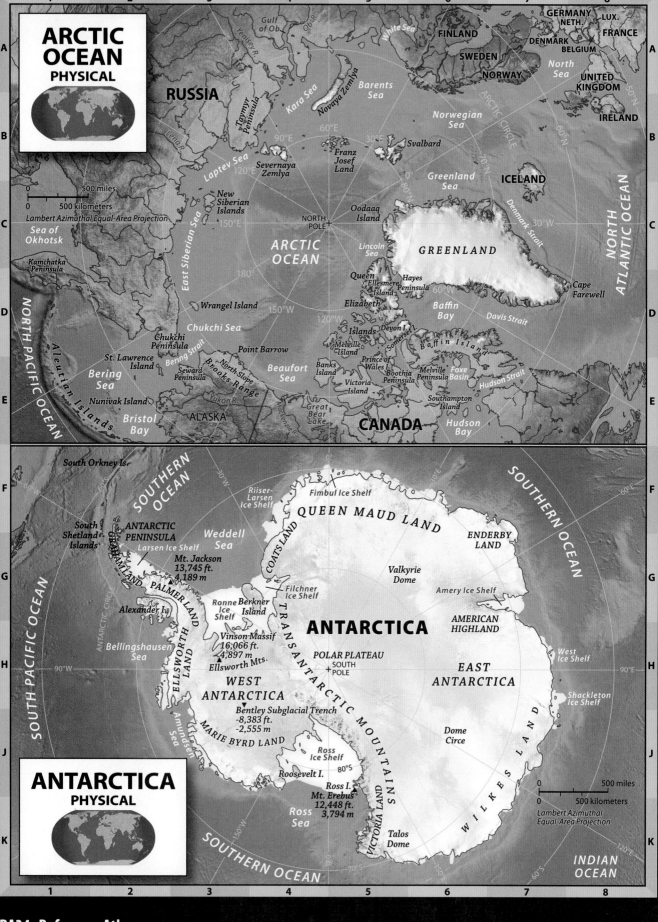

ARCTIC OCEAN
PHYSICAL

RUSSIA

Taymyr Peninsula
Kara Sea
Novaya Zemlya
Barents Sea
Gulf of Ob
Ob R.
Yenisey R.
White Sea
FINLAND
SWEDEN
NORWAY
GERMANY
NETH.
DENMARK
BELGIUM
LUX.
FRANCE
North Sea
UNITED KINGDOM
IRELAND

Severnaya Zemlya
Franz Josef Land
Svalbard
Norwegian Sea
Greenland Sea
ICELAND
ARCTIC CIRCLE

Laptev Sea
Lena R.
90°E
60°E
30°E
0°
70°N
60°N
NORTH ATLANTIC OCEAN

New Siberian Islands
120°E
150°E
NORTH POLE
Oodaaq Island
Denmark Strait
30°W
Cape Farewell

Sea of Okhotsk
ARCTIC OCEAN
Lincoln Sea
GREENLAND

Kamchatka Peninsula
180°
Queen
Hayes Peninsula
Baffin Bay
Davis Strait

East Siberian Sea
Wrangel Island
150°W
Ellesmere Island
Elizabeth
Islands
Devon I.
Somerset I.
Baffin Island

NORTH PACIFIC OCEAN
Chukchi Sea
120°W
Melville Island
Prince of Wales I.
Boothia Peninsula
Melville Peninsula
Foxe Basin

Chukchi Peninsula
Bering Strait
Point Barrow
North Slope
Beaufort Sea
Banks Island
Victoria Island
Southampton Island
Hudson Strait

St. Lawrence Island
Seward Peninsula
Brooks Range
Yukon R.
Mackenzie R.
Great Bear Lake
Hudson Bay

Aleutian Islands
Bering Sea
Nunivak Island
Bristol Bay
ALASKA
CANADA

0   500 miles
0   500 kilometers
Lambert Azimuthal Equal-Area Projection

South Orkney Is.
SOUTHERN OCEAN
60°S
Riiser-Larsen Ice Shelf
Fimbul Ice Shelf
QUEEN MAUD LAND
30°E
ENDERBY LAND
60°E
SOUTHERN OCEAN

South Shetland Islands
ANTARCTIC PENINSULA
Weddell Sea
COATS LAND
Valkyrie Dome
Amery Ice Shelf

ANTARCTIC CIRCLE
GRAHAM LAND
PALMER LAND
Larsen Ice Shelf
Mt. Jackson 13,745 ft. 4,189 m
Ronne Ice Shelf
Filchner Ice Shelf
Berkner Island
ANTARCTICA
AMERICAN HIGHLAND
West Ice Shelf

Alexander I.
ELLSWORTH LAND
Vinson Massif 16,066 ft. 4,897 m
Ellsworth Mts.
POLAR PLATEAU
SOUTH POLE
EAST ANTARCTICA
Shackleton Ice Shelf

SOUTH PACIFIC OCEAN
Bellingshausen Sea
90°W
WEST ANTARCTICA
TRANSANTARCTIC MOUNTAINS
90°E

Amundsen Sea
Bentley Subglacial Trench -8,383 ft. -2,555 m
MARIE BYRD LAND
Dome Circe
WILKES LAND

Ross Ice Shelf
Roosevelt I.
80°S
Ross I.
Mt. Erebus 12,448 ft. 3,794 m
Ross Sea
VICTORIA LAND
Talos Dome

150°W
120°E
SOUTHERN OCEAN
60°S
INDIAN OCEAN

ANTARCTICA
PHYSICAL

0   500 miles
0   500 kilometers
Lambert Azimuthal Equal-Area Projection

A WORLD OF EXTREMES

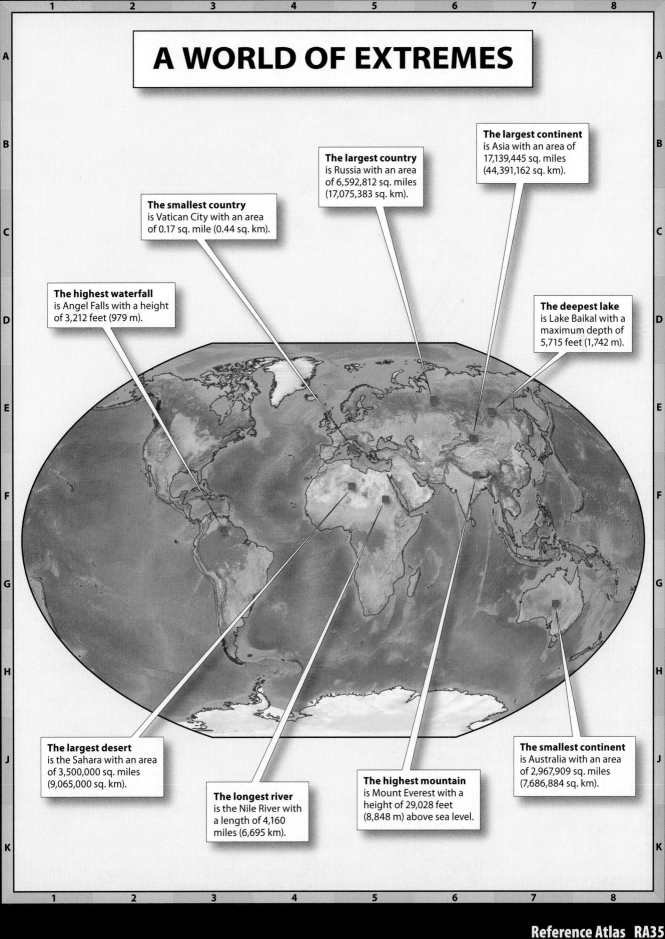

The largest continent
is Asia with an area of
17,139,445 sq. miles
(44,391,162 sq. km).

The largest country
is Russia with an area
of 6,592,812 sq. miles
(17,075,383 sq. km).

The smallest country
is Vatican City with an area
of 0.17 sq. mile (0.44 sq. km).

The deepest lake
is Lake Baikal with a
maximum depth of
5,715 feet (1,742 m).

The highest waterfall
is Angel Falls with a height
of 3,212 feet (979 m).

The largest desert
is the Sahara with an area
of 3,500,000 sq. miles
(9,065,000 sq. km).

The longest river
is the Nile River with
a length of 4,160
miles (6,695 km).

The highest mountain
is Mount Everest with a
height of 29,028 feet
(8,848 m) above sea level.

The smallest continent
is Australia with an area
of 2,967,909 sq. miles
(7,686,884 sq. km).

GEOGRAPHIC DICTIONARY

Ocean · Archipelago · Gulf · Reservoir · Volcano · Isthmus · Plateau · Highlands · Canyon · Cliff · Cape · Bay · Harbor · Reef · Island · Channel · Peninsula

archipelago a group of islands

basin area of land drained by a given river and its branches; area of land surrounded by lands of higher elevations

bay part of a large body of water that extends into a shoreline, generally smaller than a gulf

canyon deep and narrow valley with steep walls

cape point of land that extends into a river, lake, or ocean

channel wide strait or waterway between two landmasses that lie close to each other; deep part of a river or other waterway

cliff steep, high wall of rock, earth, or ice

continent one of the seven large landmasses on the Earth

delta flat, low-lying land built up from soil carried downstream by a river and deposited at its mouth

divide stretch of high land that separates river systems

downstream direction in which a river or stream flows from its source to its mouth

escarpment steep cliff or slope between a higher and lower land surface

glacier large, thick body of slowly moving ice

gulf part of a large body of water that extends into a shoreline, generally larger and more deeply indented than a bay

harbor a sheltered place along a shoreline where ships can anchor safely

highland elevated land area such as a hill, mountain, or plateau

hill elevated land with sloping sides and rounded summit; generally smaller than a mountain

island land area, smaller than a continent, completely surrounded by water

isthmus narrow stretch of land connecting two larger land areas

lake a sizable inland body of water

lowland land, usually level, at a low elevation

mesa broad, flat-topped landform with steep sides; smaller than a plateau

mountain land with steep sides that rises sharply (1,000 feet or more) from surrounding land; generally larger and more rugged than a hill

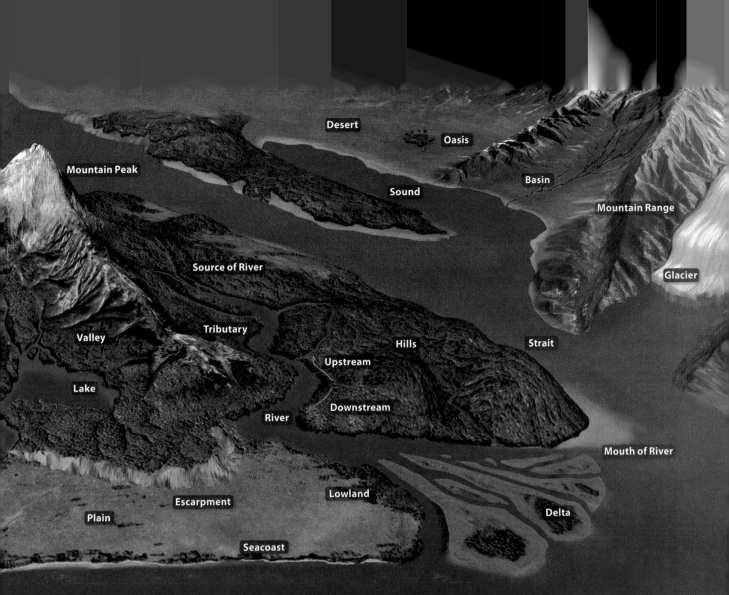

Desert

Oasis

Mountain Peak

Basin

Sound

Mountain Range

Source of River

Glacier

Valley

Tributary

Hills

Strait

Upstream

Lake

Downstream

River

Mouth of River

Escarpment

Lowland

Plain

Delta

Seacoast

mountain peak pointed top of a mountain

mountain range a series of connected mountains

mouth (of a river) place where a stream or river flows into a larger body of water

oasis small area in a desert where water and vegetation are found

ocean one of the four major bodies of salt water that surround the continents

ocean current stream of either cold or warm water that moves in a definite direction through an ocean

peninsula body of land jutting into a lake or ocean, surrounded on three sides by water

physical feature characteristic of a place occurring naturally, such as a landform, body of water, climate pattern, or resource

plain area of level land, usually at low elevation and often covered with grasses

plateau area of flat or rolling land at a high elevation, about 300 to 3,000 feet (90 to 900 m) high

reef a chain of rocks, coral or sand at or near the surface

river large natural stream of water that runs through the land

sea large body of water completely or partly surrounded by land

seacoast land lying next to a sea or an ocean

sound broad inland body of water, often between a coastline and one or more islands off the coast

source (of a river) place where a river or stream begins, often in highlands

strait narrow stretch of water joining two larger bodies of water

tributary small river or stream that flows into a large river or stream; a branch of the river

upstream direction opposite the flow of a river; toward the source of a river or stream

valley area of low land usually between hills or mountains

volcano mountain or hill created as liquid rock and ash erupt from inside the Earth

SCAVENGER HUNT

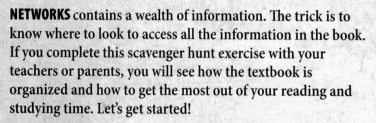

NETWORKS contains a wealth of information. The trick is to know where to look to access all the information in the book. If you complete this scavenger hunt exercise with your teachers or parents, you will see how the textbook is organized and how to get the most out of your reading and studying time. Let's get started!

1 How many lessons are in Chapter 2?

2 What does Unit 1 cover?

3 Where can you find the Essential Questions for each lesson?

4 In what three places can you find information on a Foldable?

5 How can you identify content vocabulary and academic vocabulary in the narrative?

6 Where do you find graphic organizers in your textbook?

7 You want to quickly find a map in the book about the world. Where do you look?

8 Where would you find the latitude and longitude for Dublin, Ireland?

9 If you needed to know the Spanish term for *earthquake*, where would you look?

10 Where can you find a list of all the charts in a unit?

McGraw-Hill
networks

OUR WORLD: THE WESTERN HEMISPHERE

UNIT **1**

Chapter 1
What Is Geography?

Chapter 2
Earth's Physical Geography

Chapter 3
Earth's People

EXPLORE the WORLD

Geography is the study of Earth and all of its variety. When you study geography, you learn about the planet's land, water, plants, and animals. Some people call Earth "the water planet." Do you know why? Water—in the form of streams, rivers, lakes, seas, and oceans—covers nearly 70 percent of Earth's surface.

1 **BODIES OF WATER** Underseas explorers can still experience the thrill of investigating uncharted territory—one of Earth's last frontiers. Almost all of Earth's water consists of a continuous body of water that circles the planet. This body of water makes up five oceans: the Pacific, the Atlantic, the Indian, the Southern, and the Arctic.

2 **LANDFORMS** Landforms are features of the land, such as mountains, valleys, and canyons. Landforms influence where people live and how they relate to their environment.

(bkgd) Reinhard Dirscherl/Waterframe/Getty Images; (t) Shanna Baker/Photographer's Choice/Getty Images

3 **NATURAL RESOURCES** Natural resources are products of Earth that people use to meet their needs. Solar energy is power produced by the heat of the sun. Sun and wind are renewable resources. These resources cannot be used up.

FAST FACT

Earth's longest mountain range is underwater.

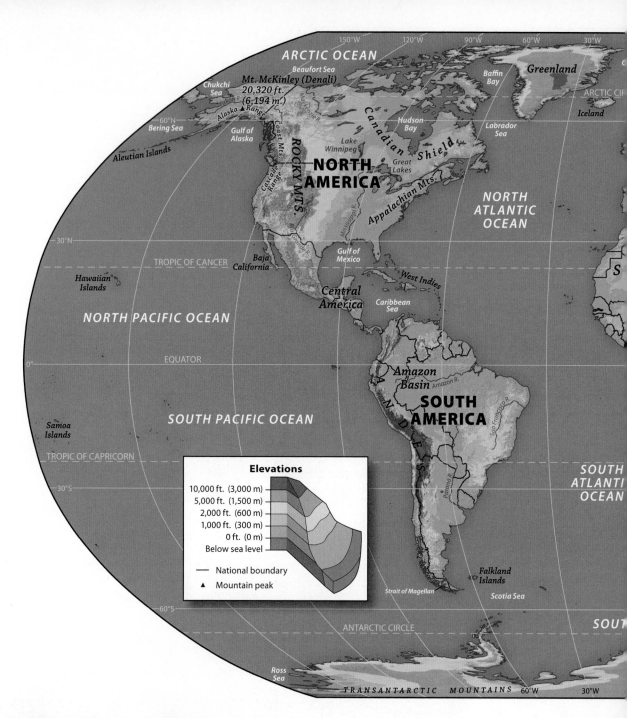

OUR WORLD

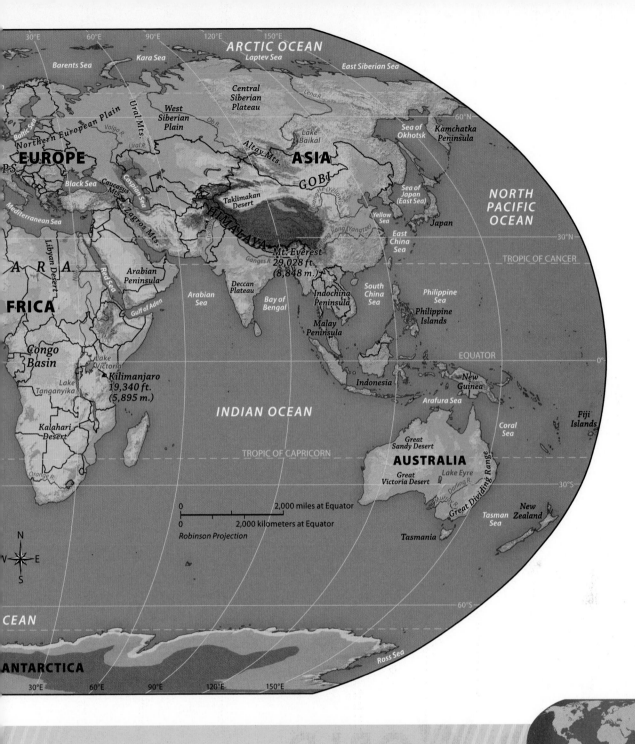

ARCTIC OCEAN

Barents Sea | Kara Sea | Laptev Sea | East Siberian Sea

Central Siberian Plateau

West Siberian Plain

Northern European Plain

Ural Mts.

Volga R.

Ural R.

Lena R.

Ob R.

Lake Baikal

Sea of Okhotsk

Kamchatka Peninsula

EUROPE

Baltic Sea

P.S

ASIA

GOBI

Altay Mts.

Caspian Sea

Black Sea

Caucasus Mts.

Zagros Mts.

Sea of Japan (East Sea)

NORTH PACIFIC OCEAN

Mediterranean Sea

Taklimakan Desert

HIMALAYA

Huang He (Yellow)

Chang Jiang (Yangtze)

Yellow Sea

Japan

A R A

Libyan Desert

Red Sea

Arabian Peninsula

Mt. Everest 29,028 ft. (8,848 m.)

East China Sea

30°N

TROPIC OF CANCER

FRICA

Gulf of Aden

Arabian Sea

Deccan Plateau

Ganges R.

Bay of Bengal

Indochina Peninsula

South China Sea

Philippine Sea

Philippine Islands

Congo Basin

Lake Victoria

Kilimanjaro 19,340 ft. (5,895 m.)

Lake Tanganyika

Malay Peninsula

EQUATOR

0°

Indonesia

New Guinea

INDIAN OCEAN

Arafura Sea

Coral Sea

Fiji Islands

Kalahari Desert

Great Sandy Desert

TROPIC OF CAPRICORN

Orange R.

AUSTRALIA

Great Victoria Desert

Lake Eyre

Murray-Darling R.

Great Dividing Range

New Zealand

Tasman Sea

30°S

0   2,000 miles at Equator

0   2,000 kilometers at Equator

Robinson Projection

Tasmania

N

W   E

S

60°S

CEAN

ANTARCTICA

Ross Sea

30°E   60°E   90°E   120°E   150°E

PHYSICAL

MAP SKILLS

1 **THE GEOGRAPHER'S WORLD** What part of North America has the highest elevation?

2 **THE GEOGRAPHER'S WORLD** Which sea is located east of Central America?

3 **PLACES AND REGIONS** How would you describe much of Europe's landscape?

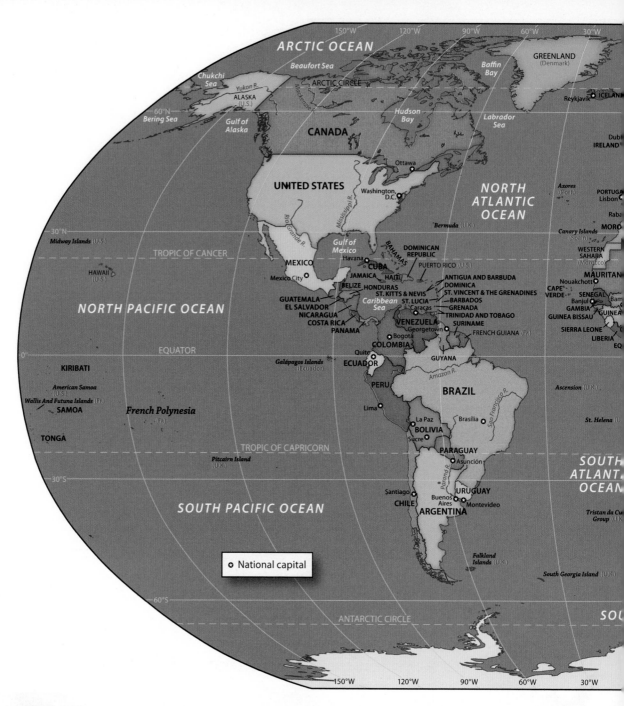

OUR WORLD

Indiana Academic Standards

6.3.1

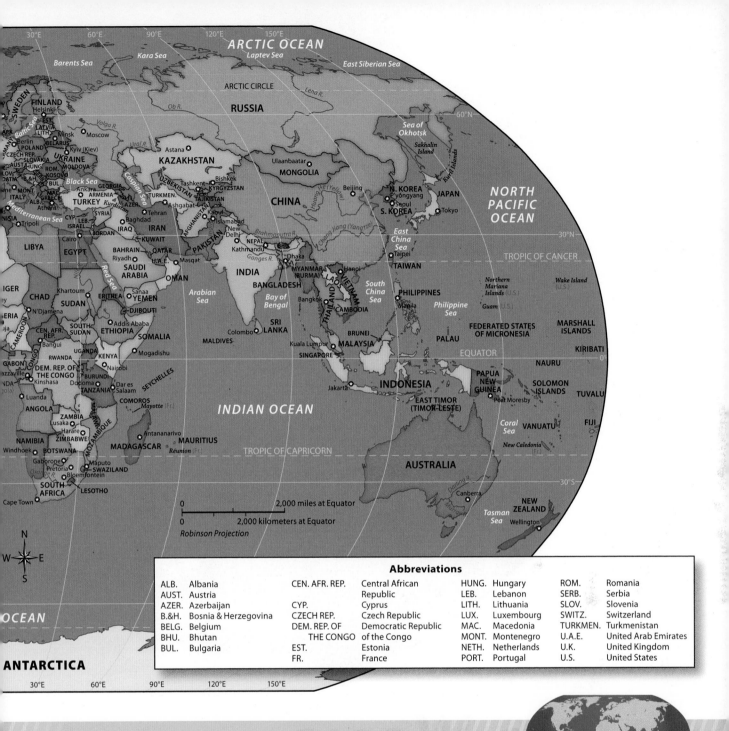

Abbreviations

ALB.	Albania	CEN. AFR. REP.	Central African Republic	HUNG.	Hungary	ROM.	Romania
AUST.	Austria			LEB.	Lebanon	SERB.	Serbia
AZER.	Azerbaijan	CYP.	Cyprus	LITH.	Lithuania	SLOV.	Slovenia
B.&H.	Bosnia & Herzegovina	CZECH REP.	Czech Republic	LUX.	Luxembourg	SWITZ.	Switzerland
BELG.	Belgium	DEM. REP. OF	Democratic Republic	MAC.	Macedonia	TURKMEN.	Turkmenistan
BHU.	Bhutan	THE CONGO	of the Congo	MONT.	Montenegro	U.A.E.	United Arab Emirates
BUL.	Bulgaria	EST.	Estonia	NETH.	Netherlands	U.K.	United Kingdom
		FR.	France	PORT.	Portugal	U.S.	United States

POLITICAL

MAP SKILLS

1 PLACES AND REGIONS Which country in South America has the largest land area?

2 THE GEOGRAPHER'S WORLD Which country is located west of Sweden?

3 PLACES AND REGIONS What is the capital of Canada?

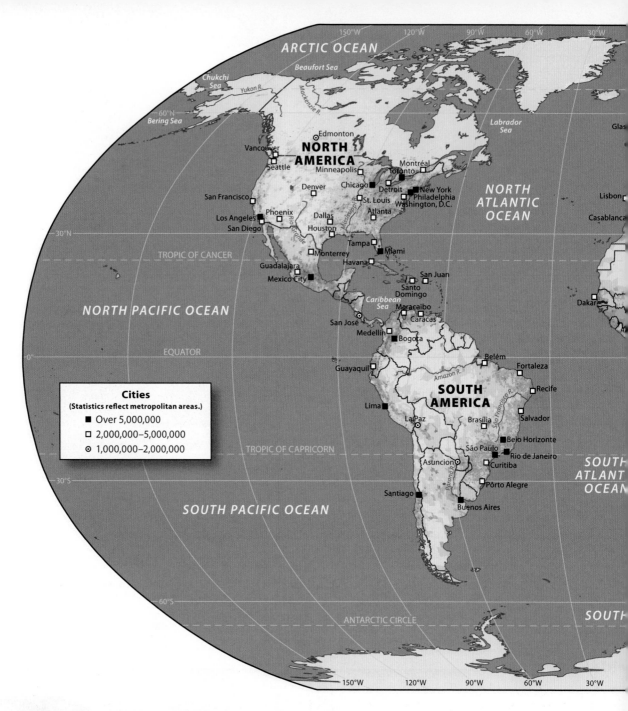

ARCTIC OCEAN

Beaufort Sea

Chukchi
Sea

Yukon R.

Mackenzie R.

Bering Sea

Labrador
Sea

Glas

⊙ Edmonton

NORTH
AMERICA

Vancouver

Seattle

Minneapolis

Montréal
Toronto

Chicago

Detroit

New York

NORTH
ATLANTIC
OCEAN

Lisbon

Denver

St. Louis

Philadelphia
Washington, D.C.

Casablanca

San Francisco

Atlanta

Los Angeles

Phoenix

Dallas

San Diego

Houston

Rio Grande

Mississippi

30°N

Tampa

TROPIC OF CANCER

Monterrey

Miami

Guadalajara

Havana

Mexico City

San Juan

Santo
Domingo

Dakar

NORTH PACIFIC OCEAN

Caribbean
Sea

Maracaibo

Caracas

San José

Medellín

Bogotá

EQUATOR

Guayaquil

Belém

Fortaleza

Amazon R.

SOUTH
AMERICA

Recife

Lima

São Francisco R.

La Paz

Brasília

Salvador

Belo Horizonte

São Paulo

Rio de Janeiro

SOUTH
ATLANT
OCEA

Asunción

Curitiba

Paraná R.

TROPIC OF CAPRICORN

Pôrto Alegre

30°S

Santiago

Buenos Aires

SOUTH PACIFIC OCEAN

60°S

ANTARCTIC CIRCLE

SOUTH

Cities
(Statistics reflect metropolitan areas.)
■ Over 5,000,000
□ 2,000,000–5,000,000
⊙ 1,000,000–2,000,000

150°W   120°W   90°W   60°W   30°W

OUR WORLD

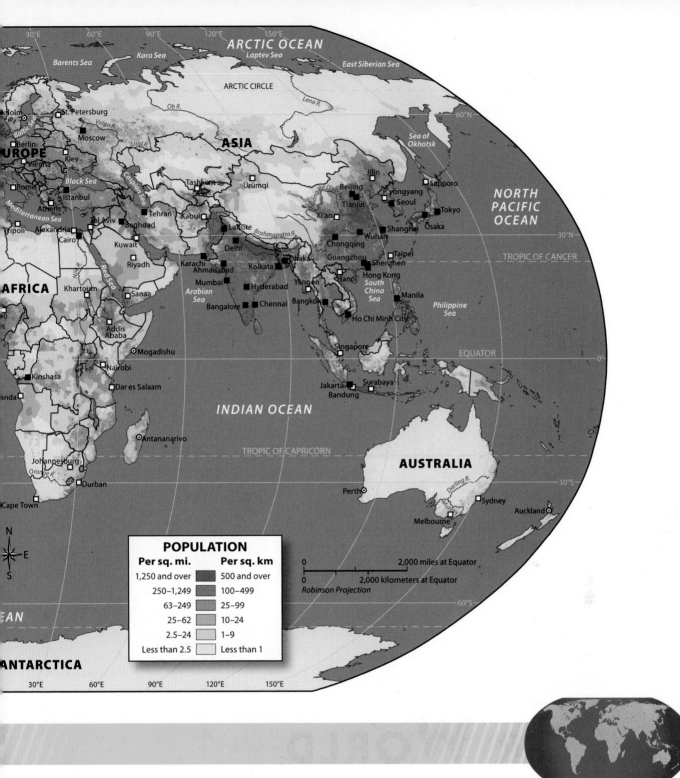

POPULATION

Per sq. mi.		Per sq. km
1,250 and over		500 and over
250–1,249		100–499
63–249		25–99
25–62		10–24
2.5–24		1–9
Less than 2.5		Less than 1

0   2,000 miles at Equator
0   2,000 kilometers at Equator
Robinson Projection

POPULATION DENSITY

MAP SKILLS

1 PLACES AND REGIONS Which parts of Europe are the least densely populated?

2 PLACES AND REGIONS Which part of the United States has the highest population density?

3 HUMAN GEOGRAPHY In general, what population pattern do you see in South America?

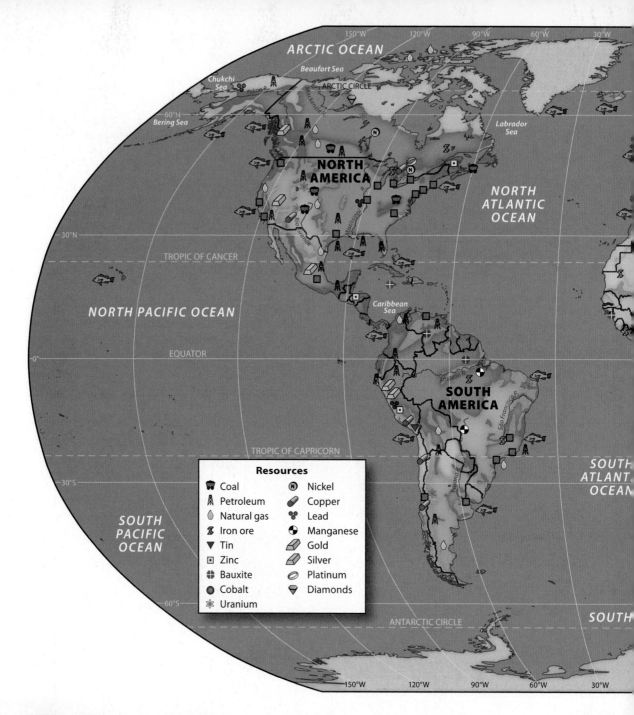

Resources

Coal		Nickel	
Petroleum		Copper	
Natural gas		Lead	
Iron ore		Manganese	
Tin		Gold	
Zinc		Silver	
Bauxite		Platinum	
Cobalt		Diamonds	
Uranium			

OUR WORLD

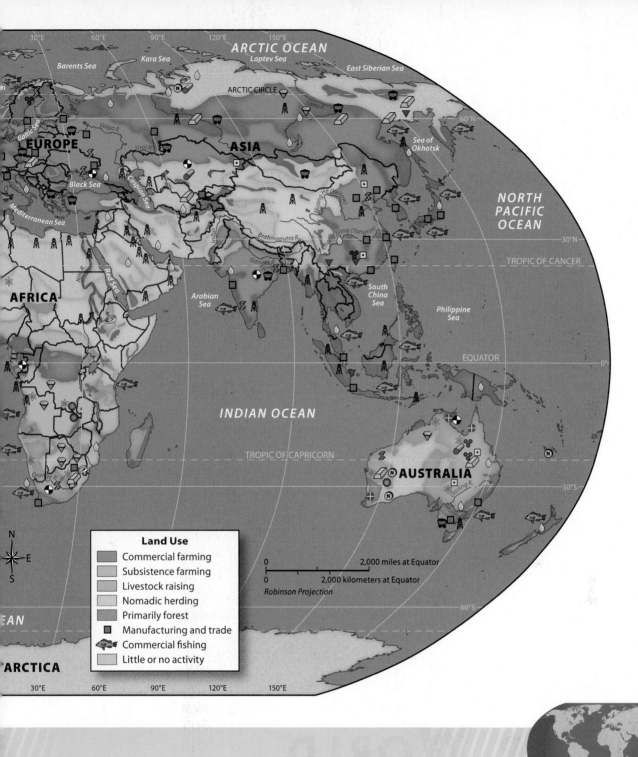

ARCTIC OCEAN

Barents Sea · Kara Sea · Laptev Sea · East Siberian Sea

ARCTIC CIRCLE

60°N

EUROPE

Baltic Sea

Black Sea · Volga R. · Ural R. · Caspian Sea

ASIA

Ob R. · Lena R.

Sea of Okhotsk

NORTH PACIFIC OCEAN

Mediterranean Sea

AFRICA

Nile R. · Red Sea · Arabian Sea · Indus R. · Brahmaputra R. · Huang He (Yellow) · Chang Jiang (Yangtze) · Ganges R.

30°N

TROPIC OF CANCER

South China Sea

Philippine Sea

EQUATOR

0°

INDIAN OCEAN

TROPIC OF CAPRICORN

AUSTRALIA

Darling R.

30°S

Land Use

▨	Commercial farming
▨	Subsistence farming
▨	Livestock raising
▨	Nomadic herding
▨	Primarily forest
■	Manufacturing and trade
🐟	Commercial fishing
▨	Little or no activity

0 ____ 2,000 miles at Equator
0 ____ 2,000 kilometers at Equator
Robinson Projection

60°S

OCEAN

ANTARCTICA

30°E · 60°E · 90°E · 120°E · 150°E

ECONOMIC RESOURCES

MAP SKILLS

1 HUMAN GEOGRAPHY Describe the general use of land in the United States.

2 HUMAN GEOGRAPHY What economic activity is found along most coastal regions?

3 PLACES AND REGIONS Which resources are located along the Amazon River?

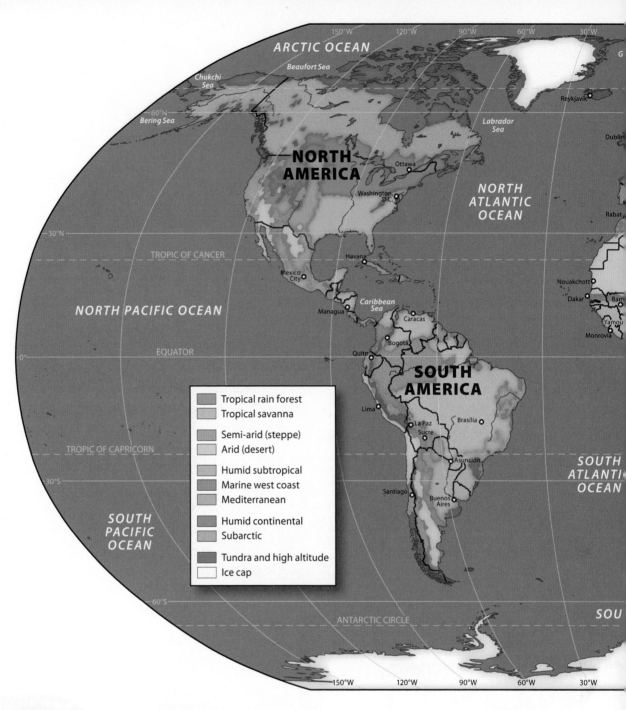

ARCTIC OCEAN

Beaufort Sea

Chukchi Sea

60°N
Bering Sea

Labrador Sea

Reykjavik

Dublin

NORTH AMERICA

Ottawa

NORTH ATLANTIC OCEAN

Washington, D.C.

Rabat

30°N

TROPIC OF CANCER

Havana

NORTH PACIFIC OCEAN

Mexico City

Nouakchott

Managua

Caribbean Sea

Caracas

Dakar

Bam

Yamou

Monrovia

Bogotá

Quito

EQUATOR

SOUTH AMERICA

Lima

Tropical rain forest
Tropical savanna

La Paz
Sucre

Brasília

TROPIC OF CAPRICORN

Semi-arid (steppe)
Arid (desert)

30°S

Humid subtropical
Marine west coast
Mediterranean

Asunción

SOUTH ATLANTIC OCEAN

SOUTH PACIFIC OCEAN

Humid continental
Subarctic

Santiago

Buenos Aires

Tundra and high altitude
Ice cap

60°S

ANTARCTIC CIRCLE

SOU

150°W   120°W   90°W   60°W   30°W

OUR WORLD

Indiana Academic Standards
6.3.1, 6.3.7

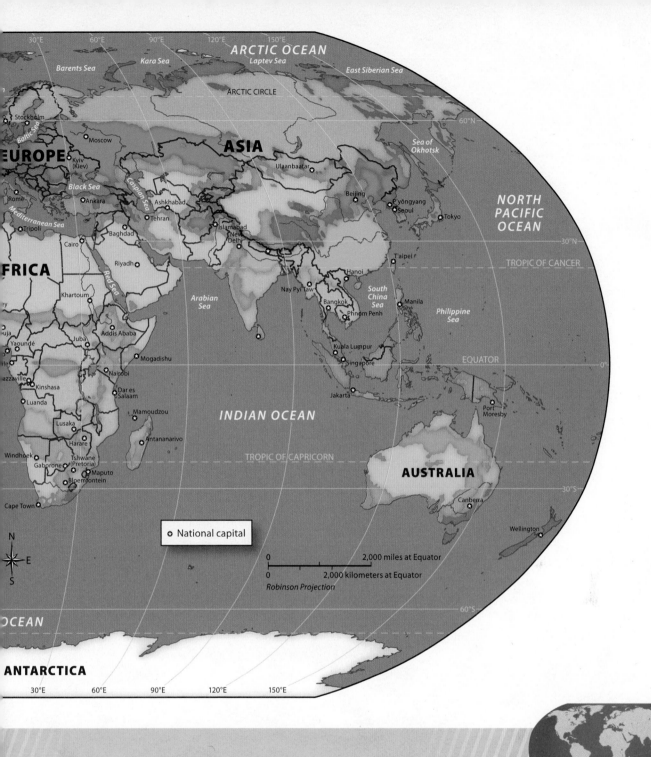

ARCTIC OCEAN

Barents Sea *Kara Sea* *Laptev Sea* *East Siberian Sea*

ARCTIC CIRCLE

Stockholm

Baltic Sea Moscow 60°N

EUROPE Kyiv (Kiev) ASIA *Sea of Okhotsk*

Rome *Black Sea* Ankara Ulaanbaatar *Caspian Sea* Ashkhabad Beijing P'yŏngyang Seoul Tokyo NORTH PACIFIC OCEAN

Mediterranean Sea Tehran Islamabad New Delhi Baghdad

Tripoli 30°N

AFRICA Cairo Riyadh *Red Sea* Taipei TROPIC OF CANCER

Khartoum *Arabian Sea* Nay Pyi Taw Hanoi *South China Sea*

Addis Ababa Juba Bangkok Manila *Philippine Sea* Phnom Penh

Yaoundé Mogadishu Kuala Lumpur EQUATOR 0°

Nairobi Singapore

Kinshasa Dar es Salaam Jakarta

Luanda Mamoudzou INDIAN OCEAN Port Moresby

Lusaka Antananarivo

Harare TROPIC OF CAPRICORN

Windhoek Tshwane (Pretoria) Gaborone Maputo AUSTRALIA

Bloemfontein Canberra 30°S

Cape Town

N E W S Wellington

○ National capital

0   2,000 miles at Equator
0   2,000 kilometers at Equator
Robinson Projection

60°S

OCEAN

ANTARCTICA

30°E   60°E   90°E   120°E   150°E

CLIMATE

MAP SKILLS

1 **PHYSICAL GEOGRAPHY** Which climate zones appear in Western Europe?

2 **PHYSICAL GEOGRAPHY** Which continent receives more rain—North America or South America? Why?

3 **PHYSICAL GEOGRAPHY** In general, how does the climate of Northern Europe differ from the climate of Southern Europe?

WHAT IS GEOGRAPHY?

ESSENTIAL QUESTION · *How does geography influence the way people live?*

Prisma/SuperStock

A geographer drills for
an ice sample.

Lesson 1
Thinking Like a Geographer

Lesson 2
Tools Used by Geographers

The Story Matters...

Since ancient times, people have drawn maps to represent Earth. As people explored, they came into contact with different places and people, which expanded their understanding of the world. Today, our understanding of the world continues to grow as geographers use the latest technology to study Earth's environments. More importantly, by understanding the connections between humans and the environment, geographers can find solutions to significant problems.

FOLDABLES
Study Organizer

Go to the Foldables® library in the back of your book to make a Foldable® that will help you take notes while reading this chapter.

Geographer's View
Geographer's Tools

WHAT IS GEOGRAPHY?

Geography is the study of Earth and all of its variety. When you study geography, you learn about the physical features and the living things—humans, plants, and animals—that inhabit Earth.

Step Into the Place

MAP FOCUS Use the map to answer the following questions.

1 **THE GEOGRAPHER'S WORLD** What are the names of the large landmasses on the maps?

2 **THE GEOGRAPHER'S WORLD** What are the names of the large bodies of water on the map?

3 **PLACES AND REGIONS** What do you think the blue lines are that appear within the landmasses?

4 **CRITICAL THINKING**
Identifying How is this world map similar to other maps you have seen?

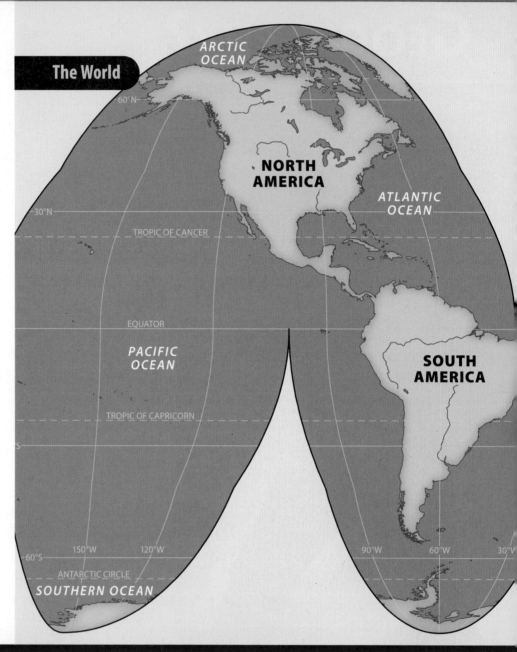

The World

ARCTIC OCEAN

60°N

NORTH AMERICA

ATLANTIC OCEAN

30°N

TROPIC OF CANCER

EQUATOR

PACIFIC OCEAN

SOUTH AMERICA

TROPIC OF CAPRICORN

150°W   120°W   90°W   60°W   30°W

60°S

ANTARCTIC CIRCLE
SOUTHERN OCEAN

Step Into the Time

DRAWING EVIDENCE Choose an event from the time line and write a paragraph describing how it might have changed the way people understood or viewed the world in which they lived.

150 A.D.
Ptolemy creates atlas of known world

Universal Images Group/Getty Images

ARCTIC OCEAN

ARCTIC CIRCLE

60°N

EUROPE

ASIA

30°N

TROPIC OF CANCER

AFRICA

PACIFIC OCEAN

EQUATOR

INDIAN OCEAN

EQUATOR

PRIME MERIDIAN

TROPIC OF CAPRICORN

AUSTRALIA

ATLANTIC OCEAN

N
W E
S

0°   30°E   60°E

90°E   120°E   150°E

60°S

ANTARCTIC CIRCLE

SOUTHERN OCEAN

ANTARCTICA

0   2,000 miles
0   2,000 kilometers
Goode's Interrupted Equal-Area projection

1953
Edmund Hillary and
Tenzing Norgay reach
the top of Mt. Everest

1969
Neil Armstrong
walks on the moon

1803 U.S. purchases Louisiana Territory

1000 A.D.

2000 A.D.

1519 Magellan sets sail on
voyage around the world

1909 Robert Peary reaches
the North Pole

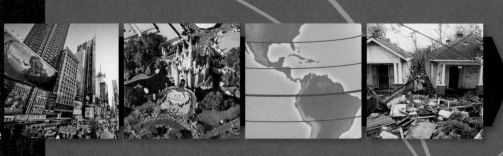

netw⊙rks

There's More Online!

☑ **CHART/GRAPH** Six Essential Elements of Geography

☑ **IMAGES** Places Change Over Time

☑ **ANIMATION** Regions of Earth

☑ **VIDEO**

Reading **HELP**DESK

Academic Vocabulary

- **dynamic**
- **component**

Content Vocabulary

- **geography**
- **spatial**
- **landscape**
- **relative location**
- **absolute location**
- **latitude**
- **Equator**
- **longitude**
- **Prime Meridian**
- **region**
- **environment**
- **landform**
- **climate**
- **resource**

TAKING NOTES: *Key Ideas and Details*

Identifying As you read, list the five themes of geography on a graphic organizer like the one below.

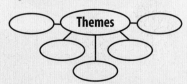

Themes

Indiana Academic Standards
6.1.18, 6.1.19, 6.3.2, 6.3.3, 6.3.11

Lesson 1
Thinking Like a Geographer

ESSENTIAL QUESTION • *How does geography influence the way people live?*

IT MATTERS BECAUSE
Thinking like a geographer helps you understand how the world works and to appreciate the world's remarkable beauty and complexity.

The World in Spatial Terms

GUIDING QUESTION *What does it mean to think like a geographer?*

We get our understanding of the world from a combination of many different sources. Biology is the study of how living things survive and relate to one another. History is the study of events that occur over time and how those events are connected. **Geography** is the study of Earth and its people, places, and environments. Geographers look at people and the world in which they live mainly in terms of space and place. They study such topics as where people live on the surface of Earth, why they live there, and how they interact with each other and the physical environment.

Viewing Earth Spatially

Geography, then, emphasizes the spatial aspects of the world. **Spatial** refers to Earth's features in terms of their locations, their shapes, and their relationships to one another.

Physical features such as mountains and lakes can be located on a map. These features can be measured in terms of height, width, and depth. Distances and directions to other features can be determined and measured. The human world also has spatial dimensions. Geographers study the size and shape of cities, states, and countries. They consider how close or far apart these human features are to one another. Geographers also think about the relationships between human features and physical features.

But thinking spatially is more than just the study of the location or size of things. It means looking at the characteristics of Earth's features. Geographers ask what mountains in different locations are made of. They examine what kinds of fish live in different lakes. The layout of cities and how easy or difficult it is for people to move around in them is also important to geographers.

Characteristics of Place

Locations on Earth are made up of different combinations of physical and human characteristics. Physical features such as climate, landforms, and vegetation combine with human features such as population, economic activity, and land use. These combinations create what geographers call places.

Places are locations on Earth that have distinctive characteristics that make them meaningful and special to people. The places where we live, work, and go to school are important to us. Our home is an important place. Even small places such as our bedroom or a classroom often have a unique and special meaning. In the same way, larger locations, such as our hometown, our country, or even Earth, are places that have meaning to people.

One way that geographers learn about places is by studying landscapes. **Landscapes** are portions of Earth's surface that can be viewed at one time and from one location. They can be as small as the view from the front porch of your home, or they can be as large as the view from a tall building that includes the city and surrounding countryside.

The geography theme of *place* describes all of the characteristics that give an area its own special quality.

▶ **CRITICAL THINKING**

Describing What are the characteristics that make a place like Times Square in New York City special?

José Fuste Raga/age fotostock

Whether we visit a landscape or look at photographs of one, the landscape can tell us much about the people who live there. Geographers look at landscapes and try to explain their unique combinations of physical and human features. As you study geography, notice the great variety in the world's landscapes.

The Perspective of Experience

Geography is not something you learn about only in school or just from books. Geography is something you experience every day.

We all live in the world. We feel the change of the seasons. We hear the sounds of birds chirping and of car horns honking. We walk on sidewalks and in forests. We ride in cars along streets and highways. We shop in malls and grocery stores. We fly in airplanes to distant places. We surf the Internet or watch TV and learn about people and events in our neighborhood, our country, and the world.

This is all geography. By learning about geography in school, we can better appreciate and understand this world in which we live.

Academic Vocabulary

dynamic always changing

A Changing World

Earth is **dynamic**, or always changing. Rivers shift course. Volcanoes erupt suddenly, forming mountains or collapsing the peaks of mountains. The pounding surf removes sand from beaches.

The things that people make change, too. Farmers shift from growing one crop to another. Cities grow larger or sometimes shrink. The borders of nations change.

Geographers, then, study how places change over time. They try to understand what impact those changes have. What factors made a city grow? What effect did a growing city have on the people who live there? What effect did the city's growth have on nearby communities and on the land and water near it? Answering questions like these is part of the field of geography.

☑ **READING PROGRESS CHECK**

Describing How is geography related to history?

Geography's Five Themes

GUIDING QUESTION *How can you make sense of a subject as large as Earth and its people?*

To organize information about the world, geographers use five themes. These themes help them view and understand Earth.

Location

Location is where something is found on Earth. There are two types of location. **Relative location** describes where a place is compared to another place. This approach often uses the cardinal directions— north, south, east, and west. A school might be on the east side of town. Relative location can also tell us about the characteristics of a

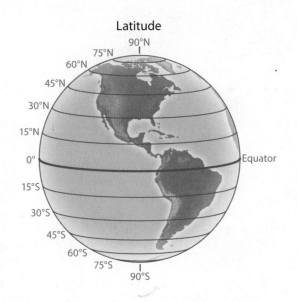

Latitude

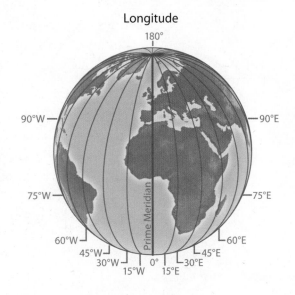

Longitude

place. For example, knowing that New Orleans is near the mouth of the Mississippi River helps us understand why the city became an important trading port.

Absolute location is the exact location of something. An address like 123 Main Street is an absolute location. Geographers identify the absolute location of places using a system of imaginary lines called latitude and longitude. Those lines form a grid for locating a place precisely.

Lines of **latitude** run east to west, but they measure distance on Earth in a north-to-south direction. One of these lines, the **Equator**, circles the middle of Earth. This line is equally distant from the North Pole and the South Pole. Other lines of latitude between the Equator and the North and South Poles are assigned a number from 1° to 90°. The higher the number, the farther the line is from the Equator. The Equator is 0° latitude. The North Pole is at 90° north latitude (90° N), and the South Pole is at 90° south latitude (90° S).

Lines of **longitude** run from north to south, but they measure distance on Earth in an east-to-west direction. They go from the North Pole to the South Pole. These lines are also called *meridians*. The **Prime Meridian** is the starting point for measuring longitude. It runs through Greenwich, England, and has the value of 0° longitude. There are 180 lines of longitude to the east of the Prime Meridian and 180 lines to the west. They meet at the meridian 180°, which is the International Date Line.

Geographers use latitude and longitude to locate anything on Earth. In stating absolute location, geographers always list latitude first. For example, the absolute location of Washington, D.C., is 38°N, 77°W.

Lines of latitude circle Earth parallel to the Equator and measure the distance north or south of the Equator in degrees. Lines of longitude circle Earth from the North Pole to the South Pole. These lines measure distances east or west of the Prime Meridian.

▶ **CRITICAL THINKING**
The Geographer's World At what degree latitude is the Equator located?

— Indiana —
CONNECTION

Locating Capital Cities

You can use latitude and longitude to locate various places discussed in this text. Using an atlas map, find the approximate absolute location of the following capital cities:
Algiers, Algeria
Bangkok, Thailand
Dakar, Senegal
Makkah, Saudi Arabia
New Delhi, India
Tokyo, Japan
Wellington, New Zealand

Disney World, located in Orlando, Florida, attracts millions of visitors every year.

▶ **CRITICAL THINKING**

Human Geography What effect do you think Disney World has on the surrounding communities?

Place

Another theme of geography is place. The features that help define a place can be physical or human.

Why is Denver called the "Mile High City"? Its location one mile above sea level gives it a special character. Why does New Orleans have the nickname "the Crescent City"? It is built on a crescent-shaped bend along the Mississippi River. That location has had a major impact on the city's growth and how its people live.

Region

Each place is unique, but two or more places can share characteristics. Places that are close to one another and share some characteristics belong to the same **region**. For example, Los Angeles and San Diego are located in southern California. They have some features in common, such as nearness to the ocean. Both cities also have mostly warm temperatures throughout the year.

In the case of those two cities, the region is defined using physical characteristics. Regions also can be defined by human characteristics. For instance, the countries of South America are part of the same region. One reason is that most of the people living in these countries follow the same religion, Roman Catholicism.

Geographers study region so they can identify the broad patterns of larger areas. They can compare and contrast the features in one region with those in another. They also examine the special features that make each place in a region distinct from the others.

Human-Environment Interaction

People and the environment interact. That is, they affect each other. The physical characteristics of a place affect how people live. Flat, rich, well-watered soil is good for farming. Mountains full of coal can be mined. The environment also can present all kinds of hazards, such as floods, droughts, earthquakes, and volcanic eruptions.

People affect the environment, too. They blast tunnels through mountains to build roadways and drain swamps to make farmland. Although these actions can improve life for some people, they can harm the environment. Exhaust from cars on the roadways can pollute the air, and turning swamps into farms destroys natural ecosystems and reduces biological diversity.

The **environment** is the natural surroundings of a place. It includes several key features. One is **landforms**, or the shape and nature of the land. Hills, mountains, and valleys are types of landforms. The environment also includes the presence or absence of a body of water. Cities located on coastlines, like New York City, have different characteristics than inland cities, like Dallas.

Weather and climate play a role in how people interact with their environment. The average weather in a place over a long period of time is called its **climate**. Alaska's climate is marked by long, cold, wet winters and short, mild summers. Hawaii's climate is warm year-round. Alaskans interact with their environment differently in December than Hawaiians do.

Another **component**, or part, of the environment is **resources**. These are materials that can be used to produce crops or other products. Forests are a resource because the trees can be used to build homes and furniture. Oil is a resource because it can be used to make energy.

Movement

Geographers also look at how people, products, ideas, and information move from one place to another. People have many reasons for moving. Some move because they find a better job.

Academic Vocabulary

component a part

In 2005 Hurricane Katrina devastated the Gulf Coast and the city of New Orleans (left). Years later, many houses remain abandoned (right).
▶ **CRITICAL THINKING**
Identifying What hazards does the environment present?

THE SIX ESSENTIAL ELEMENTS

Element	Definition
The World in Spatial Terms	Geography studies the location and spatial relationships among people, places, and environments. Maps reveal the complex spatial interactions.
Places and Regions	The identities of individuals and peoples are rooted in places and regions. Distinctive combinations of human and physical characteristics define places and regions.
Physical Systems	Physical processes, like wind and ocean currents, plate tectonics, and the water cycle, shape Earth's surface and change ecosystems.
Human Systems	Human systems are things like language, religion, and ways of life. They also include how groups of people govern themselves and how they make and trade products and ideas.
Environment and Society	Geography studies how the environment of a place helps shape people's lives. Geography also looks at how people affect the environment in positive and negative ways.
The Uses of Geography	Understanding geography and knowing how to use its tools and technologies helps people make good decisions about the world and prepares people for rewarding careers.

Being aware of the six essential elements will help you sort out what you are learning about geography.
▶ **CRITICAL THINKING**
Identifying The study of volcanoes, ocean currents, and climate is part of which essential element?

Sometimes, people are forced to move because of war, famine, or religious or racial prejudice. Movement by large numbers of people can have important effects. People may face shortages of housing and other services. If new arrivals to an area cannot find jobs, poverty levels can rise.

In our interconnected world, a vast number of products move from place to place. Apples from Washington State move to supermarkets in Texas. Clothes produced in Thailand end up in American shopping malls. Oil from Saudi Arabia powers cars and trucks across the United States. All of this movement relies on transportation systems that use ships, railroads, airplanes, and trucks.

Ideas can move at an even faster pace than people and products can. Communications systems, such as telephone, television, radio, and the Internet, carry ideas and information all around Earth. Remote villagers on the island of Borneo watch American television shows and learn about life in the United States. Political protestors in Greece and Spain use social networking sites to coordinate their activities. The geography of movement affects everyone.

The Six Essential Elements

The five themes are one way of thinking about geography. Geographers also divide the study of geography into six essential elements. Elements are the topics that make up a subject. Calling them *essential* means they are necessary to understanding geography.

☑ **READING PROGRESS CHECK**

Determining Central Ideas How is the theme of location related to the theme of place?

Geographer's Skills

GUIDING QUESTION *How will studying geography help you develop skills for everyday life?*

Have you ever used the GPS feature on a phone or in a car when you were lost? What about using a Web browser to find a route from your home to another place? If you have used these tools, your search took you to a Web site that provides maps. If you followed one of those maps to your destination, you were using a geography skill.

Interpreting Visuals

Maps are one tool geographers use to picture the world. They use other visual images, as well. These other visuals include graphs, charts, diagrams, and photographs.

Graphs are visual displays of numerical information. They can help you compare information. Charts display information in columns and rows. Diagrams are drawings that use pictures to represent an idea. A diagram might show the steps in a process or the parts that make up something.

Critical Thinking

Geographers ask analytical questions. For example, they might want to know why earthquakes are more likely in some places than in others. That question looks at causes. They might ask, How does climate affect the ways people live? Such questions examine effects.

Geographers might ask how the characteristics of a place have changed over time. That is a question of analysis. Or they could ask why people in different nations use their resources differently. That question calls on them to compare and contrast.

Learning how to ask—and answer—questions like these will help sharpen your mind. In addition to understanding geography better, you will also be able to use these skills in other subjects.

☑ **READING PROGRESS CHECK**

Determining Central Ideas How do geographers use visuals?

Indiana CONNECTION

Tools of Discovery

Scientists in other fields assist geographers in understanding societies in the present and the past. For example, archaeologists hunt for evidence in the ground where settlements might once have been. They dig up and study artifacts—weapons, tools, and other things made by humans. They also look for fossils—traces of plants or animals that have been preserved in rock. Anthropologists focus on human society. They study how humans developed and how they related to one another.

FOLDABLES Study Organizer — Geographer's View / Geographer's Tools

Include this lesson's information in your Foldable®.

LESSON 1 REVIEW

Reviewing Vocabulary

1. How do *resources* affect the types of homes people typically build in a region?

Answering the Guiding Questions

2. *Identifying* What is a place within a place that you have been in today?

3. *Determining Central Ideas* What information would you use to give your exact location on Earth?

4. *Determining Word Meanings* What is the difference between environment and landforms?

5. *Identifying* Name an innovation that has greatly affected people's movement.

6. *Informative/Explanatory Writing* Describe how you have experienced geography today.

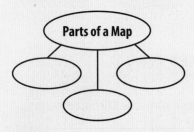

networks

There's More Online!

- ☑ **SLIDE SHOW** History of Mapmaking
- ☑ **ANIMATION** Elements of a Globe
- ☑ **VIDEO**

Reading HELPDESK

Academic Vocabulary

- sphere
- convert
- distort

Content Vocabulary

- hemisphere
- key
- scale bar
- compass rose
- map projection
- scale
- elevation
- relief
- thematic map
- technology
- remote sensing

TAKING NOTES: *Key Ideas and Details*

Describing As you read the lesson, identify three parts of a map on a graphic organizer. Then, explain what each part shows.

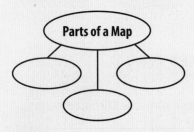

Indiana Academic Standards
6.1.17, 6.1.20, 6.3.2

Lesson 2
Tools Used by Geographers

ESSENTIAL QUESTION • *How does geography influence the way people live?*

IT MATTERS BECAUSE
The tools of geography help you understand the world.

Globes and Maps

GUIDING QUESTION *What is the difference between globes and maps?*

If you close your eyes and imagine your neighborhood, you are using a mental map. You are forming a picture of the buildings and other places and where each is in relation to the others.

Making and using maps is a big part of geography. Of course, geographers make maps that have many parts. Their maps are more detailed than your mental map. Still, their maps are essentially the same as your mental map. Both are a way to picture the world and show where things are located.

Globes

The most accurate way to show places on Earth is with a globe. Globes are the most accurate because globes, like Earth, are **spheres**; that is, they are shaped like a ball. As a result, globes represent the correct shapes of land and bodies of water. They show distances and directions between places more correctly than flat images of Earth can.

The Equator and the Prime Meridian each divide Earth in half. Each half of Earth is called a **hemisphere**. The Equator divides Earth into sections called the Northern and Southern Hemispheres. The Prime Meridian, together with the International Date Line, splits Earth into the Eastern and Western Hemispheres.

(l to r) Lana Sundman/Alamy; Antenna Audio - Inc./Getty Images; Kathy Collins/Getty Images; spacephotos.com/age fotostock; Chris Wallace/Alamy

Maps

Maps are not round like globes. Instead, maps are flat representations of the round Earth. They might be sketched on a piece of paper, printed in a book, or displayed on a computer screen. Wherever they appear, maps are always flat.

Maps **convert**, or change, a round space into a flat space. As a result, maps **distort** physical reality, or show it incorrectly. This is why maps are not as accurate as globes are, especially maps that show large areas or the whole world.

Despite this distortion problem, maps have several advantages over globes. Globes have to show the whole planet. Maps, though, can show only a part of it, such as one country, one city, or one mountain range. As a result, they can provide more detail than globes can. Think how large a globe would have to be to show the streets of a city. You could certainly never carry such a globe around with you. Maps make more sense if you want to study a small area. They can focus on just that area, and they are easy to store and carry.

Maps tend to show more kinds of information than globes. Globes generally show major physical and political features, such as landmasses, bodies of water, the countries of the world, and the largest cities. They cannot show much else without becoming too difficult to read or too large. However, some maps show these same features. But maps also can be specialized. One map might illustrate a large mountain range. Another might display the results of an election. Yet another could show the locations of all the schools in a city.

☑ **READING PROGRESS CHECK**

Analyzing What is a disadvantage of using maps?

Academic Vocabulary

sphere a round shape like a ball

convert to change from one thing to another

distort to present in a manner that is misleading

A set of imaginary lines divides Earth into hemispheres.
▶ **CRITICAL THINKING**
The Geographer's World
What line divides Earth into Eastern and Western Hemispheres?

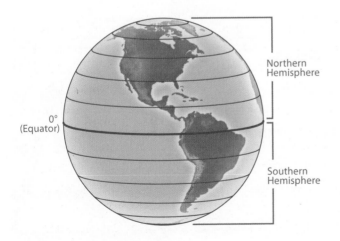

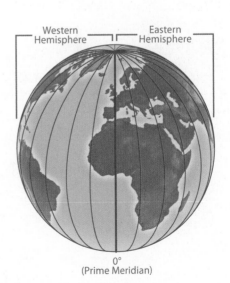

A Closer Look at Maps

GUIDING QUESTION *How do maps work?*

Maps are located in many different places. You can see them in a subway station. Subway maps indicate the routes each train takes. In a textbook, a map might show new areas that were added to the United States at different times. On a company's Web site, a map can locate all its stores in a city. The map of a state park would tell visitors what activities they can enjoy in each area of the park. Each of these maps is different from the others, but they have some traits in common.

Elements of a Map

Maps have several important elements, or features. These features are the tools the map uses to convey information.

The map title tells what area the map will cover. It also identifies what kind of information the map presents about that area. The **key** unlocks the meaning of the map by explaining the symbols, colors, and lines. The **scale bar** is an important part of the map. It tells how a measured space on the map corresponds to actual distances on Earth. For example, by using the scale bar, you can determine how many miles in the real world each inch on the map represents. The **compass rose** shows direction. This map feature points out north, south, east, and west. Some maps include insets that show more detail for smaller areas, such as cities on a state map. Many maps show latitude and longitude lines to help you locate places.

Map Projections

To convert the round Earth to a flat map, geographers use **map projections**. A map projection distorts some aspects of Earth in order to represent other aspects as accurately as possible on a flat

Many different kinds of maps are available because maps are useful for showing a wide range of information.

▶ CRITICAL THINKING
Describing What is the difference between a large-scale and a small-scale map?

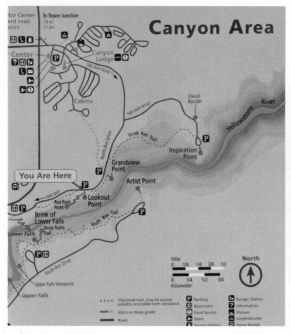

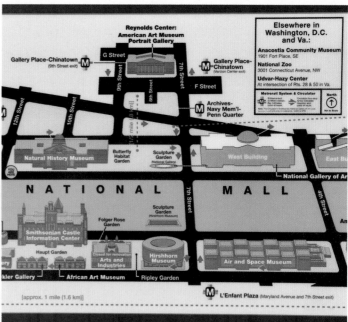

(l to r) Chris Wallace/Alamy; Lana Sundman/Alamy

map. Some projections show the correct size of areas in relation to one another. Other map projections emphasize making the shapes of areas as correct as possible.

Some projections break apart the world's oceans. By doing so, these maps show land areas more accurately. They clearly do not show the oceans accurately, though.

Mapmakers, known as cartographers, choose which projection to use based on the purpose of the map. Each projection distorts some parts of the globe more or less than other parts. Finally, mapmakers think about what part of Earth they are drawing and how large an area they want to cover.

Map Scale

Scale is another important feature of maps. As you learned, the scale bar connects distances on the map to actual distances on Earth. The scale bar is based on the scale at which the map is drawn. **Scale** is the relationship between distances on the map and on Earth.

Maps are either *large scale* or *small scale*. A large-scale map focuses on a smaller area. An inch on the map might correspond to 10 miles (16 km) on the ground. A small-scale map shows a relatively larger area. An inch on a small-scale map might be the same as 1,000 miles (1,609 km).

Each type of scale has benefits and drawbacks. Which scale to use depends on the map's purpose. Do you want to map your school and the streets and buildings near it? Then you need a large-scale map to show this small area in great detail. Do you want to show the entire United States? In that case, you need a small-scale map that shows the larger area but with less detail.

Two Types of Maps

The two types of maps are general purpose and thematic. The type depends on what kind of information is drawn on the map. General-purpose maps show a wide range of information about an area. They generally show either the human-made features of an area or its natural features, but not both.

Political maps are one common type of general-purpose map that shows human-made features. They show the boundaries of countries or divisions within them, like the states of the United States. They also show the locations and names of cities.

Physical maps display natural features such as mountains and valleys, rivers, and lakes. They picture the location, size, and shape of these features. Many physical maps show **elevation**, or how much above or below sea level a feature is. Maps often use colors to present this information. A key on the map explains what height above or below sea level each color represents.

Physical maps usually show **relief**, or the difference between the elevation of one feature and the elevation of another feature near it.

Thinking Like a Geographer

Relief

Relief is the height of a landform compared to other nearby landforms. If a mountain 10,000 feet (3,048 m) high rises above a flat area at sea level, the relief of the mountain equals its elevation: 10,000 feet. If the 10,000-foot-high mountain is in a highland region that is 4,000 feet (1,219 m) above sea level, its relief is *less than* its elevation— only 6,000 feet (1,829 m). The difference in height between it and the land around it is much less than its absolute height. *What would be the relief of a mountain 7,500 feet (2,286 m) high compared to its highest foothill, at 3,000 feet (914 m) high?*

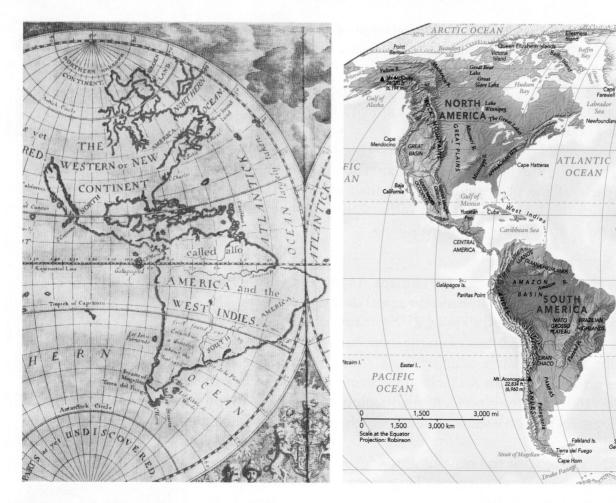

Cartography is the science of making maps. As knowledge of Earth grew, maps became increasingly accurate.

▶ **CRITICAL THINKING**
The Geographer's World
Describe two ways in which the historical map differs from the present-day map.

Elevation is an absolute number, but relief is relative. It depends on other landforms that are nearby. The width of the colors on a physical map usually shows the relief. Colors that are narrow show steep places, and colors that are wide show gently sloping land.

Thematic maps show more specialized information. A thematic map might indicate the kinds of plants that grow in different areas. That kind of map is a vegetation map. Another could show where farming, ranching, or mining takes place. That kind of map is called a land-use map. Road maps show people how to travel from one place to another by car. Just about any physical or human feature can be displayed on a thematic map.

☑ **READING PROGRESS CHECK**

Describing What are the two main types of maps?

Mapping Technology

GUIDING QUESTION *How do geographers use geospatial technologies?*

Electronic maps, such as those on cell phones and mapping devices in cars, are an example of geospatial technologies. **Technology** is any way that scientific discoveries are applied to practical use.

Geospatial technologies help us think spatially and provide practical information about the locations of physical and human features.

Global Positioning System

GPS devices work with a network called the Global Positioning System (GPS). This network was built by the U.S. government. Parts of it can be used only by the U.S. armed forces. Parts of it, though, can be used by ordinary people all over the world. The GPS has three elements.

The first element of this network is a set of more than 30 satellites that orbit Earth constantly. The U.S. government launched the satellites into space and maintains them. The satellites send out radio signals. Almost any spot on Earth can be reached by signals from at least four satellites at all times.

The second part of the network is the control system. Workers around the world track the satellites to make sure they are working properly and are on course. The workers reset the clocks on the satellites when needed.

The third part of the GPS system consists of GPS devices on Earth. These devices receive the signals sent by the satellites. By combining the signals from different satellites, a device calculates its location on Earth in terms of latitude and longitude. The more satellite signals the device receives at any time, the more accurately it can determine its location. Because satellites have accurate clocks, the GPS device also displays the correct time.

GPS is used in many ways. It is used to track the exact location and course of airplanes. That information helps ensure the safety of air travel. Farmers use it to help them work their fields. Businesses use it to guide truck drivers. Cell phone companies use GPS to provide services. GPS in cars helps guide us to our destinations.

Geographic Information Systems

Another important geospatial technology is known as a geographic information system (GIS). These systems consist of computer hardware and software that gather, store, and analyze geographic information. The information is then shown on a computer screen. Sometimes it is displayed as maps. Sometimes the information is shown in other ways. Companies and governments around the world use this tool.

A GIS is a powerful tool because it links data about all kinds of physical and human features with the locations of those features. Because computers can store and process so much data, the GIS can be accurate and detailed.

People select what features they want to study using the GIS. Then they can combine different features on the same map and analyze the patterns.

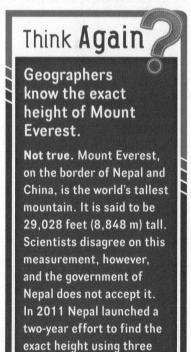

Think Again

Geographers know the exact height of Mount Everest.

Not true. Mount Everest, on the border of Nepal and China, is the world's tallest mountain. It is said to be 29,028 feet (8,848 m) tall. Scientists disagree on this measurement, however, and the government of Nepal does not accept it. In 2011 Nepal launched a two-year effort to find the exact height using three GPS devices.

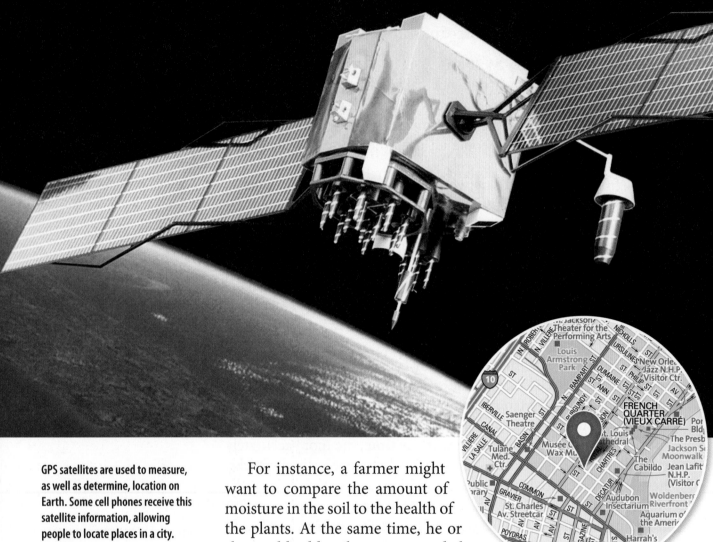

(t) spacephotos.com/age fotostock; (inset) Antenna Audio - Inc./Getty Images

GPS satellites are used to measure, as well as determine, location on Earth. Some cell phones receive this satellite information, allowing people to locate places in a city.

▶ **CRITICAL THINKING**

Analyzing Why is it important for geographers to know exactly where places are located on Earth?

For instance, a farmer might want to compare the amount of moisture in the soil to the health of the plants. At the same time, he or she could add soil types around the farm to the comparison. The farmer could then use the results of the analysis to answer all kinds of questions. What plants should I plant in different locations? How much irrigation water should I use? How can I drive the tractor most efficiently?

Satellites and Sensors

Since the 1970s, satellites have gathered data about Earth's surface. They do so using remote sensing. **Remote sensing** simply means getting information from far away. Most early satellite sensors were used to gather information about the weather. Weather satellites helped save lives during disasters by providing warnings about approaching storms. Before satellites, tropical storms were often missed because they could not be tracked over open water.

Satellites gather information in different ways. They may use powerful cameras to take pictures of the land. They can also pick up other kinds of information, such as the amount of moisture in the

soil, the amount of heat the soil holds, or the types of vegetation that are present. In the early 2000s, scientists used GIS technology to help conserve the plants and animals that lived in the Amazon rain forest. Using the technology, scientists can compare data gathered from the ground to data taken from satellite pictures. Land use planners use this information to help local people make good decisions about how to use the land. These activities help prevent the rain forest from being destroyed.

Some satellites gather information regularly on every spot in the world. That way, scientists can compare the information from one year to another. They look for changes in the shape of the land or in its makeup, spot problems, and take steps to fix them.

Limits of Technology

Geospatial technologies allow access to a wealth of information about the features and objects in the world and where those features and objects are located. This information can be helpful for identifying and navigating. By itself, however, the information does not answer questions about why features are located where they are. These questions lie at the heart of understanding our world. The answers are crucial for making decisions about this world in which we live. But it is important to go beyond the information provided by geospatial technologies. We must build understanding of people, places, and environments and the connections among them.

☑ **READING PROGRESS CHECK**

Determining Central Ideas How could remote sensing be used as part of a GIS?

Include this lesson's information in your Foldable®.

LESSON 2 REVIEW

Reviewing Vocabulary
1. Which *hemispheres* describe your location?

Answering the Guiding Questions
2. *Identifying* To plan a cross-country trip, would a large-scale or a small-scale map be more helpful? Explain.

3. *Analyzing* If half of the 30 satellites that orbit Earth were unusable, how might GPS be affected?

4. *Integrating Visual Information* Create a map of your school building and grounds. Include the key features of a map.

5. *Identifying* What are the main differences between a map and a globe?

6. *Informative/Explanatory Writing* Can geospatial technology be used to explain all aspects of geography? Explain.

What Do You Think?

Are Street-Mapping Technologies an Invasion of People's Privacy?

Suppose you are curious about a place you have never visited. Instead of going in person, you might be able to get a 360-degree view from your computer. Services like Google Street View and Bing Streetside display panoramic images of public roadways and buildings. The photos are taken by cameras attached to roving vehicles. They capture whatever is happening at the time, which means they sometimes capture random bystanders, too. Some people argue that street-level mapping programs violate the right to privacy. They point out that individuals' pictures can appear on the mapping Web sites without their knowledge or consent. Do tools like Street View intrude too much on people's privacy?

TEXT: Andrew Lavoie, "THE ONLINE ZOOM LENS: WHY INTERNET STREET-LEVEL MAPPING TECHNOLOGIES DEMAND RECONSIDERATION OF THE MODERN-DAY TORT NOTION OF 'PUBLIC PRIVACY.'" Georgia Law

Yes!

PRIMARY SOURCE

" Privacy encompasses the right to control information disseminated [spread] about oneself. ... Personal behavior disclosures that occur as a result of Internet street-level mapping technologies almost certainly violate this personal right to choose which face to display to the world ... A person may not mind that their friends and family know of their participation in certain socially stigmatizing [disapproved of] activities; an entirely new issue arises, however, should the entire public suddenly discover that the person is [doing something questionable.] ... Internet street-level mapping scenes depart from being simply a record of what a member of the public could have seen on the street [because] on the Internet, images can be—and often are—saved onto users' hard drives for later dissemination. Thus, compromising [reputation-damaging] images, even if removed by Google after the fact, can be released to the public in an ever-widening wake [path]. "

—Andrew Lavoie, Georgia attorney

Users can view high-resolution imagery from Google Earth or Street View on their screens.

A camera mounted on a car provides the technology for Google Street View.

No !

PRIMARY SOURCE

" At Google we take privacy very seriously. Street View only features imagery taken on public property and is not in real time. This imagery is no different from what any person can readily capture or see walking down the street. Imagery of this kind is available in a wide variety of formats for cities all around the world. While the Street View feature enables people to easily find, discover, and plan activities relevant to a location, we respect the fact that people may not want imagery they feel is objectionable featured on the service. We provide easily accessible tools for flagging inappropriate or sensitive imagery . . . [U]sers can report objectionable images. Objectionable imagery includes nudity, certain types of locations (for example, domestic violence shelters) and clearly identifiable individuals . . . We routinely review takedown requests and act quickly to remove objectionable imagery. "

—Stephen Chau, product manager for Google Maps

What Do You Think? DBQ

1. *Identifying* What types of images does Google consider inappropriate?

2. *Analyzing* What points does Andrew Lavoie make to argue that Street View invades people's privacy?

Critical Thinking

3. *Identifying Point of View* Describe a situation when Street View could be useful and one when it could embarrass or endanger someone. Do you think the benefits of street-mapping tools outweigh the privacy concerns?

Directions: Write your answers on a separate piece of paper.

❶ Use your FOLDABLES to explore the Essential Question.

INFORMATIVE/EXPLANATORY WRITING Research the lifestyles of early Native American people and how their activities were related to the geography of their land. Create a poster explaining what their homes looked like, what foods they ate, and what activities they engaged in. Include images on your poster.

❷ 21st Century Skills

ANALYZING Select one major geographic change or event that has occurred in the last century, such as the eruption of Mount Saint Helens or Hurricane Katrina. Write an essay that presents facts about the changes that occurred and their effects on the human population.

❸ Thinking Like a Geographer

IDENTIFYING Create a graphic organizer using the five themes of geography. Fill in information about your city for each of the five themes.

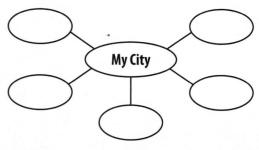

❹ GEOGRAPHY ACTIVITY

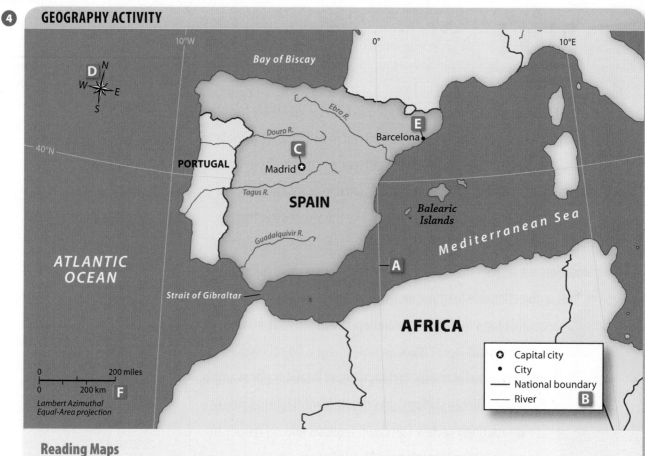

Reading Maps

Match the letters on the map with the numbered items listed below.

1. map key   **2.** compass rose   **3.** longitude line   **4.** scale bar   **5.** city   **6.** capital

REVIEW THE GUIDING QUESTIONS

Directions: Choose the best answer for each question.

1 Children think spatially when they use

    A. snap-together blocks.

    B. crayons.

    C. dress-up clothes.

    D. books.

2 When the appearance of a tree changes from season to season, it is showing that nature is

    F. static.

    G. spatial.

    H. dynamic.

    I. steady.

3 What is an address such as 414 West Third Street giving?

    A. longitude

    B. latitude

    C. absolute location

    D. relative location

4 On a map, which element tells direction?

    F. compass rose

    G. scale bar

    H. relief

    I. the dotted lines

5 A relief map shows differences in

    A. direction.

    B. elevation.

    C. theme.

    D. scale.

6 The Global Positioning System was originally created by

    F. a small committee of mapmaking companies.

    G. the Russians.

    H. a for-profit private business.

    I. the U.S. government.

DBQ ANALYZING DOCUMENTS

7 CITING TEXT EVIDENCE In the excerpt below, two geographers discuss the content of their geography book.

"*In this book we... investigate the world's great geographic realms [areas]. We will find that each of these realms possesses a special combination of cultural... and environmental properties [characteristics].*"

—from H.J. de Blij and Peter O. Muller, *Geography*

A geographic realm

A. includes only physical land characteristics.

B. involves the interaction of cultural and physical properties.

C. divides land/water into two categories.

D. makes geographic changes unlikely.

8 DETERMINING WORD MEANINGS What is one example of a geographic realm?

F. growing oranges in Florida

G. a river

H. going to school for 12 years

I. South America

SHORT RESPONSE

"*Hurricanes, wildfires, floods, earthquakes, and other natural events affect the Nation's economy,... property, and lives.... The USGS gathers and disseminates [gives out] real-time hazard data to relief workers, conducts long-term monitoring and forecasting to help minimize the impacts of future events, and evaluates conditions in the aftermath of disasters.*"

—from United States Geological Service, *The National Map—Hazards and Disasters*

9 ANALYZING After a natural disaster, some businesses may flourish, while others are destroyed. Explain.

10 DESCRIBING Select one type of natural event, and imagine that your community has just suffered from such an occurrence. What steps need to be taken right away, and who should be in charge?

EXTENDED RESPONSE

Write your answer on a separate piece of paper.

11 INFORMATIVE/EXPLANATORY WRITING Select a product such as an apple or toothpaste. Trace the path of this product from its start as raw materials through production and shipping until it appears on a store shelf. Explore the concept of buying locally, and explain what advantages it offers.

Need Extra Help?

If You've Missed Question	1	2	3	4	5	6	7	8	9	10	11
Review Lesson	1	1	1	2	2	2	1	1	1	1	1

GEOGRAPHY: REALMS, REGIONS, AND CONCEPTS, Twelfth Edition, by H. J. de Blij and Peter O. Muller. Published by John Wiley & Sons, Inc.; From "THE NATIONAL MAP—HAZARDS AND DISASTERS." National Geospatial Program Office, Fact Sheet 2009-3010. U.S. Geological Survey, Department of the Interior/USGS. The USGS home page is http://www.usgs.gov.

EARTH'S PHYSICAL GEOGRAPHY

ESSENTIAL QUESTION · *How does geography influence the way people live?*

Lesson 1
Earth and the Sun

Lesson 2
Forces of Change

Lesson 3
Landforms and Bodies of Water

The Story Matters...

Earth is part of a larger physical system called the solar system. Earth's position in the solar system makes life on our planet possible. The planet Earth has air, land, and water, which make it suitable for plant, animal, and human life. Major natural forces inside and outside of our planet shape its surface. Some of these forces can move suddenly and violently, such as earthquakes and volcanos, causing disasters that can dramatically affect life on Earth.

Carsten Peter/National Geographic/Getty Images

A geologist prepares to enter the crater of Ambrim Island volcano.

FOLDABLES
Study Organizer

Go to the Foldables® library in the back of your book to make a Foldable® that will help you take notes while reading this chapter.

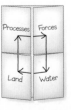

PHYSICAL GEOGRAPHY

Continents sit on large bases called plates. As these plates move on top of Earth's fluid mantle, the continents move. Sometimes, the plates collide with each other or slide under each other, creating earthquakes or volcanoes.

Step Into the Place

MAP FOCUS Use the map to answer the following questions.

1 **PHYSICAL GEOGRAPHY** Where are most of the world's volcanoes located?

2 **THE GEOGRAPHER'S WORLD** How many plates are underneath Australia?

3 **CRITICAL THINKING** **Integrating Visual Information** Why do you think the edge of the Pacific Ocean is often called the Ring of Fire?

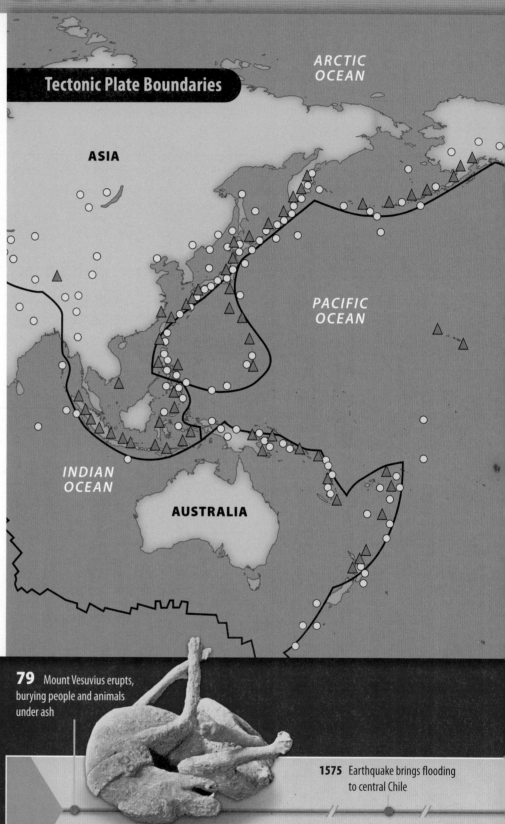

Tectonic Plate Boundaries

ARCTIC OCEAN

ASIA

PACIFIC OCEAN

INDIAN OCEAN

AUSTRALIA

Step Into the Time

DRAWING EVIDENCE Choose one event from the time line and explain how the natural forces that shape the physical geography of a particular place can have a worldwide impact.

79 Mount Vesuvius erupts, burying people and animals under ash

1575 Earthquake brings flooding to central Chile

©Sean Sexton Collection/Corbis

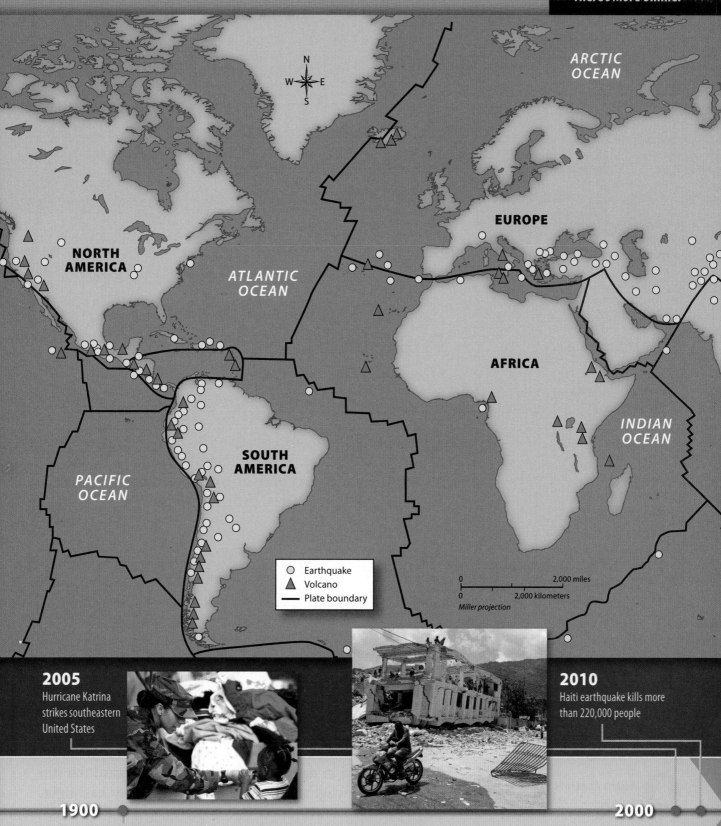

networks
There's More Online!

ARCTIC OCEAN

NORTH AMERICA

EUROPE

ATLANTIC OCEAN

AFRICA

INDIAN OCEAN

SOUTH AMERICA

PACIFIC OCEAN

○ Earthquake
△ Volcano
— Plate boundary

0 ___ 2,000 miles
0 ___ 2,000 kilometers
Miller projection

2005
Hurricane Katrina strikes southeastern United States

2010
Haiti earthquake kills more than 220,000 people

1900

1906 Earthquake, fire devastate San Francisco

2000

There's More Online!

☑ **CHART/GRAPH** Climate Zones

☑ **ANIMATION** Earth's Rotation

☑ **IMAGES** Rain Shadow

☑ **SLIDE SHOW** Effects of Climate Change

☑ **VIDEO**

Reading **HELP**DESK

Academic Vocabulary

• **accurate**

Content Vocabulary

• **orbit**
• **axis**
• **revolution**
• **atmosphere**
• **solstice**
• **equinox**
• **climate**
• **precipitation**
• **rain shadow**

TAKING NOTES: *Key Ideas and Details*

Summarize As you read, complete a graphic organizer about Earth's physical system.

Element	Description
Hydrosphere	
Lithosphere	
Atmosphere	
Biosphere	

Indiana Academic Standards
6.3.3, 6.3.6, 6.3.7, 6.3.8

Lesson 1
Earth and the Sun

ESSENTIAL QUESTION • *How does geography influence the way people live?*

IT MATTERS BECAUSE
The processes that change Earth can have sudden and lasting impacts on humans.

Planet Earth

GUIDING QUESTION *What is the structure of Earth?*

The sun is the center of the solar system. Earth is a member of the solar system—planets and the other bodies that revolve around our sun. The sun is just one of hundreds of millions of stars in our galaxy. Because the sun is so large, its gravity causes the planets to constantly **orbit**, or move around, it. Each planet follows its own path around the sun.

Earth and the Sun

Life on Earth could not exist without heat and light from the sun. Earth's orbit holds it close enough to the sun—about 93 million miles (150 million km)—to receive a constant supply of light and heat energy. The sun, in fact, is the source of all energy on Earth. Every plant and animal on the planet needs the sun's energy to survive. Without the sun, Earth would be a cold, dark, lifeless rock floating in space.

As Earth orbits the sun, it rotates, or spins, on its axis. The **axis** is an imaginary line that runs through Earth's center from the North Pole to the South Pole. Earth completes one rotation every 24 hours. As Earth rotates, different areas are in sunlight and in darkness. The part facing toward the sun experiences daylight, while the part facing away has night. Earth makes one **revolution**, or complete trip around the sun, in 365¼ days. This is what we define as one year. Every four years, the extra fourths of a day are combined and added to the calendar as February 29th. A year that contains one of these extra days is called a leap year.

Inside Earth

Thousands of miles beneath your feet, Earth's heat has turned metal into liquid. You do not feel these forces, but what lies inside affects what lies on top. Mountains, deserts, and other landscapes were formed over time by forces acting below Earth's surface—and those forces are still changing the landscape.

If you cut an onion in half, you will see that it is made up of many layers. Earth is also made up of layers. An onion's layers are all made of onion, but Earth's layers are made up of many different materials.

Layers of Earth

The inside of Earth is made up of three layers: the core, the mantle, and the crust. The center of Earth—the core—is divided into a solid inner core and an outer core of melted, liquid metal. Surrounding the outer core is a thick layer of hot, dense rock called the mantle. Scientists calculate that the mantle is about 1,800 miles (2,897 km) thick. The mantle also has two parts. When volcanoes erupt, the glowing-hot lava that flows from the mouth of the volcano is magma from Earth's outer mantle. The inner mantle is solid, like the inner core. Magma is melted rock. The outer layer is the crust, a rocky shell forming the surface of Earth. The crust is thin, ranging from about 2 miles (3.2 km) thick under oceans to about 75 miles (121 km) thick under mountains.

DIAGRAM SKILLS >

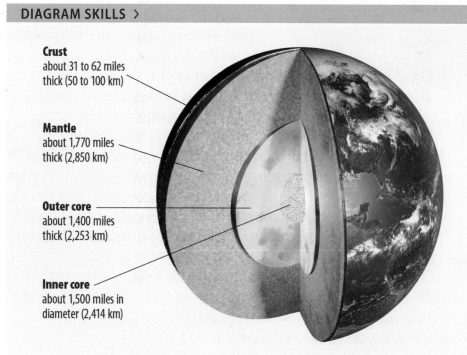

Crust
about 31 to 62 miles thick (50 to 100 km)

Mantle
about 1,770 miles thick (2,850 km)

Outer core
about 1,400 miles thick (2,253 km)

Inner core
about 1,500 miles in diameter (2,414 km)

EARTH'S LAYERS
Earth is comprised of several layers.

Identifying About how many miles thick is the mantle?

The deepest hole ever drilled into Earth was about 8 miles (13 km) deep. That is still within Earth's crust. The farthest any human has traveled down into Earth's crust is about 2.5 miles (4 km). Still, scientists have developed an **accurate** picture of the layers in Earth's structure. One way that scientists do this is to study vibrations that take place deep within Earth. The vibrations are caused by earthquakes and explosions underground. From their observations, scientists have learned what materials are inside Earth and estimated the thickness and temperature of Earth's layers.

Academic Vocabulary

accurate free from error

Earth's Physical Systems

Powerful processes operate below Earth's surface. Processes are also at work in the physical systems on the surface of Earth. Earth's physical systems consist of four major subsystems: the hydrosphere, the lithosphere, the atmosphere, and the biosphere.

About 71 percent of the earth's surface is water. The hydrosphere is the subsystem that consists of Earth's water. Water is found in the oceans, seas, lakes, ponds, rivers, ice, and groundwater. Only 3 percent of the Earth's water is freshwater.

About 29 percent of Earth's surface is land. Land makes up the part of Earth called the lithosphere. This is Earth's crust. Landforms are the shapes that occur on Earth's surface. They include mountains, hills, plateaus, plains, and ocean basins, the land beneath the ocean.

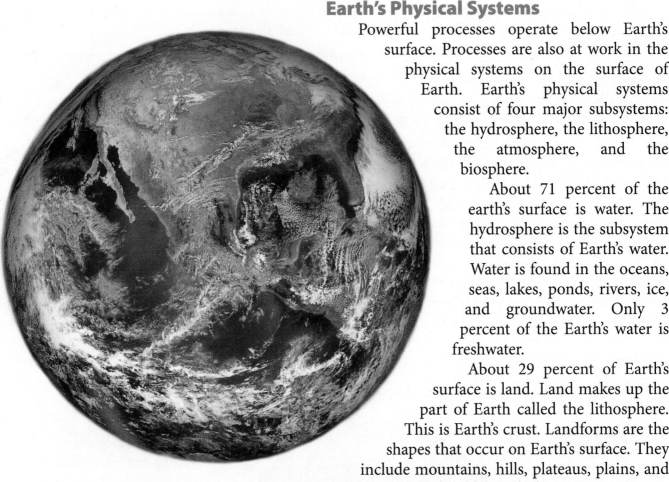

Earth is sometimes called the "water planet" because about 71 percent of it is covered with water. Almost 97 percent of this water, however, is salt water.

▶ CRITICAL THINKING

Analyzing What causes the cycle from day to night on Earth?

The air we breathe is part of the **atmosphere**, the thin layer of gases that envelop Earth. The atmosphere is made up of about 78 percent nitrogen, 21 percent oxygen, and small amounts of other gases. The atmosphere is thickest at Earth's surface and gets thinner higher up. Ninety-eight percent of the atmosphere is found within 16 miles (26 km) of Earth's surface. Outer space begins at 100 miles (161 km) above Earth, where the atmosphere ends.

The biosphere is made up of all that is living on the surface of Earth, close to the surface, and in the atmosphere. All people, animals, and plants live in the biosphere.

☑ READING PROGRESS CHECK

Identifying Which of Earth's four major subsystems consists of water?

Sun and Seasons

GUIDING QUESTION *How does Earth's orbit around the sun cause the seasons?*

Fruits such as strawberries, grapes, and bananas cannot grow in cold, icy weather. Yet grocery stores across America sell these ripe, colorful fruits all year, even in the middle of winter. Where in the world is it warm enough to grow fruit in January? To find the answer, we start with the tilt of Earth.

Earth is tilted 23.5 degrees on its axis. If you look at a globe that is attached to a stand, you will see what the tilt looks like. Because of the tilt, not all places on Earth receive the same amount of direct sunlight at the same time.

As Earth orbits the sun, it stays in its tilted position. This means that half of the planet is always tilted toward the sun, while the other half is tilted away. As a result, Earth's Northern and Southern Hemispheres experience seasons at different times.

On about June 21, the North Pole is tilted toward the sun. The Northern Hemisphere is receiving the direct rays of the sun. The sun appears directly overhead at the line of latitude called the Tropic of Cancer. This day is the summer **solstice**, or beginning of summer, in the Northern Hemisphere. It is the day of the year that has the most hours of sunlight during Earth's 24-hour rotation.

Six months later—about December 22—the North Pole is tilted away from the sun. The sun's direct rays strike the line of latitude known as the Tropic of Capricorn. This is the winter solstice—when winter occurs in the Northern Hemisphere and summer begins in the Southern Hemisphere. The days are short in the Northern Hemisphere but long in the Southern Hemisphere.

Midway between the two solstices, about September 23 and March 21, the rays of the sun are directly overhead at the Equator.

DIAGRAM SKILLS >

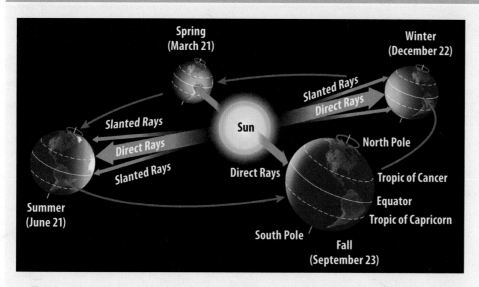

Spring (March 21)
Winter (December 22)
Slanted Rays
Direct Rays
Slanted Rays
Sun
Direct Rays
North Pole
Slanted Rays
Direct Rays
Tropic of Cancer
Equator
Tropic of Capricorn
Summer (June 21)
South Pole
Fall (September 23)

SEASONS
The tilt of Earth as it revolves around the sun causes the seasons to change.

▶ **CRITICAL THINKING**
Integrating Visual Information
When does the Northern Hemisphere receive direct rays from the sun?

These are **equinoxes**, when day and night in both hemispheres are of equal length—12 hours of daylight and 12 hours of nighttime everywhere on Earth.

✓ **READING PROGRESS CHECK**

Identifying When it is winter in the Southern Hemisphere, what season is it in the Northern Hemisphere?

Climate

GUIDING QUESTION *How do elevation, wind and ocean currents, weather, and landforms influence climate?*

Low latitudes near the Equator receive the direct rays of the sun year-round. This area, known as the Tropics, lies mainly between the Tropic of Cancer and the Tropic of Capricorn. The Tropics circle the globe like a belt. If you lived in the Tropics, you would experience hot, sunny weather most of the year because of the direct sunlight. Outside the Tropics, the sun is never directly overhead. Even when these high-latitude areas are tilted toward the sun, the sun's rays hit Earth indirectly at a slant. This means that no sunlight at all shines on the high-latitude regions around the North and South Poles for as much as six months each year. Thus, climate in these regions is always cool or cold.

A farmer harvests pineapples in Costa Rica.
▶ **CRITICAL THINKING**
Analyzing How is climate affected by elevation?

John Coletti/Photographer's Choice/Getty Images

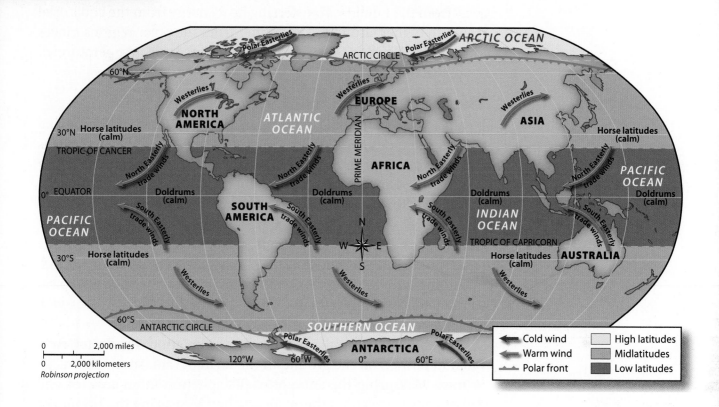

Legend:
- ← Cold wind
- ← Warm wind
- ⟋⟍ Polar front
- High latitudes
- Midlatitudes
- Low latitudes

0 — 2,000 miles
0 — 2,000 kilometers
Robinson projection

Elevation and Climate

At all latitudes, elevation influences climate. This is because Earth's atmosphere thins as altitude increases. Thinner air retains less heat. As elevation increases, temperatures decrease by about 3.5°F (1.9°C) for every 1,000 feet (305 m). For example, if the temperature averages 70°F (21.1°C) at sea level, the average temperature at 5,000 feet (1,524 m) is only 53°F (11.7°C). A high elevation will be colder than lower elevations at the same latitude.

Wind and Ocean Currents

In addition to latitude and elevation, the movement of air and water helps create Earth's climates. Moving air and water help circulate the sun's heat around the globe.

Movements of air are called winds. Winds are the result of changes in air pressure caused by uneven heating of Earth's surface. Winds follow prevailing, or typical, patterns. Warmer, low-pressure air rises higher into the atmosphere. Winds are created as air is drawn across the surface of Earth toward the low-pressure areas. The Equator is constantly warmed by the sun, so warm air masses tend to form near the Equator. This warm, low-pressure air rises, and then cooler, high-pressure air rushes in under the warm air, causing wind. This helps balance Earth's temperature.

MAP SKILLS

1. **PHYSICAL GEOGRAPHY** In what general direction does the wind blow over North America?

2. **PHYSICAL GEOGRAPHY** What two areas in the Pacific have calm winds?

Just as winds move in patterns, cold and warm streams of water, known as currents, circulate through the oceans. Warm water moves away from the Equator, transferring heat energy from the equatorial region to higher latitudes. Cold water from the polar regions moves toward the Equator, also helping to balance the temperature of the planet.

Weather and Climate

Weather is the state of the atmosphere at a given time, such as during a week, a day, or an afternoon. Weather refers to conditions such as hot or cold, wet or dry, calm or stormy, or cloudy or clear. Weather is what you can observe any time by going outside or looking out a window. **Climate** is the average weather conditions in a region or an area over a longer period. One useful measure for comparing climates is the average daily temperature. This is the average of the highest and lowest temperatures that occur in a 24-hour period. In addition to the average temperature, climate includes typical wind conditions and rainfall or snowfall that occur in an area year after year.

Rainfall and snowfall are types of precipitation. **Precipitation** is water deposited on the ground in the form of rain, snow, hail, sleet, or mist. Measuring the amount of precipitation in an area for one day provides data about the area's weather. Measuring the amount of precipitation for one full year provides data about the area's climate.

Landforms

It might seem strange to think that landforms such as mountains can affect weather and climate, but landforms and landmasses change the strength, speed, and direction of wind and ocean

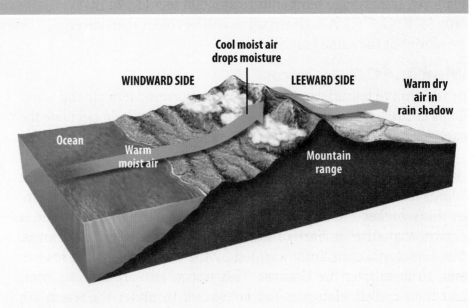

DIAGRAM SKILLS >

RAIN SHADOW
A rain shadow affects the amount of rain a region receives.

▶ **CRITICAL THINKING**

1. *Determining Word Meanings*
   What is precipitation?

2. *Describing* What happens to the air as it passes over the peaks and moves to the other side of the mountain?

Cool moist air drops moisture

WINDWARD SIDE

LEEWARD SIDE

Warm dry air in rain shadow

Ocean

Warm moist air

Mountain range

currents. Wind and ocean currents carry heat and precipitation, which shape weather and climate. The sun warms the land and the surface of the world's oceans at different rates, causing differences in air pressure. As winds blow inland from the oceans, they carry moist air with them. As the land rises in elevation, the atmosphere cools. When masses of moist air approach mountains, the air rises and cools, causing rain to fall on the side of the mountain facing the ocean. The other side of the mountain receives little rain because of the rain shadow effect. A **rain shadow** is a region of reduced rainfall on one side of a high mountain; the rain shadow occurs on the side of the mountain facing away from the ocean.

The climate in a zone affects how people live and work.

Identifying What three things are used to define a climate zone?

☑ **READING PROGRESS CHECK**

Determining Word Meanings Do the terms *weather* and *climate* mean the same thing? Explain.

Climate Regions

GUIDING QUESTION *What are the characteristics of Earth's climate zones?*

Why do Florida and California attract so many visitors? These places have cold or stormy weather at times, but their climates are generally warm, sunny, and mild, so people can enjoy the outdoors all year.

The Zones

In the year 1900, German scientist Wladimir Köppen invented a system that divides Earth into five basic climate zones. Climate zones are regions of Earth classified by temperature, precipitation, and distance from the Equator. Köppen used names and capital letters to label the climate zones as follows: Tropical (A); Desert (B); Humid Temperate (C); Cold Temperate (D); and Polar (E). Years later, a sixth climate zone was added: High Mountain (F).

Each climate zone also can be divided into smaller subzones, but the areas within each zone have many similarities. Tropical areas are hot and rainy, and they often are covered with dense forests. Desert areas are always dry, but they can be cold or hot, depending on their latitude. Humid temperate areas experience all types of weather with changing seasons. Cold temperate climates have a short summer season but are generally cold and windy. Polar climates are very cold, with ice and snow covering the ground most of the year. High mountain climates are found only at the tops of high mountain ranges such as the Rockies, the Alps, and the Himalaya. High mountain climates have variable conditions because the atmosphere cools with increasing elevation. Some of the highest mountaintops are cold and windy and stay white with snow all year.

Different types of plants grow best in different climates, so each climate zone has its own unique types of vegetation and animal life. These unique combinations form ecosystems of plants and animals that are adapted to environments within the climate zone. A biome is a type of large ecosystem with similar life-forms and climates. Earth's biomes include rain forest, desert, grassland, and tundra. All life is adapted to survive in its native climate zone and biome.

Animals that live in that environment have unique adaptations that help them survive.

Ryan Hagerty/USFWS

Changes to Climate

Many scientists say climates are changing around the world. If this is true, the world could experience new weather patterns. These changes might mean more extreme weather in some places and milder weather in others. Human activities can affect weather and climate. For example, people have cut down millions of square miles of rain forests in Central and South America. As a result, fewer trees are available to release moisture into the air. The result is a drier climate in the region.

Metal, asphalt, and concrete surfaces in cities absorb a huge amount of heat from the sun. Because of this, an enormous mass of warmer air builds up in and around the city, which affects local weather.

In recent years, scientists have become aware of a problem called global warming. Global warming is an increase in the average temperature of Earth's atmosphere. Industries created by humans dump polluting chemicals into the atmosphere. Many scientists say that a buildup of this pollution is contributing to the increasing temperature of Earth's atmosphere.

If the temperature of the atmosphere continues to rise, all of Earth's climates could be affected. Changes in climate could alter many natural ecosystems. Another consequence is that the survival of some plant and animal species will be threatened. It is also likely to be expensive and difficult for humans to adapt to these changes.

☑ **READING PROGRESS CHECK**

Identifying Which climate zone experiences all types of weather with changing seasons?

cold   cool   warm   hot

A thermal image shows heat escaping from the roofs of buildings.

▶ **CRITICAL THINKING**

Analyzing How do structures in cities affect local weather?

Study Organizer

Include this lesson's information in your Foldable®.

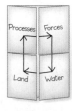

LESSON 1 REVIEW

Reviewing Vocabulary

1. Describe the difference between weather and *climate*.

Answering the Guiding Questions

2. ***Describing*** Explain the role of Earth's axis as it relates to rotation.

3. ***Describing*** Describe what happens before, during, and after the equinoxes.

4. ***Analyzing*** If you were climbing a 10,000-foot (3,048 m) mountain, what would you wear and why?

5. ***Identifying*** What has happened in Central and South America that many scientists believe is impacting the world?

6. ***Narrative Writing*** Imagine you are climbing a very tall mountain such as Mont Blanc in the Alps during the summer. Describe how the vegetation would change as you climb higher.

networks

There's More Online!

☑ **ANIMATION** Magnitude of Earthquakes

☑ **IMAGES** Human Impact on Earth

☑ **MAP** Risk of Earthquakes in the United States

☑ **VIDEO**

Lesson 2
Forces of Change

ESSENTIAL QUESTION · *How does geography influence the way people live?*

Reading **HELP**DESK

Academic Vocabulary

• **intense**

Content Vocabulary

• **continent**
• **tectonic plate**
• **fault**
• **earthquake**
• **Ring of Fire**
• **tsunami**
• **weathering**
• **erosion**
• **glacier**

TAKING NOTES: *Key Ideas and Details*

Identify As you read, use a graphic organizer like this one to describe the external forces that have shaped Earth.

External Forces
_____
_____
_____
_____

Indiana Academic Standards
6.3.13

IT MATTERS BECAUSE
Internal and external forces change Earth, the setting for human life.

Shaping Earth

GUIDING QUESTION *How did the surface of Earth form?*

Earth's surface has been moving continually since the formation of the planet. Landmasses have shifted and moved over time. Landforms have been created and destroyed. The way Earth looks from space has changed many times because of the movement of continents.

Earth's Surface

A **continent** is a large, continuous mass of land. Continents are part of Earth's crust. Earth has seven continents: Asia, Africa, North America, South America, Europe, Antarctica, and Australia. The region around the North Pole is not a continent because it is made of a huge mass of dense ice, not land. Greenland might seem as big as a continent, but it is classified as the world's largest island. Each of the seven continents has features that make it unique. Some of the most interesting features on the continents are landforms.

Even though you usually cannot feel it, the land beneath you is moving. This is because Earth's crust is not a solid sheet of rock. Earth's surface is like many massive puzzle pieces pushed close together and floating on a sea of boiling rock. The movement of these pieces is one of the major forces that create Earth's land features. Old mountains are worn down, while new mountains grow taller. Even the continents move.

(l to r) ©Jeff Vanuga/Corbis; MARCELLO PATERNOSTRO/AFP/Getty Images; MARTIN BERNETTI/AFP/Getty Images; Stockbyte/Getty Images; Science & Society Picture Library/SSPL/Getty Images

Plate Movements

Earth's rigid crust is made up of 16 enormous pieces called **tectonic plates**. These plates vary in size and shape. They also vary in the amount they move over the more flexible layer of the mantle below them. Heat from deep within the planet causes plates to move. This movement happens so slowly that humans do not feel it. But some of Earth's plates move as much as a few inches each year. This might not seem like much, but over millions of years, it causes the plates to move thousands of miles.

Movement of surface plates changes Earth's surface features very, very slowly. It takes millions of years for plates to move enough to create landforms. Some land features form when plates are crushed together. At times, forces within Earth push the edge of one plate up over the edge of a plate beside it. This dramatic movement can create mountains, volcanoes, and deep trenches in the ocean floor.

At other times, plates are crushed together in a way that causes the edges of both plates to crumble and break. This event can form jagged mountain ranges. If plates on the ocean floor move apart, the space between them widens into a giant crack in Earth's crust. Magma rises through the crack and forms new crust as it hardens and cools. If enough cooled magma builds up that it reaches the surface of the ocean, an island will begin to form.

Powerful forces within Earth cause the Old Faithful geyser in Yellowstone National Park (left) to erupt. Those forces also cause lava to flow from Mount Etna, a volcano in Italy (right).

▶ **CRITICAL THINKING**

Describing What can happen if plates on the ocean floor move apart?

A road collapses as a result of a powerful earthquake in Chile.

▶ **CRITICAL THINKING**

Describing What natural forces cause earthquakes?

MARTIN BERNETTI/AFP/Getty Images

Sudden Changes

Changes to Earth's surface also can happen quickly. Events such as earthquakes and volcanoes can destroy entire areas within minutes. Earthquakes and volcanoes are caused by plate movement. When two plates grind against each other, faults form. A **fault** results when the rocks on one side or both sides of a crack in Earth's crust have been moved by forces within Earth. **Earthquakes** are caused by plate movement along fault lines. Earthquakes also can be caused by the force of erupting volcanoes.

Various plates lie at the bottom of the Pacific Ocean. These include the huge Pacific Plate along with several smaller plates. Over time, the edges of these plates were forced under the edges of the plates surrounding the Pacific Ocean. This plate movement created a long, narrow band of volcanoes called the **Ring of Fire**. The Ring of Fire stretches for more than 24,000 miles (38,624 km) around the Pacific Ocean.

The **intense** vibrations caused by earthquakes and erupting volcanoes can transfer energy to Earth's surface. When this energy travels through ocean waters, it can cause enormous waves to form on the water's surface. A **tsunami** is a giant ocean wave caused by volcanic eruptions or movement of the earth under the ocean floor. Tsunamis have caused terrible flooding and damage to coastal areas. The forces of these mighty waves can level entire coastlines.

☑ **READING PROGRESS CHECK**

Determining Central Ideas Earth's surface plates are moving. Why don't we feel the ground moving under us?

Academic Vocabulary

intense great or strong

External Forces

GUIDING QUESTION *How can wind, water, and human actions change Earth's surface?*

What happens when a stormy ocean tide hits the shore and sweeps over a sand dune? The water wears down the sand dune. Similar changes take place on a larger scale across Earth's lithosphere. These changes happen much slower—over hundreds, thousands, or even millions of years.

Weathering

Some landforms are created when materials such as rocks and soil build up on Earth's surface. Other landforms take shape as rocks and soil break down and wear away over time. **Weathering** is a process by which Earth's surface is worn away by forces such as wind, rain, chemicals, and the movement of ice and flowing water. Even plants can cause weathering. Plant roots and small seeds can grow into tiny cracks in rock, gradually splitting the rock apart as the roots expand.

You might have seen the effects of weathering on an old building or statue. The edges become chipped and worn, and features such as raised lettering are smoothed down. Landforms such as mountains are affected by weathering, too. The Appalachian Mountains in the eastern United States have become rounded and crumbled after millions of years of weathering by natural forces.

Erosion

Erosion is a process that works with weathering to change surface features of Earth. **Erosion** is a process by which weathered bits of rock are moved elsewhere by water, wind, or ice. Rain and moving water can erode even the hardest stone over time. When material is broken down by weathering, it can easily be carried away by the action of erosion. For example, the Grand Canyon was formed by weathering and erosion caused by flowing water and blowing winds. Water flowed over the region for millions of years, weakening the surface of the rock. The moving water carried away tiny bits of rock. Over time, weathering and erosion carved a deep canyon into the rock. Erosion by wind and chemicals caused the Grand Canyon to widen until it became the amazing landform we see today.

Weathering and erosion cause different materials to break down at different speeds. Soft, porous rocks, such as sandstone and limestone, wear away faster than dense rocks like granite. The spectacular rock formations in Utah's Bryce Canyon were formed as different types of minerals within the rocks were worn away by erosion, some more quickly than others. The result is landforms with jagged, rough surfaces and unusual shapes.

Erosion created this rock formation, named Thor's Hammer, in Bryce Canyon National Park, Utah.

▶ **CRITICAL THINKING**

Explaining Why do materials affected by erosion break down at different speeds?

Huge, tunnel-boring machines were used to dig the Channel Tunnel, or Chunnel. The Chunnel is an underseas rail tunnel connecting the United Kingdom and France.

▶ **CRITICAL THINKING**
Determining Central Ideas Why have humans changed the environment faster and more broadly in the past 50 years than at any other time in history?

Buildup and Movement

The buildup of materials creates landforms such as beaches, islands, and plains. Ocean waves pound coastal rocks into smaller and smaller pieces until they are tiny grains of sand. Over time, waves and ocean currents deposit sand along coastlines, forming sandy beaches. Sand and other materials carried by ocean currents build up on mounds of volcanic rock in the ocean, forming islands. Rivers deposit soil where they empty into larger bodies of water, creating coastal plains and wetland ecosystems.

Entire valleys and plains can be formed by the incredible force and weight of large masses of ice and snow. These masses are often classified by size as glaciers, polar ice caps, or ice sheets. A **glacier**, the smallest of the ice masses, moves slowly over time, sometimes spreading outward on a land surface. Although glaciers are usually thought of as existing during the Ice Age, glaciers can still be found on Earth today.

Ice caps are high-altitude ice masses. Ice sheets, extending more than 20,000 square miles (51,800 sq. km), are the largest ice masses. Ice sheets cover most of Greenland and Antarctica.

Human Actions

Natural forces are awesome in their power to change the surface of Earth. Human actions, however, have also changed Earth in many ways. Activities such as coal mining have leveled entire mountains. Humans use explosives such as dynamite to blast tunnels through mountain ranges when building highways and railroads. Canals dug by humans change the natural course of waterways. Humans have cut down so many millions of acres of forests that deadly landslides and terrible erosion occur on the deforested lands.

Pollution caused by humans can change Earth, as well. When people burn gasoline and other fossil fuels, toxic chemicals are released into the air. These chemicals settle onto the surfaces of mountains, buildings, oceans, rivers, grasslands, and forests. The chemicals poison waterways, kill plants and animals, and cause erosion. The buildings in many cities show signs of being worn down by chemical erosion.

Studies show that humans have changed the Earth's environment faster and more broadly in the last 50 years than at any time in history. A major cause of all the change is the demand for food and natural resources is greater than ever before, and it continues to grow.

Changes to Earth's surface caused by natural weathering and erosion happen slowly. These forces create different kinds of landforms that make our planet unique. Erosion and other changes caused by humans, however, can damage Earth's surface quickly. Their effects threaten our safety and survival. We need to protect our environment to ensure that our quality of life improves for future generations.

☑ **READING PROGRESS CHECK**

Identifying Categorize each of the following events as a change caused by nature or a change caused by humans: coal mining, glaciers melting, air pollution, volcano eruption, acid rain, water erosion, and plate movement.

FOLDABLES
Study Organizer

Include this lesson's information in your Foldable®.

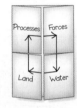

LESSON 2 REVIEW

Reviewing Vocabulary
1. Why can *tsunamis* cause such massive destruction?

Answering the Guiding Questions
2. *Describing* Select one country in Europe. Write about two special or unique landforms of that country. Include photos.

3. *Describing* Describe the importance of tectonic plates.

4. *Determining Central Ideas* How can land features be created by natural forces?

5. *Identifying* How can chemicals contribute to weathering?

6. *Distinguishing Fact from Opinion* Is the following statement a fact or an opinion? Explain.
Although coal mining can damage the environment, it is worth it because the coal is needed to produce electricity.

7. *Describing* Explain the process by which sharp mountain peaks become rounded.

8. *Informative/Explanatory Writing* What human actions have changed the landscape in your area?

Reading **HELP**DESK

Academic Vocabulary

• **transform**

Content Vocabulary

• **plateau**
• **plain**
• **isthmus**
• **continental shelf**
• **trench**
• **desalinization**
• **groundwater**
• **delta**
• **water cycle**
• **evaporation**
• **condensation**
• **acid rain**

TAKING NOTES: *Key Ideas and Details*

Describe Using a chart like this one, describe two kinds of landforms and two bodies of water.

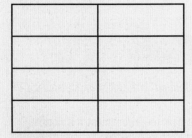

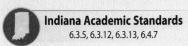
Indiana Academic Standards
6.3.5, 6.3.12, 6.3.13, 6.4.7

Lesson 3
Landforms and Bodies of Water

ESSENTIAL QUESTION • *How does geography influence the way people live?*

IT MATTERS BECAUSE
Earth's landforms and bodies of water influence our ways of life.

Types of Landforms

GUIDING QUESTION *What kinds of landforms cover Earth's surface?*

What is the land like in mountainous areas? Are unique landforms located along a seacoast? Have you ever wondered how different kinds of landforms developed? The surface of Earth is covered with landforms and bodies of water. Our planet is filled with variety on land and under water.

Surface Features on Land

Earth has many different landforms. When scientists study landforms, they find it useful to group them by characteristics. One characteristic that is often used is elevation.

Elevation describes how far above sea level a landform or a location is. Low-lying areas, such as ocean coasts and deep valleys, may be just a few feet above sea level. Mountains and highland areas can be thousands of feet above sea level. Even flat areas of land can have high elevations, especially when they are located far inland from ocean shores.

Plateaus and plains are flat, but a **plateau** rises above the surrounding land. A steep cliff often forms at least one side of a plateau. **Plains** can be flat or have a gentle roll and can be found along coastlines or far inland. Some plains are home to grazing animals, such as horses and antelope. Farmers and ranchers use plains areas to raise crops and livestock for food. A valley is a lowland area between two higher sides. Some

(l to r) Kei Uesugi/Getty Images; Anders Ryman/Alamy; Ingram Publishing/SuperStock; Edwin Remsberg/Taxi/Getty Images

valleys are small, level places surrounded by hills or mountains. Other valleys are huge expanses of land with highlands or mountain ranges on either side. Because they are often supplied with water runoff and topsoil from the higher lands around them, many valleys have rich soil and are used for farming and grazing livestock.

Another way to classify some landforms is to describe them in relation to bodies of water. Some types of landforms are surrounded by water. Continents are the largest of all landmasses. Most continents are bordered by land and water. Only Australia and Antarctica are completely surrounded by water. Islands are landmasses that are surrounded by water, but they are much smaller than continents.

A peninsula is a long, narrow area that extends into a river, a lake, or an ocean. Peninsulas at one end are connected to a larger landmass. An **isthmus** is a narrow strip of land connecting two larger land areas. One well-known isthmus is the Central American country of Panama. Panama connects two massive continents: North America and South America. Because it is the narrowest place in the Americas, the Isthmus of Panama is the location of the Panama Canal, a human-made canal connecting the Atlantic and Pacific oceans.

Under the Oceans

The ocean floor is also covered by different landforms. The ocean floor, like the ground we walk on, is part of Earth's crust. In many ways, the ocean floor and land are similar. If you could see an ocean without its water, you would see a huge expanse of plains, valleys, mountains, hills, and plateaus. Some of the landforms were shaped by the same forces that created the features we see on land.

If you were to explore the oceans, you would see landforms under the water that are similar to those on land.

Identifying What landforms on the map are classified by elevations?

Kei Uesugi/Getty Images

A DROP IN THE OCEAN
SALT WATER VS FRESHWATER

Earth's surface is about 70 percent water. That seems like a lot of water, but how much can we humans actually use? Hint: probably less than you think.

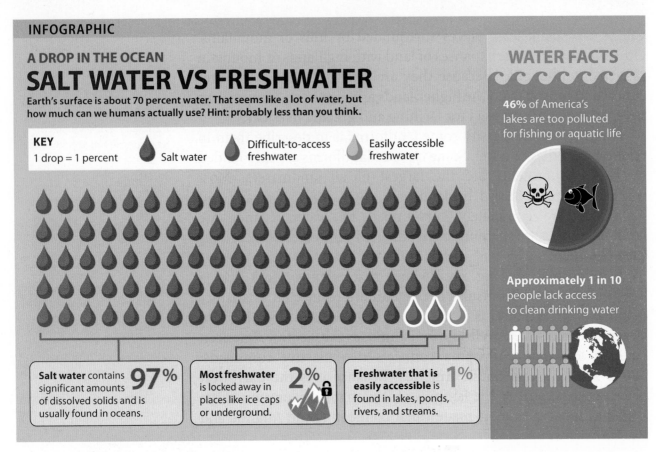

KEY
1 drop = 1 percent — Salt water — Difficult-to-access freshwater — Easily accessible freshwater

Salt water contains **97%** significant amounts of dissolved solids and is usually found in oceans.

Most freshwater 2% is locked away in places like ice caps or underground.

Freshwater that is easily accessible 1% is found in lakes, ponds, rivers, and streams.

WATER FACTS

46% of America's lakes are too polluted for fishing or aquatic life

Approximately 1 in 10 people lack access to clean drinking water

The surface of Earth is made up of water and land. Oceans, lakes, rivers, and other bodies of water make up a large part of Earth.

▶ **CRITICAL THINKING**

Citing Text Evidence How much of Earth's surface is water?

One type of ocean landform is the continental shelf. A **continental shelf** is an underwater plain that borders a continent. Continental shelves usually end at cliffs or downward slopes to the ocean floor.

When divers explore oceans, they sometimes find enormous underwater cliffs that drop off into total darkness. These cliffs extend downward for hundreds or even thousands of feet. The water below is so deep that it is beyond the reach of the sun's light. The deepest location on Earth is the Mariana Trench in the Pacific Ocean. A **trench** is a long, narrow, steep-sided cut in the ground or on the ocean floor. At its deepest point, the Mariana Trench is more than 35,000 feet (10,668 m) below the surface of the ocean.

Other landforms on the ocean floor include volcanoes and mountains. When underwater volcanoes erupt, islands can form because layers of lava build up until they reach the ocean's surface. Mountains on the ocean floor can be as tall as Mount Everest. Undersea mountains can also form ranges. The Mid-Atlantic Ridge, the longest underwater mountain range, is longer than any mountain range on land.

☑ **READING PROGRESS CHECK**

Determining Word Meanings How is an ocean trench like a canyon?

The Water Planet

GUIDING QUESTION *What types of water are found on Earth's surface?*

You can find water in many different forms. Water exists in three states of matter—solid, liquid, and gas—can be found all over the world. Glaciers, polar ice caps, and ice sheets are large masses of water in solid form. Rivers, lakes, and oceans contain liquid water. The atmosphere contains water vapor, which is water in the form of a gas.

A resident in a village in South America pumps water from a well.

▶ **CRITICAL THINKING**

Describing Why is groundwater important to people?

Two Kinds of Water

Water at Earth's surface can be freshwater or salt water. Salt water is water that contains a large percentage of salt and other dissolved minerals. About 97 percent of the planet's water is salt water. Salt water makes up the world's oceans and also a few lakes and seas, such as the Great Salt Lake and the Dead Sea.

Salt water supports a huge variety of plant and animal life, such as whales, fish, and other sea creatures. Because of its high concentration of minerals, humans and most animals can not drink salt water. However, there is a way to remove minerals from salt water to produce water that is safe to drink. **Desalinization** is a process that separates most of the dissolved chemical elements. People who live in dry regions of the world use desalinization to process seawater into drinking water. But this process is expensive.

Freshwater makes up the remaining 3 percent of water on Earth. Most freshwater stays frozen in the ice caps of the Arctic and Antarctic. Only about 1 percent of all water on Earth is the liquid freshwater that humans and other living organisms use. Liquid freshwater is found in lakes, rivers, ponds, swamps, and marshes, and in the rocks and soil underground.

Water contained inside Earth's crust is called **groundwater**. Groundwater is an important source of drinking water, and it is used to irrigate crops. Groundwater often gathers in aquifers. These are underground layers of rock through which water flows. When humans dig wells down into rocks and soil, groundwater flows from the surrounding area and fills the well. Groundwater also flows naturally into rivers, lakes, and oceans.

Bodies of Water

You are probably familiar with some of the different kinds of bodies of water. Some bodies of water contain salt water, and others hold freshwater. The world's largest bodies of water are its five vast saltwater oceans.

Anders Ryman/Alamy

Humans use bodies of water for recreational activities and to earn a living.

▶ **CRITICAL THINKING**
Describing How do people rely on bodies of water to meet their need for fuel?

From largest to smallest, the oceans are the Pacific, Atlantic, Indian, Southern, and Arctic. The Pacific Ocean covers more area than all of Earth's land combined. The Southern Ocean surrounds the continent of Antarctica. Although it is convenient to name the different oceans, it is important to remember that these bodies of water are actually connected and form one global ocean. Things that happen in one part of the ocean can affect the ocean all around the world.

When oceans meet landmasses, unique land features and bodies of water form. A coastal area where ocean waters are partially surrounded by land is called a bay. Bays are protected from ocean waves by the surrounding land, making them useful for fishing and boating. Larger areas of ocean waters partially surrounded by landmasses are called gulfs. The Gulf of Mexico is an example of ocean waters surrounded by continents and islands. Gulfs have many of the features of oceans but are smaller and are affected by the landmasses around them.

Bodies of water such as lakes, rivers, streams, and ponds usually hold freshwater. Freshwater contains some dissolved minerals, but only a small percentage. The fish, plants, and other life-forms that live in freshwater cannot live in salty ocean water.

Freshwater rivers are found all over the world. Rivers begin at a source where water feeds into them. Some rivers begin where two other rivers meet; their waters flow together to form a larger river. Other rivers are fed by sources such as lakes, natural springs, and melting snow flowing down from higher ground.

A river's end point is called the mouth of the river. Rivers end where they empty into other bodies of water. A river can empty into a lake, another river, or an ocean. A **delta** is an area where sand, silt, clay, or gravel is deposited at the mouth of a river. Some deltas flow onto land, enriching the soil with the nutrients they deposit. River deltas can be huge areas with their own ecosystems.

Bodies of water of all kinds affect the lives of people who live near them. Water provides food, work, transportation, and recreation to people in many parts of the world. People get food and earn a

(t) Ingram Publishing/SuperStock; (b) Edwin Remsberg/Taxi/Getty Images living by fishing in rivers, lakes, and oceans. The ocean floor is mined for minerals and drilled for oil. All types of waters have been used for transportation for thousands of years. People also use water for sports and recreation, such as swimming, sailing, fishing, and scuba diving. Water is vital to human culture and survival.

☑ **READING PROGRESS CHECK**

Describing In what ways do you depend on water?

The Water Supply

GUIDING QUESTION *What is the water cycle?*

Water is a basic need of all living things. Humans and other mammals, birds, reptiles, insects, fish, green plants, fungi, and bacteria must have water to survive. Water is essential for all life on Earth. To provide for the trillions of living organisms that use water every day, the planet needs a constant supply of fresh, clean water. Fortunately, water is recycled and renewed continually through Earth's natural systems of atmosphere, hydrosphere, lithosphere, and biosphere.

A Cycle of Balance

When it rains, puddles of water form on the ground. Have you noticed that after a day or two, puddles dry up and vanish? Where does the water go? It might seem as if water disappears and then new water is created, but this is not true. Water is not made or destroyed; it only changes form. When a puddle dries, the liquid water has turned into gas vapor that we cannot see. In time, the vapor will become liquid again, and perhaps it will fill another puddle someday.

Scientists believe the total amount of water on Earth has not changed since our planet formed billions of years ago. How can this be true? It is possible because the same water is being recycled. At all times, water is moving over, under, and above Earth's surface and changing form as it is recycled. Earth's water recycling system is called the **water cycle**. The water cycle keeps Earth's water supply in balance.

Water Changes Form

The sun's energy warms the surface of Earth, including the surface of oceans and lakes. Heat energy from the sun causes liquid water on Earth's surface to change into water vapor in a process called **evaporation**. Evaporation is happening all around us, at all times. Water in oceans, lakes, rivers, and swimming pools is constantly evaporating into the air. Even small amounts of water—in the soil, in the leaves of plants, and in the breath we exhale—evaporate to become part of the atmosphere.

Air that contains water vapor is less dense than dry air. This means that moist air tends to rise. As water evaporates, tiny droplets of water vapor rise into the atmosphere. Water vapor gathers into clouds of varying shapes and sizes. Sometimes clouds continue to build until they are saturated with water vapor and can hold no more. A process called **condensation** occurs, in which water vapor **transforms** into a denser liquid or solid state.

Condensation causes water to fall back to Earth's surface as rain, hail, or snow. Hail and snow either build up and stay solid or melt into liquid water. Snow stays solid when it falls in cold climates or on frozen mountaintops. When snow melts, it flows into rivers and lakes or melts directly into the ground.

Liquid rainwater returns water to rivers, lakes, and oceans. Rainwater also soaks into the ground, supplying moisture to plants and refilling underground water supplies to wells and natural springs. Much of the rainwater that soaks into the ground filters through soil and rocks and trickles back into rivers, lakes, and oceans. In this way, water taken from Earth's surface during evaporation returns in the form of precipitation. This cycle repeats all over the world, recycling the water every living organism needs to survive.

DIAGRAM SKILLS >

Condensation

Clouds

Surface runoff

Precipitation (snow, sleet, hail, rain)

Evaporation from lakes and streams

Evaporation from ocean

Groundwater to rivers and oceans

Water is constantly moving—from the oceans to the air to the ground and finally back to the oceans. The water cycle is the name given to this regular movement of water.

▶ **CRITICAL THINKING**

Analyzing In which part of the water cycle is water vapor changed into a denser liquid or a solid state? Why is this change important?

Human actions have damaged the world's water supply. Waste from factories and runoff from toxic chemicals used on lawns and farm fields has polluted rivers, lakes, oceans, and groundwater. Chemicals such as pesticides and fertilizers seep into wells that hold drinking water, poisoning the water and causing deadly diseases.

The fossil fuels we burn release poisonous gases into the atmosphere. These gases combine with water vapor in the air to create toxic acids. These acids then fall to Earth as a deadly mixture called **acid rain**. Acid rain damages the environment in several ways. It pollutes the water humans and animals drink. The acids damage trees and other plants. As acid rain flows over the land and into waterways, it kills plant and animal life in bodies of water. This upsets the balance of the ecosystem.

☑ **READING PROGRESS CHECK**

Determining Central Ideas Explain how water that evaporates from a puddle can end up in a lake hundreds of miles away.

Some human activities pollute our rivers and oceans.
▶ **CRITICAL THINKING**
Explaining What causes acid rain?

Include this lesson's information in your Foldable®.

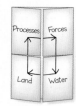

LESSON 3 REVIEW

Reviewing Vocabulary

1. What is the role of *evaporation* in the *water cycle*?

Answering the Guiding Questions

2. ***Analyzing*** Why might life-forms change significantly in the area surrounding a continental shelf?

3. ***Identifying*** How are aquifers useful to humans?

4. ***Determining Central Ideas*** Explain why the water that you drink and bathe in today is the same water that the dinosaurs used millions of years ago.

5. ***Analyzing*** Why is a country that has little or no plains at a disadvantage?

6. ***Narrative Writing*** Imagine that you are a diver exploring the ocean floor. Describe various features you might see.

Directions: Write your answers on a separate piece of paper.

1 Use your **FOLDABLES** to explore the Essential Question.

INFORMATIVE/EXPLANATORY WRITING Starting with the solar system and ending with your bedroom, describe your place in the universe in progressively more specific terms.

2 **21st Century Skills**

INTEGRATING VISUAL INFORMATION Work with a partner to research the Isthmus of Panama. Create a PowerPoint presentation that shows its location and special characteristics. Include information about the history of the Panama Canal and its value to the people of the area.

3 **Thinking Like a Geographer**

ANALYZING Use various resources to research the history and evolution of the Appalachian and the Rocky Mountains, including their formation and their current conditions. Make a two-column poster chart describing their similarities and differences in detail.

4 **GEOGRAPHY ACTIVITY**

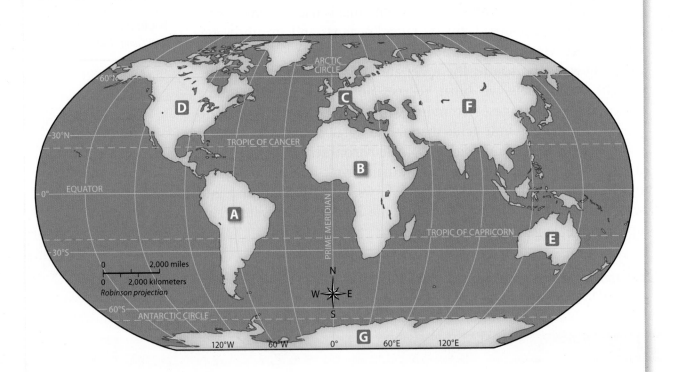

Locating Places

Match the letters on the map to the numbered continents below.

1. Africa
2. Australia
3. Europe
4. North America
5. Asia
6. South America
7. Antarctica

REVIEW THE GUIDING QUESTIONS

Directions: Choose the best answer for each question.

1 We walk on Earth's
  A. atmosphere.
  B. biosphere.
  C. hydrosphere.
  D. lithosphere.

2 When Earth makes a complete trip around the sun, it makes a
  F. revolution.
  G. rotation.
  H. spin.
  I. month.

3 The Ring of Fire was created by
  A. an ancient tsunami.
  B. movement of Earth's plates.
  C. glaciers moving over the surface of Earth.
  D. erosion.

4 Damage to buildings from air pollution is an example of
  F. human forces.
  G. natural forces.
  H. global warming.
  I. erosion.

5 The Gulf of Mexico is filled with
  A. salt water.
  B. freshwater.
  C. a mixture of freshwater and saltwater fish species.
  D. currently inactive volcanoes.

6 How many saltwater oceans does the world have?
  F. two
  G. three
  H. five
  I. six

DBQ ANALYZING DOCUMENTS

7 **ANALYZING** In this paragraph, a science writer describes the two types of planets in our solar system.

"*The planets can be divided quite easily into two categories—the inner planets, which are small and rocky, and the gas giants that circle through the outer reaches of the solar system. Within each class, the planets bear a striking resemblance to each other, but the two classes themselves are very different.*"

—from James S. Trefil, *Space, Time, Infinity*

The author infers that

A. the planets are all similar.

B. the order of the planets changes.

C. the planets that are in the same category are located together as a group.

D. planets can switch categories as their structure changes.

8 **IDENTIFYING** Which feature of Earth would the author use to classify the type of planet we live on?

F. the presence of oceans

G. a hard surface

H. the existence of human beings

I. our moon

SHORT RESPONSE

"*There was a great rattle and jar. . . . [Then] there came a really terrific shock; the ground seemed to roll under me in waves, interrupted by a violent joggling up and down, and there was a heavy grinding noise as of brick houses rubbing together.*"

—from Mark Twain, *Roughing It*

9 **DETERMINING WORD MEANINGS** What do you think Twain meant by "great rattle and jar"?

10 **DETERMINING CENTRAL IDEAS** What do you think Twain just experienced? Explain how you know.

EXTENDED RESPONSE

11 **INFORMATIVE/EXPLANATORY WRITING** Considering the characteristics of various places in North America, South America, and Europe, where would you most like to live? Support your answer with detailed information about that area's climate, climate zone, landforms, and proximity to areas that suffer from potential natural disasters such as floods or earthquakes.

Need Extra Help?

If You've Missed Question	**1**	**2**	**3**	**4**	**5**	**6**	**7**	**8**	**9**	**10**	**11**
Review Lesson	1	1	2	2	3	3	1	1	2	2	1

EARTH'S PEOPLE

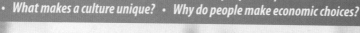

ESSENTIAL QUESTIONS
- How do people adapt to their environment?
- What makes a culture unique?
- Why do people make economic choices?

©Hugh Sitton/Corbis

Young girl from South America

networks

There's More Online about Earth's People.

CHAPTER 3

Lesson 1
A Changing Population

Lesson 2
Global Cultures

Lesson 3
Economic Systems

The Story Matters...

In our study of geography, we study culture, which is the way of life of people who share similar beliefs and customs. A particular culture can be understood by looking at the languages the people speak, what beliefs they hold, and what smaller groups form as parts of their society.

FOLDABLES
Study Organizer

Go to the Foldables® library in the back of your book to make a Foldable® that will help you take notes while reading this chapter.

Adaptations Cultural Views Basic Needs

EARTH'S PEOPLE

Earth's People

This image shows the world at night. Although the world's population is increasing, people still live on only a small part of Earth's surface. Some people live in highly urbanized areas, such as cities. Others, however, live in areas where they may not have access to electiricty or even running water.

Step Into the Place

MAP FOCUS Use the image to answer the following questions.

1 HUMAN GEOGRAPHY
What do you think the brightly lit areas on the map represent?

2 HUMAN GEOGRAPHY
Why might some areas be brighter than others?

3 ENVIRONMENT AND SOCIETY Are the lights evenly distributed across the land?

4 CRITICAL THINKING
ANALYZING Why do some areas have fewer lights?

Step Into the Time

IDENTIFYING POINT OF VIEW Research one event from the time line. Write a journal entry describing the daily life of the time.

c. 200–550
Maya civilization uses hillside terracing and irrigation

c. 35,000–20,000 B.C. Early peoples come to North America

3500–2180 B.C. First Egyptian dynasties build trade network with West Asia

B.C.   A.D.

c. 4500–4000 B.C. Sumer civilization develops between Tigris and Euphrates rivers

1492
Christopher Columbus
lands in Americas

1892–1924
Millions of immigrants
arrive at Ellis Island

1000

1200s West African kingdom
of Benin trade center

1845–1849 Ireland's Great
Famine leads to mass emigration

2000

2011 South Sudan secedes,
forming new nation

networks

There's More Online!

☑ **IMAGES** Refugee Camps

☑ **MAP** Landforms, Waterways, and Population

☑ **SLIDE SHOW** Places around the World

☑ **VIDEO**

Reading **HELP**DESK

Academic Vocabulary

- **mature**

Content Vocabulary

- **death rate**
- **birthrate**
- **doubling time**
- **population distribution**
- **population density**
- **urban**
- **rural**
- **emigrate**
- **immigrate**
- **refugee**
- **urbanization**
- **megalopolis**

TAKING NOTES: *Key Ideas and Details*

Determining Cause and Effect
As you read, use a graphic organizer like this one to take notes about the causes of population growth and migration.

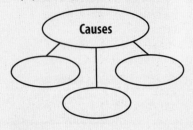

Causes

Indiana Academic Standards
6.1.15, 6.1.18, 6.1.19, 6.3.9, 6.3.13, 6.3.14

72

Lesson 1
A Changing Population

ESSENTIAL QUESTION • *How do people adapt to their environment?*

IT MATTERS BECAUSE
Billions of people share Earth. They have many different ways of life.

Population Growth

GUIDING QUESTION *What factors contribute to Earth's constantly rising population?*

The world's population is growing fast. In 1800 about 860 million people lived in the world. During the next 100 years, the population doubled to nearly 1.7 billion. By 2012, the total passed 7 billion.

Causes of Population Growth

How has Earth's population become so large? What has caused our population to grow so quickly? Many factors cause populations to increase. One major cause of population growth is a falling death rate. The death rate is the number of deaths compared to the total number of individuals in a population at a given time. On average, about 154,080 people die every day worldwide. The **death rate** has decreased for many reasons. Better health care, more food, and cleaner water have helped more people—young and old—live longer, healthier lives.

Another major cause of population growth is the global birthrate. The **birthrate** is the number of babies born compared to the total number of individuals in a population at a given time. On average, about 215,120 babies are born each day worldwide. In time, the babies born today will **mature** and have children and grandchildren of their own.

This is how more and more people join the human population with each passing day.

However, during the past 60 years, the world's human birthrate has been decreasing slowly, although the global birthrate is still higher than the global death rate. This means that at any given time, such as a day or a year, more births than deaths occur. This results in population growth.

Growth Rates

In some countries, a high number of births has combined with a low death rate to greatly increase population growth. As a result, **doubling time**, or the number of years it takes a population to double in size based on its current growth rate, is relatively short. In some parts of Asia and Africa, for example, the doubling time is relatively short—25 years or less. In contrast, the average doubling time of countries with slow growth rates, such as Canada, can be more than 75 years.

Despite the fact that the global population is growing, the rate of growth is gradually slowing. The United Nations Department of Economic and Social Affairs predicts that the world's population will peak at 9 billion by the year 2050. After that, the population will begin to decrease. This means that for the next few decades, Earth's population will continue to grow. In time, however, this growth trend might stop.

Academic Vocabulary

mature to become fully grown and developed as an adult; also refers to older adults

GRAPH SKILLS ›

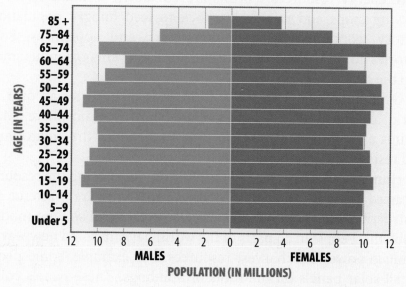

Source: U.S. Census Bureau, Statistical Abstract of the United States: 2012

POPULATION PYRAMID

A population pyramid shows a country's population by age and gender. Geographers use population pyramids to plan for a country's future needs. A population pyramid is two bar graphs. These bar graphs show the number of males and females living in a region. The number of males and females is given as a percentage along the bottom of the graph. The age range for each group is listed along the left side of the graph. This graph shows the population of the U.S.

▶ **CRITICAL THINKING**

1. *Analyzing* Which age group has the most females?

2. *Analyzing* About how many males are in the 10–14 age group?

Farmers in many parts of the world clear land by cutting and burning forests.

▶ **CRITICAL THINKING**

Analyzing Why do some groups practice slash-and-burn agriculture?

Population Challenges

When human populations grow, the places people inhabit can become crowded. In many parts of the world, cities, towns, and villages have grown and expanded beyond a comfortable capacity. Some cities are now so filled with people that they are becoming dangerously overcrowded.

When the population of an already crowded area continues to grow, serious problems can arise. For example, diseases spread quickly in crowded environments. Sometimes there is not enough work for everyone, and many households live in ongoing poverty. Where many people share tight living spaces, crime can be a serious problem and pollution can increase.

Effects on the Environment

On a global scale, rapid population growth can harm the environment. Each year, more people are sharing the same amount of space. People demand fuel for their cars and power for their homes. Miners drill and dig into the earth in a constant search for more energy resources. Forests are cut down to make farms for growing crops and raising livestock to feed hungry populations. Factory workers build cars, computers, and appliances. Some factories dump chemicals into waterways and vent poisonous smoke into the air.

Over time, and with many thousands of factories all over the world, chemical wastes have polluted Earth's atmosphere. Many groups and individuals are working to clean up the environment and restore once-polluted areas.

Humans have many methods of finding and using the resources we need for survival. Some of these methods are wasteful and destructive. However, humans are also creative in solving modern problems. People in all parts of the world have invented new ways to produce power and harvest resources. For example, some people install solar panels on the roofs of buildings. These panels collect energy from the sun, which can be used to produce heat and electric energy. Wind, solar, and geothermal energy are resources that do not pollute the environment. Humans are rising to the challenge of finding new methods of using these natural resources.

Population Growth Rates

Human populations grow at different rates in different areas of the world for many reasons. Often, the number of children each family will have is influenced by the family's culture and religion. In some cultures, families are encouraged to have as many children as they can. Although birthrates have fallen greatly in many parts of Asia, Africa, and South America in recent decades, the rates are still higher than in industrialized nations such as the United States.

In locations with the largest and fastest-growing populations, the need for resources, jobs, health care, and education is great. When millions of people living in a small area need food, water, and housing, there is sometimes not enough for everyone. People in many parts of the world go hungry or die of starvation. Water supplies in crowded cities are often polluted with wastes that can cause diseases. Some areas do not have enough land resources and materials for people to build safe, sturdy homes.

Children in areas affected by extreme poverty often do not receive an education. In areas with job shortages, people are forced to live on low incomes. People living in poverty often live in crowded neighborhoods called *slums*. Slums surround many of the world's cities. These places are often dirty and unsafe. Governments and organizations such as the United Nations are working to make these areas safer, healthier places to live. Because populations grow at different rates some areas experience more severe problems.

✓ **READING PROGRESS CHECK**

Analyzing How can rapid population growth harm the environment?

Solar panels collect energy from the sun.

▶ **CRITICAL THINKING**
Identifying Why do some people prefer solar energy?

Much of the land surrounding this village is used for farming.

▶ **CRITICAL THINKING**

Describing What are rural areas?

Where People Live

GUIDING QUESTION *Why do more people live in some parts of the world than in others?*

Throughout history, people have moved from one place to another. People continue to move today, and the population continues to grow.

Population Distribution

Population growth rates vary among Earth's regions. The **population distribution**, or the geographic pattern of where people live on Earth, is uneven as well. One reason people live in a certain place is work. During the industrial age, for example, people moved to places that had important resources such as coal or iron ore to make and operate machinery. People gather in other places because these places hold religious significance or because they are government or transportation centers.

Population Density

One way to look at population is by measuring **population density**—the average number of people living within a square mile or a square kilometer. To say that an area is *densely populated* means the area has a large number of people living within it.

Keep in mind that a country's population density is the average for the entire country. But population is not distributed evenly throughout the country. As a result, some areas are more densely populated than their country's average indicates. In Egypt, for example, the population is concentrated along the Nile River; in China, along its eastern seaboard; and in Mexico, on the Central Plateau.

Derek Trask/Photolibrary/Getty Images

Where People Are Located

Urban areas are densely populated. **Rural** areas, in contrast, are sparsely populated. People inhabit only a small part of Earth. Remember that land covers about 30 percent of Earth's surface, and half of this land is not useful to humans. This means that only about 15 percent of Earth's surface is inhabitable. Large cities have dense populations, while deserts, oceans, and mountaintops are uninhabited.

The main reason people settle in some areas and not in others is the need for resources. People live where their basic needs can be met. People need shelter, food, water, and a way to earn a living. Some people live in cities, which have many places to live and work. Other people make their homes on open grasslands where they build their own shelters, grow their own food, and raise livestock.

☑ **READING PROGRESS CHECK**

Identifying What is the main reason people settle in some areas instead of others?

Changing Populations

GUIDING QUESTION *What are the causes and effects of human migration?*

When people move, either as individuals or in large groups, areas change. When many people leave an area, that area's population decreases.

Guanajuato is a town in central Mexico that began as a major mining center in the mid-1500s.

▶ **CRITICAL THINKING**
Describing What is the main difference between rural areas and urban areas?

People in rural areas often obtain food in oudoor markets such as this one in northern Michigan.

When large numbers of people move into an area, the population of that area increases. Moving from one place to another is called *migration*.

Causes of Migration

What causes people to leave their homelands and migrate to different parts of the world? To **emigrate** means to leave one's home "to live in another place." Emigration can happen within the same nation, such as when people move from a village to a city inside the same country. Often, emigration happens when people move from one nation to another. For example, millions of people have emigrated from countries in Europe, Asia, and Africa to start new lives in the United States. The term *immigrate* is closely related to *emigrate*, but it does not mean the same thing. To **immigrate** means "to enter and live in a new country."

The reasons for leaving one area and going to another are called push-pull factors. *Push* factors drive people from an area. For example, when a war breaks out in a country or a region, people emigrate from that place to escape danger. People who flee a country because of violence, war, or persecution are called **refugees**. Sometimes people emigrate from an area after a natural disaster such as a flood, an earthquake, or a tsunami has destroyed their homes and land. If the economy of a place becomes so weak that little or no work is available, people emigrate to seek new opportunities.

Think **Again**?

Early immigrants came to the United States for both push and pull factors.

True. Some came to avoid war or escape persecution. In addition to these "push" factors, others were "pulled" to gain freedom, education, and economic opportunity.

Pull factors attract people to an area. Some people move to new places to be with friends or family members. Many young people move to cities or countries to attend universities or other schools. Some relocate in search of better jobs. Families sometimes move to places where their children will be able to attend good schools.

Effects of Migration

The movement of people to and from different parts of the world can affect the land, resources, culture, and economy of an area. Some of these effects are positive, but others can be harmful.

One positive effect of migration is cultural blending. As people from diverse cultures migrate to the same place and live close together, their cultures become mixed and blended. This blending creates new, unique cultures and ways of life. Artwork and music created in diverse urban areas is often an interesting mixture of styles and rhythms from around the world. Food, clothing styles, and languages spoken in urban areas change when people migrate into that area and bring new influences.

Some families and cultural groups work to preserve their original culture. These people want to keep their cultural traditions alive so they can be passed down to future generations. For example, the traditional Chinese New Year is an important celebration for many Chinese American families. Chinese Americans can be part of a blended American culture but still enjoy traditional Chinese foods, music, and arts, and celebrate Chinese holidays. It is possible to adapt to a local culture yet maintain strong ties to a home culture.

Some people migrate by choice. Others, such as the Libyan refugees shown here, are forced to flee to another country to live.

▶ **CRITICAL THINKING**
Identifying What are examples of "pull" causes of migration?

Carlos Spottorno/Getty Images News/ Getty Images

Nearly half the world's people live in urban areas. Many live in very large cities such as New York City.

▶ **CRITICAL THINKING**

Determining Word Meanings What is a megalopolis?

Causes and Effects of Urbanization

Another effect of migration is the growth of urban areas. **Urbanization** happens when cities grow larger and spread into surrounding areas. Migration is a primary reason that urbanization occurs.

People move to cities for many reasons. The most common reason is to find jobs. Transportation and trade centers draw people primarily by creating new opportunities for business. As the businesses grow and people move into an area, the need for services also grows. Workers fill positions in medical services, education, entertainment, housing, and food sectors.

As more people migrate to cities, urban areas become increasingly crowded. When populations within urban areas increase, cities grow and expand. Farmland is bought by developers to build homes, apartment buildings, factories, offices, schools, and stores to provide for the growing number of people. The loss of farmland means that food must be grown farther from cities, resulting in additional shipping and related pollution.

Urbanization is happening in cities all over the world. In some places, cities have grown so vast that they have reached the outer edges of other cities. The result is massive clusters of urban areas that continue for miles. A huge city or cluster of cities with an extremely large population is called a **megalopolis**. These huge cities are growing larger every day, and they face the challenges that come with population growth and urbanization.

Robert Glusic/Photodisc/Getty Images

Delhi, one of India's largest cities, is a megalopolis. Its sprawling land area takes in a section called the Old City, dating from the mid-1600s. It also encompasses New Delhi, the modern capital city built by British colonial rulers in the early 1900s.

The largest megalopolis in the Americas is Mexico City. Because of its size and influence, Mexico City is a primate city, an urban area that dominates its country's economy and political affairs. Primate cities include Cairo, Egypt, in Africa, Amman, Jordan, in Asia, and Paris, France, in Europe.

Examples of Urbanization

Urbanization takes place around the world but for different reasons and at different rates. Paraguay, for example, has long been one of the least urbanized regions in the Americas. In 1965, about two of every three Paraguayans lived in rural areas. Most worked in agriculture. Over time, Paraguayans began leaving their rural villages in order to find better job opportunities. Even those who continued to work in agriculture moved closer to urban areas. About 70 percent of all citizens live within 120 miles (193 km) of the capital, Asunción.

Europe is highly urbanized. Beginning in the late 1700s, the Industrial Revolution transformed Europe from a rural, agricultural society to an urban, industrial society. The growth of industries and cities began first in Western Europe. Later, after World War II, the process spread to Eastern Europe.

Include this lesson's information in your Foldable®.

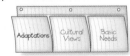

Adaptations | Cultural Views | Basic Needs

✔ **READING PROGRESS CHECK**

Describing Explain why Mexico City is a primate city and a megalopolis.

LESSON 1 REVIEW

Reviewing Vocabulary
1. Why does *population distribution* vary?

Answering the Guiding Questions
2. *Analyzing* If Canada's population is doubling about every 75 years, is the death rate or the birthrate higher? Explain.

3. *Identifying* Which parts of North America are less densely populated than most other parts? Explain.

4. *Determining Central Ideas* Explain factors that might have led to the formation of the megalopolis of New York City.

5. *Informative/Explanatory Writing* Imagine that you are planning to leave the area where you currently live. Write a paragraph explaining the factors you would consider in selecting a new location. Select a city that is at least 100 miles from where you live now and analyze whether that city meets the criteria you selected. Write another paragraph that includes this analysis.

networks

There's More Online!

☑ **IMAGE** Cultural Change

☑ **MAP** World Culture Regions

☑ **SLIDE SHOW** Different Languages

☑ **VIDEO**

Reading **HELP**DESK

Academic Vocabulary

- **behalf**

Content Vocabulary

- **culture**
- **ethnic group**
- **dialect**
- **cultural region**
- **democracy**
- **representative democracy**
- **monarchy**
- **dictatorship**
- **human rights**
- **globalization**

TAKING NOTES: *Key Ideas and Details*

Organize On a graphic organizer like this one, take notes about the different forms of government.

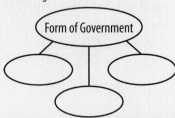

Form of Government

Indiana Academic Standards
6.1.16, 6.1.17, 6.1.20, 6.2.1, 6.2.2, 6.2.4, 6.2.5, 6.2.7, 6.3.4, 6.3.10, 6.4.9

Lesson 2
Global Cultures

ESSENTIAL QUESTION • *What makes a culture unique?*

IT MATTERS BECAUSE
Culture shapes the way people live and how they view the world.

What Is Culture?

GUIDING QUESTION *How is culture part of your life?*

Think about the clothes you wear, the music you listen to, and the foods you eat. These things are all part of your culture. Do you like pizza, rice and beans, pasta, or *samosas*? Have you ever thought about the people and cultures that invented the foods you enjoy eating? Foods are created, grown, or developed by people of different cultures.

Culture is the set of beliefs, behaviors, and traits shared by a group of people. The term *culture* can also refer to the people of a certain culture. For example, saying "the Hindu culture" can mean the Hindu cultural traditions, the people who follow these traditions, or both.

You might be part of more than one culture. If your family has strong ties to a culture, such as that of a religion or a nation, you might follow this cultural tradition at home. You also might be part of a more mainstream American culture while at school and with friends.

If your family emigrated from Somalia to the United States, for example, you might speak the Somali language, wear traditional Somali clothing, and eat Somali foods. Your family might celebrate holidays observed in Somalia as well as American holidays, such as Thanksgiving and Independence Day. When you are with your friends, you might speak English, listen to American music, and watch American sports.

Different Groups

We can look at members of a culture in terms of age, gender, or ethnic group. An **ethnic group** is a group of people with a common racial, national, tribal, religious, or cultural background. Members of the same Native American nation are an example of people of the same ethnic group. Other examples include the Maori of New Zealand and the Han Chinese. Large countries such as China, can be home to hundreds of different ethnic groups. Some ethnic groups in a country are minority groups—people whose race or ethnic origin is different from that of the majority group. The largest ethnic minority groups in the United States are Hispanic Americans and African Americans.

Members of a culture might have special roles or positions as part of their cultural traditions. In some cultures, women are expected to care for and educate children. Most cultures expect men to earn money to support their families or to provide in other ways, such as by hunting and farming. Many cultures respect the elderly and value their wisdom. The leaders of older, traditional cultures are often elderly men or women who have leadership experience. Most cultures have clearly defined roles for their members. From an early age, young people learn what their culture expects of them. It is possible, too, to be part of more than one culture.

Language

Language serves as a powerful form of communication. Through language, people communicate information and experience and pass on cultural beliefs and traditions. Thousands of different languages are spoken in the world. Some languages have become world languages, or languages that are commonly spoken in many parts of the world. Some languages are spoken differently in different regions or by different ethnic groups. A **dialect** is a regional variety of a language with unique features, such as vocabulary, grammar, or pronunciation. People who speak the same language can sometimes understand other dialects, but at times, the pronunciation, or accent, of a dialect can be nearly impossible for others to understand.

Cities often have communities within them that share a distinct and common culture, language, and customs.

▶ **CRITICAL THINKING**

Describing What is a dialect?

MAJOR WORLD RELIGIONS

Religion	Major Leader	Beliefs
Buddhism	Siddhārtha Gautama, the Buddha	Suffering comes from attachment to earthly things, which are not lasting. People become free by following the Eightfold Path, rules of right thought and conduct. People who follow the Path achieve nirvana—a state of endless peace and joy.
Christianity	Jesus Christ	The one God is Father, Son, and Holy Spirit. God the Son became human as Jesus Christ. Jesus died and rose again to bring God's forgiving love to sinful humanity. Those who trust in Jesus and follow his teachings of love for God and neighbor receive eternal life with God.
Hinduism	No one founder	One eternal spirit, Brahman, is represented as many deities. Every living thing has a soul that passes through many successive lives. Each soul's condition in a specific life is based on how the previous life was lived. When a soul reaches purity, it finally joins permanently with Brahman.
Islam	Muhammad	The one God sent a series of prophets, including the final prophet Muhammad, to teach humanity. Islam's laws are based on the Quran, the holy book, and the Sunnah, examples from Muhammad's life. Believers practice the five pillars—belief, prayer, charity, fasting, and pilgrimage—to go to an eternal paradise.
Judaism	Abraham	The one God made an agreement through Abraham and later Moses with the people of Israel. God would bless them, and they would follow God's laws, applying God's will in all parts of their lives. The main laws and practices of Judaism are stated in the Torah, the first five books of the Hebrew Bible.
Sikhism	Guru Nanak	The one God made truth known through 10 successive gurus, or teachers. God's will is that people should live honestly, work hard, and treat others fairly. The Sikh community, or Khalsa, bases its decisions on the principles of a sacred text, the Guru Granth Sahib.

Religion

Religion has a major influence on how people of a culture see the world. Religious beliefs are powerful. Some individuals see their religion as merely a tradition to follow during special occasions or holidays. Others view religion as the foundation and most important part of their life. Religious practices vary widely. Many cultures base their way of life on the spiritual teachings and laws of holy books. Religion is a central part of many of the world's cultures. Throughout history, religious stories and symbols have influenced painting, architecture, and music.

Customs

Customs are also an important outward display of culture. In many traditional cultures, a woman is not permitted to touch a man other than her husband, even for a handshake. In modern European cultures, polite greetings include kissing on the cheeks. People of many cultures bow to others as a sign of greeting, respect, and goodwill. The world's many cultures have countless fascinating customs. Some are used only formally, and others are viewed as good manners and respectful, professional behavior.

History

History shapes how we view the world. We often celebrate holidays to honor the heroes and heroines who brought about successes. Stories about heroes reveal the personal characteristics that people think are important. Groups also remember the dark periods of history when they met with disaster or defeat. These experiences, too, influence how groups of people see themselves.

The Arts and Sports

Dance, music, visual arts, and literature are important elements of culture. Nearly all cultures have unique art forms that celebrate their history and enrich people's lives. Some art forms, such as singing and dancing, are serious parts of religious ceremonies or other cultural events. Aboriginal peoples of the Pacific Islands have songs, dances, and chants that are vital parts of their cultural traditions. Art can be forms of personal expression or worship, entertainment, or even ways of retelling and preserving a culture's history.

In sports, as in many other aspects of culture, activities are adopted, modified, and shared. Many sports that we play today originated with different culture groups in the past. Athletes in ancient Japan, China, Greece, and Rome played a game similar to soccer. Scholars believe that the Maya of Mexico and Central America developed "ballgame," the first organized team sport. Playing on a 40- to 50-foot long (12m to 15m) recessed court, the athletes' goal was to kick a rubber ball through a goal.

Features of Government

Government is another element of culture. Governments share certain features. They maintain order within an area and provide protection from outside dangers. Governments also provide services to citizens, such as education. Different cultures have different ways of distributing power and making rules.

Soccer is one of the most popular international sports.

▶ **CRITICAL THINKING**
Identifying Which culture first played the game of soccer?

Economy

Economies control the use of natural resources and define how goods are produced and distributed to meet human needs. Some cultures have their own type of economy, but most follow the economy of the country or area where they live. This allows people of different cultures living in an area to trade and conduct other types of business with one another. For example, many people in Benin, West Africa, sell goods in open-air markets. Some people bring items to the markets to trade for the goods they need, but others pay for goods using paper money and coins.

Cultural Regions

A **cultural region** is a geographic area in which people have certain traits in common. People in a cultural region often live close to one another to share resources, for social reasons, and to keep their cultures and communities strong. Cultural regions can be large or relatively small. For example, one of the world's largest cultural areas stretches across northern Africa and Southwest Asia. This cultural region is home to millions of people of the Islamic, or Muslim, culture. A much smaller cultural region is Spanish Harlem in New York City. This cultural region is home to a large and growing Hispanic culture.

✓ READING PROGRESS CHECK

Identifying Point of View What cultural traditions do you practice? Make a list of the beliefs, behaviors, languages, foods, art, music, clothing, and other elements of culture that are part of your daily life.

On July 4, 1776, the Second Continental Congress approved the Declaration of Independence, establishing the United States as an independent country.

▶ **CRITICAL THINKING**

Describing What is a representative democracy?

Government

GUIDING QUESTION *How does government affect way of life?*

All nations need some type of formal leadership. What differs among countries is how leaders are chosen, who makes the rules, how much freedom people have, and how much control governments have over people's lives. Many different kinds of government systems operate in the world today. Three of the most common are democracy, monarchy, and dictatorship.

Democracy

In a democracy, the people hold the power. Citizens of a nation make the decisions themselves. A **democracy** is a system of government that is run by the people. In

©PoodlesRock/Corbis

democratic systems of government, people are free to propose laws and policies. Citizens then vote to decide which laws and policies will be set in place. When people run the government, citizens' rights and freedoms are protected.

In some democracies, the people elect leaders to make and carry out laws. A **representative democracy** is a form of democracy in which citizens elect government officials to represent the people; the government representatives make and carry out laws and policies on **behalf** of the people. The United States in an example of a representative democracy.

The queen is the symbolic head of the United Kingdom, but elected leaders hold the power to rule.

▶ **CRITICAL THINKING**
Describing What is a monarchy?

Monarchy

A **monarchy** is ruled by a king or a queen. In a monarchy, power and leadership are passed down from older to younger generations through heredity. The ruler of a monarchy, called a *monarch*, is usually a king, a queen, a prince, or a princess. In the past, monarchs had absolute power, or total power. Today, most monarchs only represent, or stand for, a country's traditions and values, while elected officials run the government. The United Kingdom is an example of a monarchy.

Dictatorship

A **dictatorship** is a form of government in which one person has absolute power to rule and control the government, the people, and the economy. People who live under a dictatorship often have few rights. With absolute power, a dictator can make laws with no concern for how just, fair, or practical the laws are. North Korea is an example of a dictatorship.

Some dictators abuse their power for personal gain. One negative consequence of abuse of power is lack of personal freedoms and human rights for the general public. **Human rights** are the rights that belong to all individuals. Those rights are the same for every human in every culture. Some basic human rights are the right to life, liberty, security, privacy, freedom from slavery, fair treatment before the law, and marry and have children.

Academic Vocabulary

behalf in the interest of; in support of; in defense of

☑ **READING PROGRESS CHECK**

Describing In your own words, define human rights and list three important human rights.

Cargo containers are stockpiled and ready to be loaded onto ships in the port of Johor, Malaysia. International trade is the exchange of goods and services between countries. When people trade, they not only exchange goods, they also exchange customs and ideas.

©Justin Guariglia/Corbis

Shifts in Culture

GUIDING QUESTION *How do cultures change over time?*

Over time, cultures change for a variety of reasons. When people relocate, they bring their cultural traditions with them. The traditions often influence or blend with the cultures of the places where they settle. Over time, as people of many cultures move to a location, the culture of that location takes on elements of all the cultures within it. Cities, such as London and New York, are examples of areas that have richly diverse cultures.

Cultural Change

Change also can occur as a result of trade, travel, war, and exchange of ideas. Trade brings people to new areas to sell and barter goods. Whenever people travel, they bring their language, customs, and ideas with them. They also bring elements of foreign cultures back with them when they return home. Throughout history, traders and explorers have brought home new foods, clothing, jewelry, and other goods. Some of these, such as gold, chocolate, gunpowder, and silk, became popular all over the world. Trade in these items changed the course of history.

Culture can also change as a result of technology. The telegraph, telephones, and e-mail have made communication increasingly faster and easier. Television and the Internet have given people in all parts of the world easy access to information and new ideas. Elements of culture such as language, clothing styles, customs, and behaviors spread quickly as people discover them by watching television and using the Internet.

Global Culture

Today's world is becoming more culturally blended every day. As cultures combine, new cultural elements and traditions are born. The spread of culture and ideas has caused our world to become globalized. **Globalization** is the process by which nations, cultures, and economies become integrated, or mixed. Globalization has had the positive effect of making people more understanding and accepting of other cultures. It also has helped spread ideas and innovations. Technology has made communication faster and easier. Travel also has become faster and easier, allowing more people to visit more places in less time. This is resulting in cultural blending on a wider scale than ever before.

The process of cultural blending through globalization is not always smooth and easy. Sometimes it produces tension and conflict as people from different cultures come into contact with one another. Some people do not want their cultures to change, or they want to control the amount of change. Sometimes the changes come too fast, and cultures can be damaged or destroyed.

Just as no one element defines a culture, no one culture can define the world. All cultures have value and add to the human experience. As the world becomes more globalized, people must continue to respect other ways of life. We have much to learn, and much to gain, from the many cultures that make our world a fascinating place.

Widespread use of technology, such as cell phones, allows us to share information with a larger audience.
▶ **CRITICAL THINKING**
Analyzing How does technology help spread new ideas?

Include this lesson's information in your Foldable®.

✓ **READING PROGRESS CHECK**

Determining Word Meanings What is globalization?

Paul J. Richards/AFP/Getty Images

LESSON 2 REVIEW

Reviewing Vocabulary

1. What is the difference between a *democracy* and a *representative democracy*?

Answering the Guiding Questions

2. *Determining Central Ideas* Give several examples of parts of a culture that might change over a period of time.

3. *Analyzing* What can be a disadvantage of a representative form of government?

4. *Describing* Describe how the Internet has changed the culture of the countries in which a large percentage of the population utilizes it.

5. *Informative/Explanatory Writing* Write a short essay explaining ways in which globalization has affected your family. Consider foods you eat, things you like to do, and people you have met.

Social Media in a Changing World

News media, such as newspapers and radio, give us information. Social media, such as Facebook and Twitter, provide information, but they help us communicate, too.

Family and Friends The Internet helps us stay in touch with family and friends. It provides a good way for members of the military and their families and friends to share what is happening in their lives. Online connections also help service members in other ways. Where soldiers have access to the Internet, many have continued taking college courses, even in Iraq and Afghanistan.

In College Nearly one-third of higher-education students take at least one course online. For several years, online enrollment has grown faster than total higher-education enrollment.

> **It makes it easy to share information.**

In an Emergency At times, social media provides the only way to communicate during an emergency like the tsunami that struck Japan in 2011. Through the use of sites such as Facebook and Twitter, people were able to contact family and friends. Social networks also posted information about shelters, medical help, and relief efforts.

Political Effects

The use of social media increases during times of political trouble or change, such as the "Arab Spring" protests. For example, the number of tweets from Egypt rose from 2,300 to 230,000 in the week leading to the resignation of Egyptian president Hosni Mubarak. The tweets spread the message of protest.

Safety Tips The Internet is a fun way to interact with friends. It makes it easy to share information. But some information should *not* be shared. Remember these safety tips:

- Keep your personal information to yourself. Don't post your full name, Social Security number, address, phone number, or bank account numbers.
- Don't accept as a friend anyone you do not know.
- Post only information that you are comfortable with others seeing—and knowing—about you.

(l) Ariel Skelley/the Agency Collection/Getty Images; (r) Daniel Berehulak/Getty Images News/Getty Images

THERE'S MORE ONLINE

HEAR about China's internet restrictions • *SEE* how people influence media • *WATCH* a video on piracy

These numbers and statistics can help you see how social media is changing the world.

840 Million

By 2012, Facebook had more than 840 million users. That's more than the population of every country in the world except China and India.

$1 Billion

In April 2012, Facebook, Inc. purchased the photo-sharing app called Instagram for $1 billion. Instagram allows users to apply a filter on a photograph and share it on any of a number of social networks. The company has 13 employees and has been in business since October 2010.

Fastest to 10 MILLION

Which social network was the quickest to reach 10 million users? That's Google+, taking only 16 days to reach that number. By contrast, Twitter reached 10 million users in 780 days. It took Facebook 852 days.

Singer from band Train sends a picture from the concert stage

190 Million
That's the number of tweets sent on an average day.

23 minutes

Social networks are taking up more and more of our time online. Users spend on average about 23 minutes of every hour of computer time on social networking sites.

62% Of online users worldwide, 62 percent use the Internet for social networking. Social media is most popular in Indonesia, where 83 percent take part. Although the use of social networking is growing, e-mail remains number one. About 85 percent of those surveyed say it remains their top online activity.

April 23, 2005

On that day, the very first video was uploaded to YouTube. It was called "Me at the Zoo." A little more than a year later, more than 65,000 videos were being uploaded every day.

55 and older | Nielsen reported that users 55 and older are the fastest-growing group on social networks. However, people between ages 18 and 34 are the most active age group.

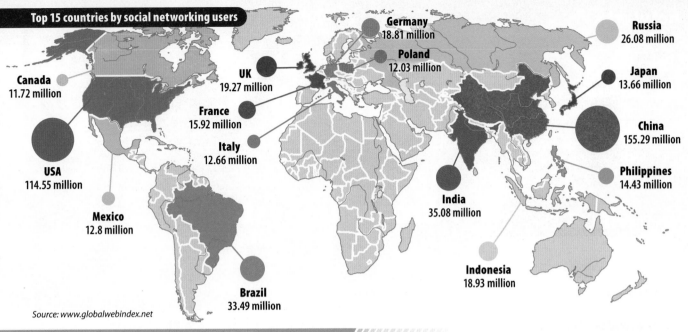

Canada
11.72 million

USA
114.55 million

Mexico
12.8 million

Brazil
33.49 million

UK
19.27 million

France
15.92 million

Italy
12.66 million

Germany
18.81 million

Poland
12.03 million

Russia
26.08 million

Japan
13.66 million

China
155.29 million

Philippines
14.43 million

India
35.08 million

Indonesia
18.93 million

Source: www.globalwebindex.net

GLOBAL IMPACT

THE GEOGRAPHY OF SOCIAL MEDIA When social media and the Internet were developed, many people said that physical location would no longer matter. People are able to communicate as easily with someone half a world away as with the person next door. However, most communication on social media is with people who are close by. They could go to see the person they are addressing, but they prefer to send a message. Messages on Facebook can be sent to many people at the same time.

Facebook Users

The graph shows the number of Facebook users worldwide from 2004 to 2011.

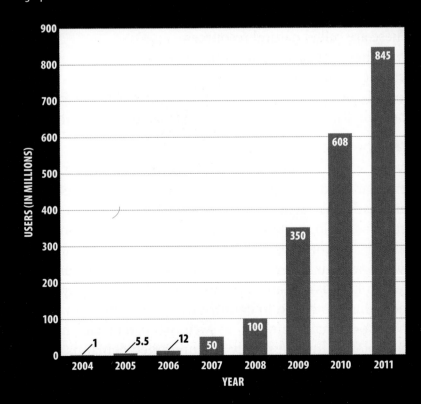

Thinking like a
Geographer

1. ***Analyzing*** Why is social media useful?

2. ***Describing*** How would you use social media to plan an event at your school?

3. ***Integrating Visual Information*** Research one country where "Arab Spring" protests took place in 2011. Prepare a PowerPoint® presentation that includes photographs, maps, and messages. Present your slide show to the class.

networks

There's More Online!

- ☑ **CHART/GRAPH** Economic Terms
- ☑ **IMAGE** Renewable Resources
- ☑ **GAME** Bartering and Trade
- ☑ **VIDEO**

Reading **HELP**DESK

Academic Vocabulary

- **currency**

Content Vocabulary

- **renewable resource**
- **nonrenewable resource**
- **opportunity cost**
- **economic system**
- **traditional economy**
- **market economy**
- **command economy**
- **mixed economy**
- **gross domestic product**
- **standard of living**
- **productivity**
- **export**
- **import**
- **free trade**
- **sustainability**

TAKING NOTES: *Key Ideas and Details*

Organize As you read, summarize the key ideas about each economic system.

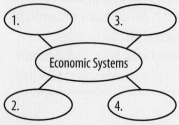

Indiana Academic Standards
6.3.1, 6.3.3, 6.3.12, 6.4.1, 6.4.2, 6.4.3, 6.4.4, 6.4.5, 6.4.7, 6.4.9

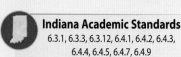

Lesson 3
Economic Systems

ESSENTIAL QUESTION • *Why do people make economic choices?*

IT MATTERS BECAUSE
People strive to meet their basic needs and their desires for a better life.

Economic Principles

GUIDING QUESTION *How do people get the things they want and need?*

All human beings have wants and needs. How do you get the things you want and the things you need? To obtain these items, people use resources. Resources are the supplies that are used to meet our wants and needs. Some types of resources, such as water, soil, plants, and animals, come from the earth. These are called natural resources.

Other resources are supplied by humans. Human resources include the labor, skills, and talents people contribute. Countries also have wants and needs. Like individuals, nations must use resources to meet their needs.

Wants and Resources

What would happen if 14 students each wanted a glass of lemonade from a pitcher that contained only 12 glasses of lemonade? What if even more students wanted a glass of lemonade? No matter how many people want lemonade, the pitcher still contains just 12 glasses. There is not enough for everyone. This is an example of a limited supply and unlimited demand. This situation is not uncommon. It happens to individuals and also to countries. You probably can think of many personal examples, as well as current and historical examples, of limited supply and unlimited demand.

One type of resource everyone needs is energy. Energy is the power to do work. Energy resources are the supplies that provide the power to do work. Many types of energy resources exist in our world. Energy resources can be renewable or nonrenewable. **Renewable resources** are resources that can be totally replaced or are always available naturally. They can be regenerated and replenished. Examples of renewable resources include water, trees, and energy from the wind and the sun.

In contrast are nonrenewable resources. **Nonrenewable resources** can not be totally replaced. Once nonrenewable resources are consumed, they are gone. Examples of nonrenewable resources include the fossil fuels oil, coal, and natural gas. These fuels received their name because they formed millions of years ago. Humans' increasing need for energy is taking its toll as supplies of nonrenewable resources shrink.

Making Choices

If the peoples of all nations have unlimited wants but face limited resources, what must happen? We must make choices. Do we continue to use nonrenewable resources? If so, at what rate should we be using them? Should we switch to renewable resources?

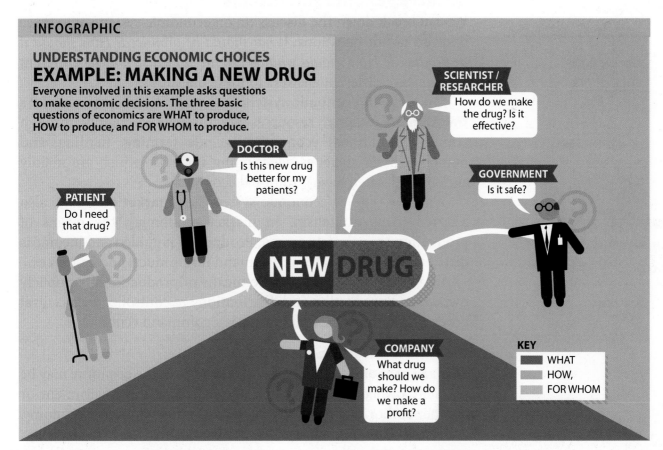

INFOGRAPHIC

UNDERSTANDING ECONOMIC CHOICES
EXAMPLE: MAKING A NEW DRUG
Everyone involved in this example asks questions to make economic decisions. The three basic questions of economics are WHAT to produce, HOW to produce, and FOR WHOM to produce.

SCIENTIST / RESEARCHER
How do we make the drug? Is it effective?

DOCTOR
Is this new drug better for my patients?

GOVERNMENT
Is it safe?

PATIENT
Do I need that drug?

NEW DRUG

COMPANY
What drug should we make? How do we make a profit?

KEY
- WHAT
- HOW,
- FOR WHOM

Every economic system must address three basic questions.

▶ **CRITICAL THINKING**
Determining Central Ideas How are the three basic economic questions related to the problem of limited supply?

No countries today rely on a command economic system.

Not true. Some countries still have planned economies, including Cuba, Saudi Arabia, Iran, and North Korea. These nations have an economic system in which supply and prices are regulated by the government, not by the market.

We must weigh the **opportunity cost**, or the value of what we must give up to acquire something else, of using renewable resources versus nonrenewable resources. We must take into account these and many more considerations as we make choices now and in the future.

☑ **READING PROGRESS CHECK**

Describing How are renewable and nonrenewable resources alike, and how are they different?

National Economies

GUIDING QUESTION *What kinds of economic systems are used in our world today?*

Economic resources are another important resource. Economic resources include the goods and services a society provides and how they are produced, distributed, and used. How a society decides on the ownership and distribution of its economic resources is its **economic system**. Do you ever stop to think about the goods and services you use in a single day? How do these goods and services become available to you?

Different Economic Systems

We can break down the discussion on economic systems into three basic economic questions: *What should be produced? How should it be produced? How should what is produced be distributed?* Different nations have different answers to these questions.

One type of economic system is the traditional economy. In a **traditional economy**, resources are distributed mainly through families. Traditional economies include farming, herding, and hunter-gatherer societies. Developing societies, which are mainly agricultural, often have traditional economies.

Another type of economic system is a **market economy**, also referred to as capitalism. In market economies, the means of production are privately owned. Production is guided and income is distributed through sales and demand for products and resources.

In a **command economy**, the means of production are publicly owned, and production and distribution are controlled by a central governing authority. Communism is a command economy.

What Is a Mixed Economy?

A **mixed economy** is just that—mixed. Parts of the economy may be privately owned, and parts may be owned by the government or another authority. The United States has a mixed market economy. Another economic system is socialism. In socialist societies, property and the distribution of goods and income are controlled by the community. How do individuals get the goods or services they need under each of these economic systems?

Buyers and sellers come together in an outdoor market in a village in the Central Highlands of Mexico.

▶ **CRITICAL THINKING**

Describing Is the seller involved in a primary, secondary, or tertiary economic activity?

Parts of the Economy

We can break down the economy into parts. In economics, *land* is a factor of production that includes natural resources. Another factor of production is *labor*, which refers to all paid workers within a system. The other factor is *capital*, the human-made resources used to produce other goods. An *industry* is a branch of a business. For example, the *agriculture industry* grows crops and raises livestock. *Service industries* provide services rather than goods. Banking, retail, food service, transportation, and communications are examples of service industries.

Types of Economic Activities

Another way to view the parts of the economy is by the type of economic activity. Economists use the terms *primary sector*, *secondary sector*, and *tertiary sector* to group these activities. The primary sector includes activities that produce raw materials and basic goods. These activities include mining, fishing, agriculture, and logging.

The secondary sector makes finished goods. This sector includes home and building construction, food processing, and aerospace manufacturing. The tertiary sector of the economy is the service industry. Service sectors include sales, restaurants, banking, information technology, and health care.

Indiana CONNECTION

Saving and Investing

In all societies, saving and investing helps increase productivity and economic growth. There are various ways that one can save or invest. One way is to open a savings account at a bank or credit union. These institutions accept people's money, pay them a fairly low rate of interest, and loan the money on deposit to other customers. You can withdraw your money at any time, and the interest you earn is added automatically to your principal, or the amount you initially deposited.

Another way to save is with certificate of deposit (CDs). CDs are a kind of time deposit, in which you agree to deposit a sum of money with a financial institution for a certain amount of time. In return, you are guaranteed a set rate of interest that will be added to your principal when the CD comes due. The rate of interest on a CD is almost always higher than that on a savings account. This is because you have less flexibility to withdraw your money. If you want to withdraw it before the stated date, you must pay a substantial penalty.

Almost all investors also invest in stocks. When you buy shares of stock, you are buying partial ownership in a company. You can sell the shares at any time, hopefully for more than you paid, and the difference is your profit. Stocks generally earn a higher return because they carry greater risk. There is no guarantee that you will make money on your stock investment. In fact, if the company goes out of business, you will lose your entire investment.

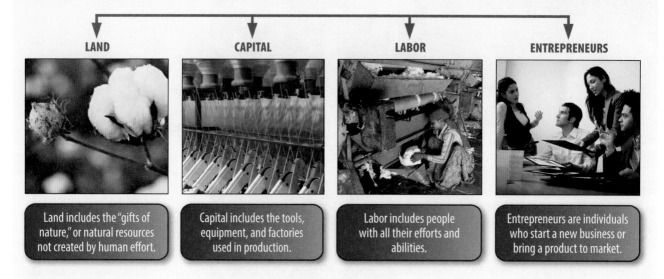

LAND

CAPITAL

LABOR

ENTREPRENEURS

Land includes the "gifts of nature," or natural resources not created by human effort.

Capital includes the tools, equipment, and factories used in production.

Labor includes people with all their efforts and abilities.

Entrepreneurs are individuals who start a new business or bring a product to market.

FACTORS OF PRODUCTION

The factors of production are broad categories of resources we need to produce the goods and services we want.

▶ **CRITICAL THINKING**

Identifying A tractor used to plow a field for crops is an example of what factor of production?

Economic Performance

Economic performance measures how well an economy meets the needs of society. Economic performance can be determined by several factors that measure economic success. The **gross domestic product (GDP)** is the total dollar value of all final goods and services produced in a country during a single year. The **standard of living** is the level at which a person, a group, or a nation lives as measured by the extent to which it meets its needs. These needs include food, shelter, clothing, education, and health care. Per capita income is the total national income divided by the number of people in the nation.

When referring to economics, **productivity** is a measurement of what is produced and what is required to produce it. Sustainable growth is the growth rate a business can maintain without having to borrow money. The employment rate is the percentage of the labor force that is employed. These factors help determine a nation's economic strength and performance.

Types of National Economies

National economies also can be classified by types. Developed countries are industrialized countries. Developing countries are less industrialized, agricultural countries that are working to become more advanced economically. Developing countries often have weak economies, and most of their population lives in poverty. Newly industrialized countries (NICs) are in the process of becoming developed and economically secure. Their economies are growing and struggling to become fully developed, but they still face many economic and social challenges.

☑ **READING PROGRESS CHECK**

Describing How is standard of living a sign of economic performance or success?

World Economy

GUIDING QUESTION *How do the world's economies interact and affect one another?*

You have read about different economic systems and different types of economies. All the world's nations can be classified into the different economic categories. All nations must find ways to interact with one another. Look at the labels in your clothes or on other products you buy. How many different country names can you find? How and why do we get goods from far across the world?

Trade

Trade is the business of buying, selling, or bartering. When you buy something at the store, you are trading money for a product. On a much bigger scale, nations trade with each other. Countries have different resources. Resources can include raw materials, such as iron ore. Even labor may be cheaper in another country where workers earn lower wages. As a result, goods can be produced in some countries more easily or efficiently than in other countries.

Trade can benefit countries. One country can **export**, or send to another country, a product that it is able to produce. Another country **imports**, or buys that product from another country.

GRAPH SKILLS >

GDP COMPARISON

GDP per capita can help us compare the economic output of a country as it relates to the standard of living of its people.

▶ **CRITICAL THINKING**

1. *Integrating Visual Information*
   The GDP of which countries surpassed $6 trillion?

2. *Integrating Visual Information*
   To reach $2 trillion, Mexico's GDP has to grow by how many dollars?

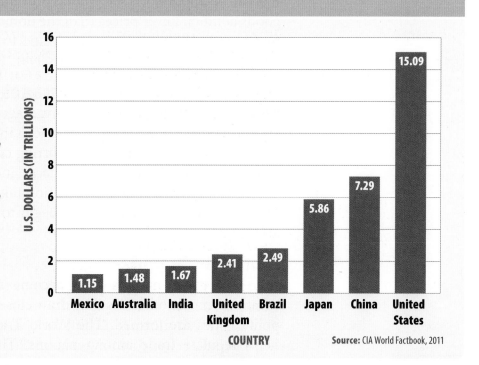

U.S. DOLLARS (IN TRILLIONS)

Mexico	Australia	India	United Kingdom	Brazil	Japan	China	United States
1.15	1.48	1.67	2.41	2.49	5.86	7.29	15.09

COUNTRY

Source: CIA World Factbook, 2011

International trade involves preparing cargo for shipping (left) and transporting goods (right).

▶ **CRITICAL THINKING**

Describing Why do countries trade?

The country that imported the product can in turn export its products to another country. In global trade, extra fees are often added to the cost of importing products by a country's government. The extra money is a type of tax called a tariff. Governments often create tariffs to persuade their people to buy products made in their own country.

Sometimes a quota, a limit on the amount of one particular good that can be imported, is set. Quotas prevent countries from exporting goods at much lower prices than the domestic market can sell them for. A group of countries may decide to set little or no tariffs or quotas when trading among themselves. This is called **free trade**.

Advantages and Disadvantages of Trade

Trade has advantages and disadvantages. Trade can help build economic growth and increase a nation's income. On the other hand, jobs might be lost because of importing certain goods and services. With its benefits and its barriers, increasing trade leads to globalization. Economic globalization takes place when businesses move past national markets and begin to trade with other nations around the world.

Economic Organizations

In recent years, nations have become more interdependent, or reliant on one another. As they draw closer together, economic and political ties are formed. The World Trade Organization (WTO) helps regulate trade among nations. The World Bank provides financing, advice, and research to developing nations to help them grow their economies. The International Monetary Fund (IMF) is a group that monitors economic development. The IMF also lends money to nations in need and provides training and technical help. One well-known policy and organization that promotes global trade is the North American Free Trade Agreement (NAFTA). NAFTA encourages free trade among the United States, Canada, and Mexico.

The European Union (EU) is a group of European countries that operate under one economic unit and one **currency**, or type of money—the euro. The Mercado Camon del Sur (formerly called MERCOSUR) is a group of South American countries that promote free trade, economic development, and globalization. The Mercado Camon del Sur helps countries make better use of their resources while preserving the environment.

The Association of Southeast Asian Nations (ASEAN) is a group of countries in Southeast Asia that promote economic, cultural, and political development. The Dominican Republic-Central America Free Trade Agreement (CAFTA-DR) is an agreement among the United States, five developing Central American countries, and the Dominican Republic. The agreement promotes free trade.

Whether a nation produces its own goods or trades, one basic principle exists: sustainability. The principle of **sustainability** is central to the discussion of resources. When a country focuses on sustainability, it works to create conditions where all the natural resources for meeting the needs of society are available.

What can countries do to ensure sustainability now and into the future? What can you do to plan for your future and the future of your community? Just as every nation is part of a global system, you are part of your community. The choices you make affect you and those around you. What can you do now to plan for a bright economic future?

☑ **READING PROGRESS CHECK**

Analyzing What are some possible disadvantages of trade?

Include this lesson's information in your Foldable®.

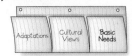

LESSON 3 REVIEW

Reviewing Vocabulary
1. List examples of *nonrenewable resources*.

Answering the Guiding Questions
2. *Determining Word Meanings* Explain the opportunity cost of buying a new pair of jeans.

3. *Determining Central Ideas* Explain why the United States has a market economy.

4. *Determining Central Ideas* Why is it advantageous to the United States to have a free trade agreement with Mexico and Canada?

5. *Informative/Explanatory Writing* Write a short essay explaining decisions that you can make that help ensure the sustainability of our resources. Consider things you do daily as well as things you do or buy less frequently.

Directions: Write your answers on a separate piece of paper.

❶ Exploring the Essential Questions

INFORMATIVE/EXPLANATORY WRITING Research the culture of a South American country. Create a poster with pictures or photos showing some of the unique features, holidays, and celebrations that you learn about.

❷ 21st Century Skills

INTEGRATING VISUAL INFORMATION Select one country in Europe to examine more closely. Create a slide show presentation that includes the following information: population, population density, population patterns, primary language, type of government, customs and other cultural information, resources, and type of economy.

❸ Thinking Like a Geographer

IDENTIFYING Look up the population density and total population figures for North America, South America, and Europe. Make a chart that shows both figures for each continent. Write a narrative explanation of your findings also.

❹ GEOGRAPHY ACTIVITY

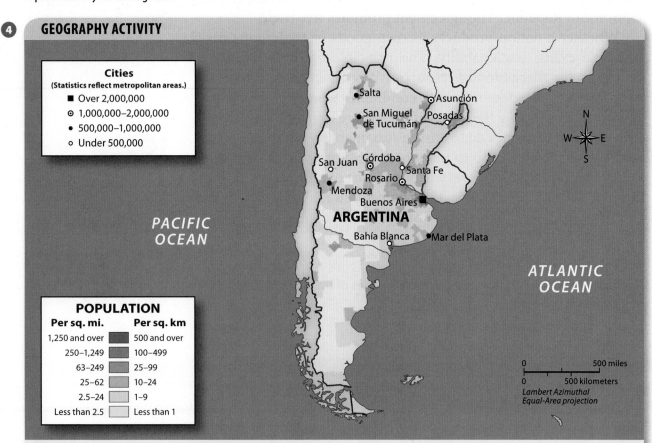

Study the map above and answer the following questions.

1. What is the population of Buenos Aires?

2. What is the population of Salta?

3. About how many people per square mile live in San Juan?

REVIEW THE GUIDING QUESTIONS

Directions: Choose the best answer for each question.

1 The doubling time refers to the amount of time it takes for

   A.   the birthrate to double.

   B.   the population to double.

   C.   the death rate to double.

   D.   a child to reach adulthood.

2 A megalopolis is

   F.   a huge urban area.

   G.   an area that will not grow any larger.

   H.   found in a rural setting.

   I.   found most often in inland areas.

3 The American culture encourages

   A.   intolerance of diverse religious beliefs.

   B.   adherence to common goals.

   C.   the same education for everyone.

   D.   a blending of lifestyles and backgrounds.

4 Because of globalization,

   F.   new languages are forming.

   G.   people now travel less.

   H.   birthrates are increasing rapidly.

   I.   international trade is increasing

5 Choosing between buying a new car or a used car is an example of considering

   A.   a quota.

   B.   an economic system.

   C.   opportunity cost.

   D.   an energy resource.

DBQ ANALYZING DOCUMENTS

6 DETERMINING WORD MEANINGS A news story reports on India's population growth.

"*India . . . will surpass China to become the world's most populous country in less than two decades. The population growth will mean a nation full of working-age youth, which economists say could allow the already booming economy to maintain momentum.*"

What does *surpass* mean in this story?

A. to move ahead of

B. to fall behind

C. to increase in speed

D. to be the same as

7 ANALYZING How will the increase in population help India?

F. It will allow the people to use more resources.

G. The farmers will produce more food.

H. India's companies will have enough workers.

I. Schools will be educating more students.

SHORT RESPONSE

"*A school that is . . . easy for students, teachers, [and] parents . . . to reach on foot or by bicycle helps reduce the air pollution from automobile use, protecting children's health. Building schools . . . in the neighborhoods they serve minimizes the amount of paved surface . . ., which can help protect water quality by reducing polluted runoff.*"

—from Environmental Protection Agency, "Smart Growth and Schools"

8 DETERMINING CENTRAL IDEAS Explain in your own words what the author would like to see change.

9 ANALYZING What are some disadvantages to building community schools?

EXTENDED RESPONSE

10 ARGUMENT WRITING From the early years of the founding of this country, debate has raged over how many immigrants to allow in and from where. More recently, discussion has focused on what to do about people who have entered the United States illegally. The debate continues today. Research various countries in Europe and North and South America to collect statistics about which countries have large numbers of immigrants including refugees and where they are primarily from. Also consider the reasons that these people might be moving from one country to another. Write a letter to your congressional representative explaining these facts and how other countries are dealing with their immigration situation. Then state your position on immigration to this country. Back up your position with firm reasons.

Need Extra Help?

If You've Missed Question	❶	❷	❸	❹	❺	❻	❼	❽	❾	❿
Review Lesson	1	1	2	3	3	1	1	2	2	1

"India Challenged to Provide Jobs, Education to Young Population," by Anjana Pasricha, October 31, 2011. *Voice of America*, http://voanews.com; "Smart Growth and Schools," The United States Environmental Protection Agency official Web site, http://www.epa.gov/dced/school.htm

NORTH AMERICA

UNIT **2**

Chapter 4
The United States East
of the Mississippi River

Chapter 5
The United States West
of the Mississippi River

Chapter 6
Canada

Chapter 7
Mexico, Central America,
and the Caribbean Islands

EXPLORE the CONTINENT

NORTH AMERICA is made up of three large countries—Canada, the United States, Mexico—plus the Caribbean Islands and the countries of Central America. North America is a cultural kaleidoscope—a mixture of the many groups who have settled in the region.

1 LANDFORMS At 800 miles long (1,287 km), the Baja Peninsula of Mexico is one of the world's longest peninsulas. Running nearly its entire length is a chain of mountain ranges. Four deserts are also physical features of Baja.

2 BODIES OF WATER Boats line the harbor of a fishing village in Nova Scotia, along Canada's Atlantic coast. In addition to the waters of the Atlantic, the coastal waters of the Pacific Ocean and the Gulf of Mexico are important fisheries.

(3) **NATURAL RESOURCES** Wheat is grown on the Great Plains of the United States, a region often called the Wheat Belt. The type of wheat grown depends on the climate. Farmers in the northern plains, with their short growing season, plant wheat in the spring and harvest it in the fall. Farther south, farmers plant winter wheat, which is harvested in early summer.

FAST FACT

Canada is the world's second largest country in area.

(tr) NOAA National Geophysical Data Center (NGDC)

NORTH AMERICA

PHYSICAL

MAP SKILLS

1 PHYSICAL GEOGRAPHY What physical features probably acted as barriers to settlement in the United States and Canada?

2 THE GEOGRAPHER'S WORLD Which body of water is located east of Panama?

3 PHYSICAL GEOGRAPHY What mountain range lies in the eastern United States?

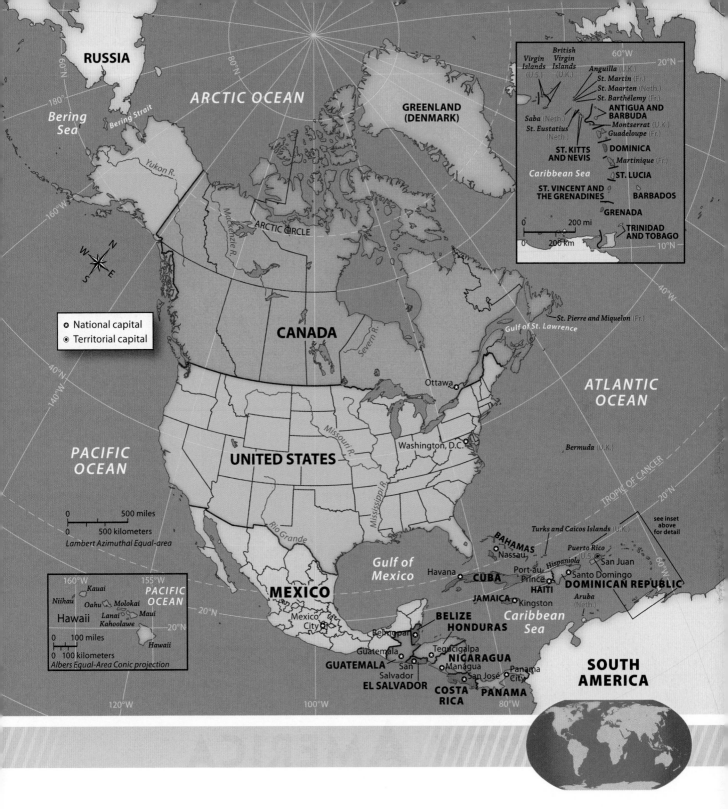

RUSSIA

ARCTIC OCEAN

Bering Sea

GREENLAND (DENMARK)

Yukon R.

ARCTIC CIRCLE

Mackenzie R.

- ✪ National capital
- ◉ Territorial capital

CANADA

Severn R.

St. Pierre and Miquelon (Fr.)

Gulf of St. Lawrence

Ottawa ✪

ATLANTIC OCEAN

PACIFIC OCEAN

Missouri R.

UNITED STATES

Washington, D.C. ✪

Bermuda (U.K.)

TROPIC OF CANCER

500 miles

500 kilometers

Lambert Azimuthal Equal-area

Mississippi R.

Rio Grande

Gulf of Mexico

MEXICO

Havana

CUBA

BAHAMAS

Nassau

Turks and Caicos Islands (U.K.)

Puerto Rico (U.S.) San Juan

Hispaniola

Port-au-Prince

HAITI

Santo Domingo

DOMINICAN REPUBLIC

160°W   155°W

PACIFIC OCEAN

Kauai

Niihau   Oahu   Molokai

Hawaii   *Lanai   Maui*

Kahoolawe

20°N

Hawaii

0   100 miles

0   100 kilometers

Albers Equal-Area Conic projection

Mexico City ✪

BELIZE

Belmopan ✪

HONDURAS

Tegucigalpa ✪

JAMAICA

Kingston

Aruba (Neth.)

Caribbean Sea

SOUTH AMERICA

Guatemala ✪

GUATEMALA

San Salvador ✪

EL SALVADOR

NICARAGUA

Managua ✪

San José ✪

COSTA RICA

Panama City ✪

PANAMA

Caribbean inset:

60°W   20°N

Virgin Islands (U.S.)

British Virgin Islands (U.K.)

Anguilla (U.K.)

St. Martin (Fr.)

St. Maarten (Neth.)

St. Barthélemy (Fr.)

ANTIGUA AND BARBUDA

Saba (Neth.)

St. Eustatius (Neth.)

Montserrat (U.K.)

Guadeloupe (Fr.)

ST. KITTS AND NEVIS

DOMINICA

Martinique (Fr.)

Caribbean Sea

ST. LUCIA

ST. VINCENT AND THE GRENADINES

BARBADOS

GRENADA

TRINIDAD AND TOBAGO

10°N

0   200 mi

0   200 km

see inset above for detail

POLITICAL

MAP SKILLS

1 **PLACES AND REGIONS** What is the capital of Canada?

2 **THE GEOGRAPHER'S WORLD** What physical feature forms the natural boundary between the United States and Mexico?

3 **PLACES AND REGIONS** What is the capital of Cuba?

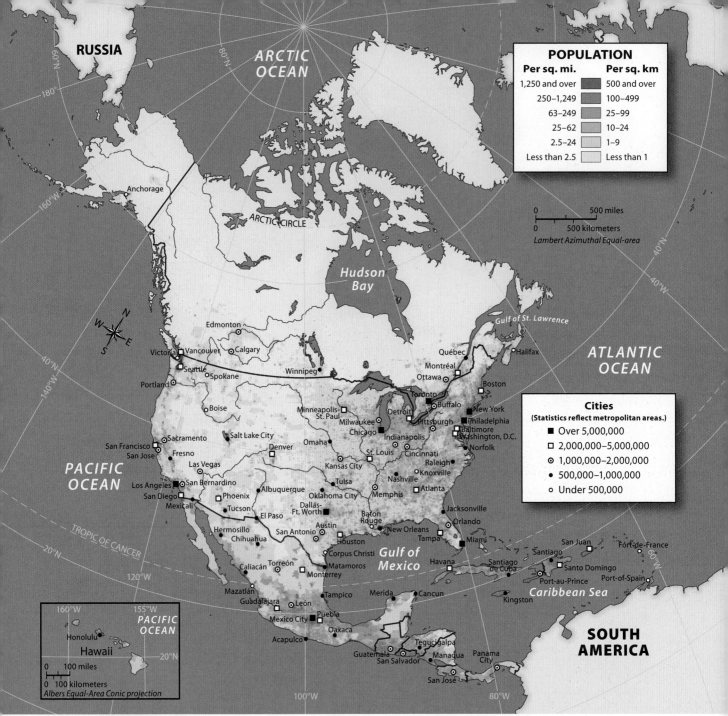

NORTH AMERICA

POPULATION DENSITY

MAP SKILLS

1. **PLACES AND REGIONS** Where is the greatest population density located in the United States?

2. **PLACES AND REGIONS** Contrast the population density of northern Mexico with southern Mexico.

3. **ENVIRONMENT AND SOCIETY** What generalizations can you make about the populations of the Caribbean islands?

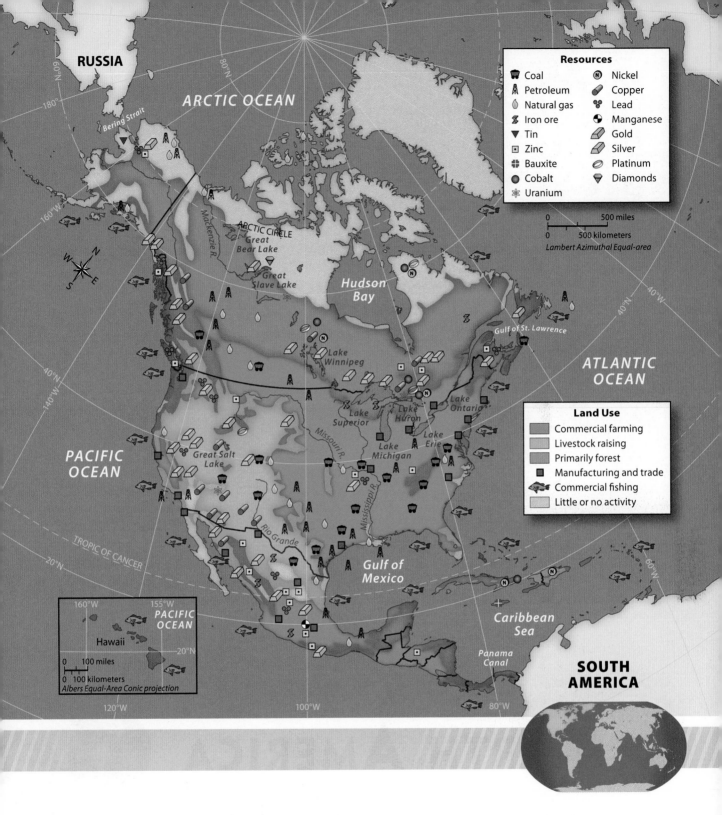

Resources

- Coal
- Petroleum
- Natural gas
- Iron ore
- Tin
- Zinc
- Bauxite
- Cobalt
- Uranium
- Nickel
- Copper
- Lead
- Manganese
- Gold
- Silver
- Platinum
- Diamonds

500 miles
500 kilometers
Lambert Azimuthal Equal-area

Land Use

- Commercial farming
- Livestock raising
- Primarily forest
- Manufacturing and trade
- Commercial fishing
- Little or no activity

RUSSIA

ARCTIC OCEAN

Bering Strait

ARCTIC CIRCLE

Great Bear Lake

Mackenzie R.

Great Slave Lake

Hudson Bay

Gulf of St. Lawrence

ATLANTIC OCEAN

PACIFIC OCEAN

Lake Winnipeg

Lake Superior

Lake Michigan

Lake Huron

Lake Ontario

Lake Erie

Missouri R.

Great Salt Lake

Mississippi R.

TROPIC OF CANCER

Rio Grande

Gulf of Mexico

Caribbean Sea

Panama Canal

SOUTH AMERICA

160°W   155°W
PACIFIC OCEAN

Hawaii

0   100 miles
0   100 kilometers
Albers Equal-Area Conic projection

120°W   100°W   80°W

ECONOMIC RESOURCES

MAP SKILLS

1 HUMAN GEOGRAPHY Describe how most land is used throughout the United States.

2 HUMAN GEOGRAPHY What is the main economic activity near Panama?

3 HUMAN GEOGRAPHY Contrast how land is used in northern Canada with Central America.

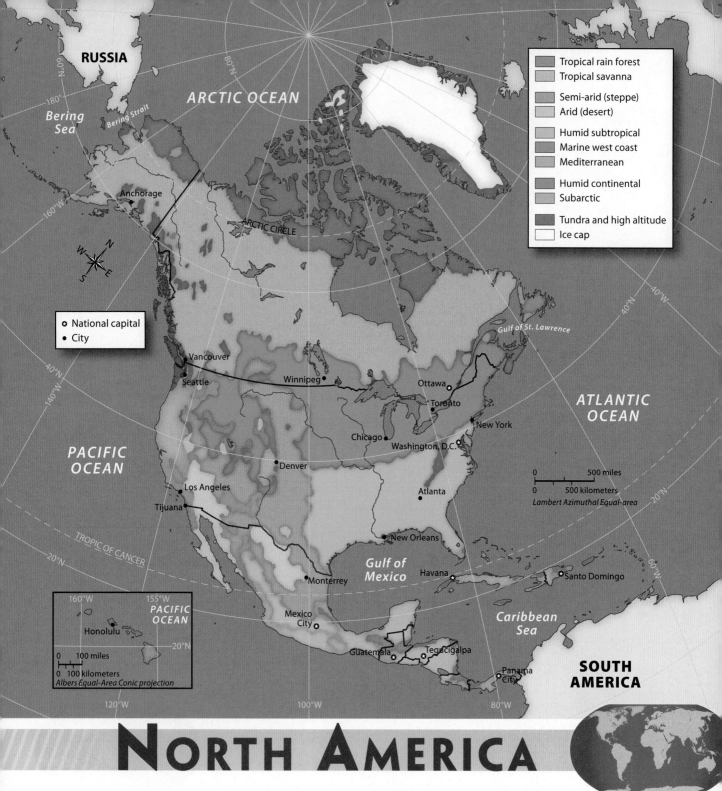

RUSSIA

ARCTIC OCEAN

Bering Sea

Bering Strait

Legend:
- Tropical rain forest
- Tropical savanna
- Semi-arid (steppe)
- Arid (desert)
- Humid subtropical
- Marine west coast
- Mediterranean
- Humid continental
- Subarctic
- Tundra and high altitude
- Ice cap

Anchorage

ARCTIC CIRCLE

- ◎ National capital
- ● City

Vancouver

Seattle

Winnipeg

Ottawa

Toronto

New York

Chicago

Washington, D.C.

Denver

Los Angeles

Atlanta

Tijuana

TROPIC OF CANCER

New Orleans

Gulf of Mexico

Havana

Monterrey

Santo Domingo

Caribbean Sea

Mexico City

Guatemala

Tegucigalpa

Panama City

SOUTH AMERICA

Gulf of St. Lawrence

ATLANTIC OCEAN

PACIFIC OCEAN

0   500 miles
0   500 kilometers
Lambert Azimuthal Equal-area

PACIFIC OCEAN

Honolulu

160°W   155°W

20°N

0   100 miles
0   100 kilometers
Albers Equal-Area Conic projection

NORTH AMERICA

CLIMATE

MAP SKILLS

1 **PHYSICAL GEOGRAPHY** What type of climate is most common throughout Canada?

2 **PLACES AND REGIONS** The Caribbean islands and central Mexico appear to have two things in common. What are these commonalities?

3 **PHYSICAL GEOGRAPHY** What is the overall climate of the southeastern United States?

THE UNITED STATES EAST OF THE MISSISSIPPI RIVER

netw⊙rks

There's More Online about The United States East of the Mississippi River.

CHAPTER 4

Lesson 1
Physical Features

Lesson 2
History of the Region

Lesson 3
Life in the Region

ESSENTIAL QUESTIONS • *How does geography influence the way people live?*
• *Why is history important?* • *What makes a culture unique?*

Christopher Morris/VII/Corbis

Worker at Michigan auto assembly plant

The Story Matters...

The eastern United States is a region of diverse physical features and many natural resources. Bordered by the Atlantic Ocean and the Mississippi River, this region is home to mighty rivers, the largest group of freshwater lakes in the world, and old-growth forests. The physical features and wealth of resources found in the eastern United States have played a key role in the region's history. They also influence where and how people live in this region today.

FOLDABLES
Study Organizer

Go to the Foldables® library in the back of your book to make a Foldable® that will help you take notes while reading this chapter.

Diversity
Geographic Barriers
East and West
The US East of the Mississippi

THE UNITED STATES EAST OF THE MISSISSIPPI RIVER

The United States east of the Mississippi River is one of the two regions that make up the United States. As you study the map, identify the geographic features of the region.

Step Into the Place

MAP FOCUS Use the map to answer the following questions.

1. **THE GEOGRAPHER'S WORLD** Which body of water lies to the east of the United States?

2. **THE GEOGRAPHER'S WORLD** Which state has the longest coastline?

3. **PLACES AND REGIONS** Name the states that use the Mississippi River as all or part of their western border.

4. **CRITICAL THINKING ANALYZING** Why do you think rivers such as the Mississippi and the Ohio were used as state borders?

A

THE CAPITOL The U.S. Capitol in Washington, D.C., is the meeting place of Congress, the nation's legislature.

B

THE GREAT SMOKY MOUNTAINS Part of the Appalachian Mountain Range, the Smoky Mountains are named for the blue-gray mist that seems to hang above the peaks and valleys.

Step Into the Time

DESCRIBING Choose an event from the time line and write a paragraph describing a social, environmental, or economic effect that event had on the region, the country, or the world.

1620 Pilgrims' ship *Mayflower* arrives in Plymouth

1800

1776 Declaration of Independence is signed

1825 Erie Canal links New York City and the Great Lakes

1861 U.S. Civil War begins

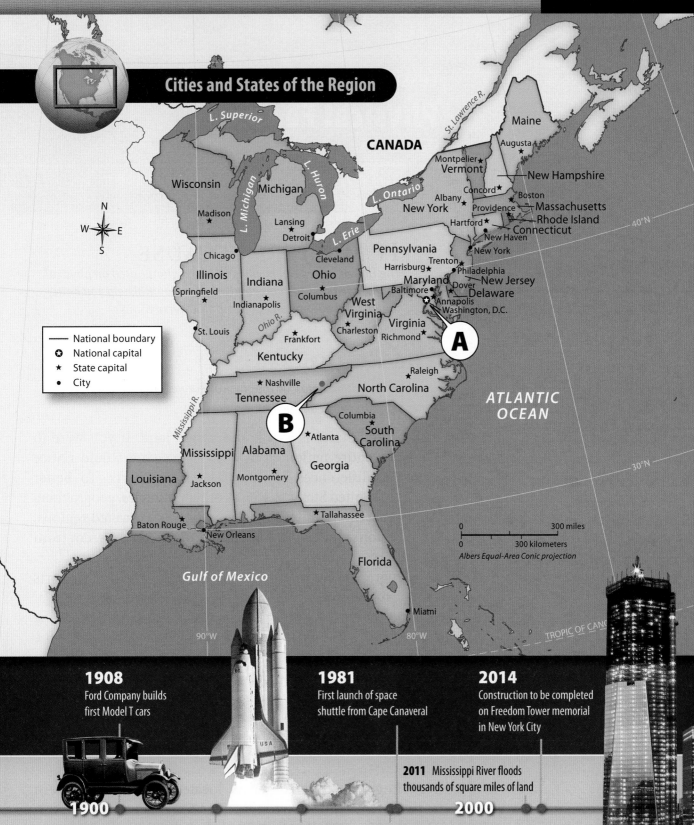

Cities and States of the Region

CANADA

L. Superior

L. Michigan

L. Huron

L. Ontario

L. Erie

St. Lawrence R.

Maine
Augusta ★

Montpelier ★
Vermont

New Hampshire

Concord ★

Albany ★
New York

Boston
Providence ●
Massachusetts
Rhode Island
Hartford ★ Connecticut
New Haven ●
New York ●

Wisconsin

Michigan

Madison ★

Lansing ★
Detroit ●

Chicago ●

Cleveland ●

Pennsylvania

Trenton ★
Harrisburg ★ ● Philadelphia
Maryland ● New Jersey
Baltimore ● Dover ★ Delaware
Annapolis ★
Washington, D.C.

Illinois

Indiana

Ohio

Springfield ★

Columbus ★

Indianapolis ★

West
Virginia

Virginia

St. Louis ●

Ohio R.

Frankfort ★

Charleston ●

Richmond ★

A

Kentucky

Raleigh ★

	National boundary
✪	National capital
★	State capital
●	City

Mississippi R.

★ Nashville

North Carolina

ATLANTIC OCEAN

Tennessee

B

Columbia
★
South
Carolina

Mississippi

Alabama

● Atlanta

Georgia

Louisiana

Jackson ★

Montgomery ★

Jackson ★

Florida

Baton Rouge ●
New Orleans ●

★ Tallahassee

Miami ●

Gulf of Mexico

0   300 miles
0   300 kilometers
Albers Equal-Area Conic projection

40°N

30°N

90°W   80°W   TROPIC OF CANCER

1908
Ford Company builds
first Model T cars

1981
First launch of space
shuttle from Cape Canaveral

2014
Construction to be completed
on Freedom Tower memorial
in New York City

2011 Mississippi River floods
thousands of square miles of land

1900

2000

1933 Tennessee Valley Authority
brings electricity to southeastern U.S.

1955 Montgomery bus
boycott begins in Alabama

1979 Nuclear accident occurs
at a Pennsylvania power plant

Reading **HELP**DESK

Academic Vocabulary

- **parallel**

Content Vocabulary

- **subregion**
- **lock**
- **tributary**
- **levee**
- **coastal plain**
- **fall line**
- **hurricane**

TAKING NOTES: *Key Ideas and Details*

Organize As you read about the region's physical landscape, take notes on a graphic organizer like this one.

U.S. East of the Mississippi

Landscape   Bodies of Water

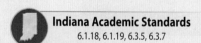

Indiana Academic Standards
6.1.18, 6.1.19, 6.3.5, 6.3.7

Lesson 1
Physical Features

ESSENTIAL QUESTION • *How does geography influence the way people live?*

IT MATTERS BECAUSE
The United States can be divided into regions based on physical characteristics. Learning about each region will help you better understand our nation's geographic diversity.

The Regions

GUIDING QUESTION *How do the physical features of the eastern United States make the region unique?*

The United States is a vast and varied land. If you were to view our entire nation from outer space, you would notice dramatic differences between its various parts. To better study the United States as well as other countries, geographers divide these parts into large geographic areas called regions. Each region's characteristics make it distinctly different from the others.

Geographers can further divide regions into smaller parts called **subregions**. Like a region, a subregion has special features that make it unique. The most basic way to divide the United States is into two regions: the United States east of the Mississippi and the United States west of the Mississippi. The Mississippi River is the dividing line between the two regions. In this lesson, you will learn about the United States east of the Mississippi and its four subregions: New England, the Mid-Atlantic, the Midwest, and the Southeast.

New England

New England is the subregion located in the northeastern corner of the United States, between Canada and the Atlantic Ocean. Many of the first English colonists who came to

America during the 1600s settled in this area. The settlers named the area New England in honor of their distant homeland. New England includes the states of Maine, New Hampshire, Vermont, Massachusetts, Rhode Island, and Connecticut.

The Mid-Atlantic

Located along the Atlantic coast, just south of New England, is the Mid-Atlantic subregion. The Mid-Atlantic includes the states of Delaware, Maryland, New Jersey, New York, and Pennsylvania. These states were part of America's original thirteen colonies. Our nation's capital, Washington, D.C., is also located in the Mid-Atlantic.

The Midwest

The states of Illinois, Indiana, Michigan, Ohio, and Wisconsin are part of the subregion called the Midwest. All five of these states share borders with one or more of the Great Lakes. The Midwest is nicknamed "the nation's breadbasket" because a large percentage of America's food crops are grown in its rich soil.

The Southeast

The Southeast is the largest subregion in the eastern United States. The Southeast is made up of 11 states: Alabama, Florida, Georgia, Kentucky, Louisiana, Mississippi, North Carolina, South Carolina, Tennessee, Virginia, and West Virginia. Some Southeastern states have long coastal borders where they meet the Atlantic Ocean or the Gulf of Mexico.

✔ **READING PROGRESS CHECK**

Determining Central Ideas Why do you think geographers divide the United States at the Mississippi River instead of dividing it through the middle into equal halves?

Bodies of Water

GUIDING QUESTION *Which of North America's major bodies of water are located east of the Mississippi?*

Oceans, lakes, and rivers have helped make this region prosperous. Oceans link the region to other countries for trade. An abundant supply of freshwater provides power for homes and industries.

MAP SKILLS

1 **PLACES AND REGIONS** Which subregion includes the state of Alabama?

2 **THE GEOGRAPHER'S WORLD** Which subregion extends the farthest north?

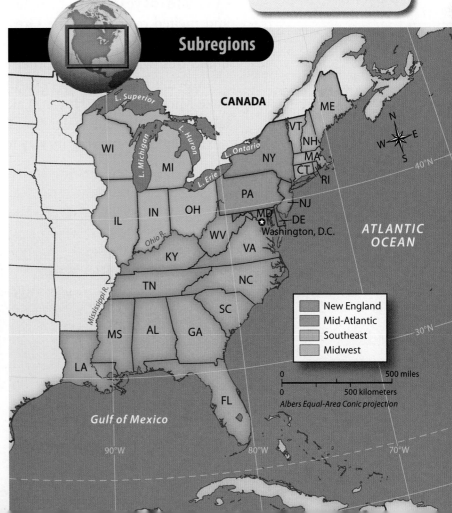

Subregions

L. Superior • CANADA • ME • VT • NH • MA • CT • RI • L. Huron • L. Ontario • NY • L. Erie • WI • L. Michigan • MI • PA • NJ • IL • IN • OH • MD • DE • Washington, D.C. • WV • VA • ATLANTIC OCEAN • Ohio R. • KY • TN • NC • Mississippi R. • SC • MS • AL • GA • LA • FL • Gulf of Mexico

New England
Mid-Atlantic
Southeast
Midwest

0   500 miles
0   500 kilometers
Albers Equal-Area Conic projection

40°N • 30°N • 90°W • 80°W • 70°W

The Atlantic Ocean and the Gulf Coast

The eastern United States is nearly surrounded by water. The largest body of water east of the Mississippi is the Atlantic Ocean. This enormous salt-water ocean borders the states along the East Coast. The East Coast is a shoreline that stretches for more than 2,000 miles (3,219 km), from Maine in the north to Florida in the south. It is jagged and rocky in New England but smooth and sandy in the Mid-Atlantic and Southeast. The Atlantic Ocean affects the region's land, weather, economy, and people in many ways.

The East Coast borders the Atlantic Ocean. The Gulf Coast borders a smaller body of water called the Gulf of Mexico. The Gulf of Mexico covers an area of about 600,000 square miles (1,550,000 sq. km) and is nearly surrounded by land. Several currents flow through the Gulf of Mexico like giant underwater rivers. One of the currents feeds into the Gulf Stream, a powerful current that flows through the Atlantic Ocean.

The Gulf Coast extends from Florida to Texas and Mexico. The land along the Gulf Coast varies from sandy beaches to marshes, bays, and lagoons. Waters in the Gulf of Mexico are warmer and generally calmer than those of the Atlantic.

The Great Lakes

The term *Great Lakes* refers to a cluster of five huge lakes located in the American Midwest and central Canada. These lakes were formed thousands of years ago when massive glaciers carved out the ground and melted over time. The Great Lakes form the largest group of freshwater lakes in the world. Together, the Great Lakes hold more liquid freshwater than any other location on Earth.

Moving from west to east, the Great Lakes are Lake Superior, Lake Michigan, Lake Huron, Lake Erie, and Lake Ontario. The five lakes are connected. Water flows west to east from one lake to the next, eventually making its way to the long St. Lawrence River.

Destin Beach is located on the Florida Panhandle along the Gulf of Mexico.
▶ **CRITICAL THINKING**
Describing How far does the Gulf Coast extend?

Karl Weatherly/The Image Bank/Getty Images

The St. Lawrence Seaway

The St. Lawrence River carries the water eastward for 750 miles (1,207 km), until it empties into the Atlantic Ocean. Because the Great Lakes border the United States and Canada, these nations work together to set up environmental programs for the region. Important goals include addressing population threats and protecting the health and safety of people living in the Great Lakes region.

During the 1950s, the United States and Canada worked together to build canals and gated passageways called **locks** between the Great Lakes and into the St. Lawrence River. The locks and canals made it possible for ships to travel the entire length of the Great Lakes and the St. Lawrence River. The final passageway, extending 2,340 miles (3,766 km) from Lake Superior to the Atlantic Ocean, is called the St. Lawrence Seaway. The St. Lawrence Seaway connects the Midwest to seaports all over the world. This has made it faster and easier for businesses in the Midwest to ship their products to buyers worldwide.

The Mississippi River

The "Mighty Mississippi" is one of the longest rivers in North America. Many people consider it the most important river in the United States. From its source in Minnesota, the Mississippi River winds its way southward for 2,350 miles (3,782 km). **Tributaries** such as the Missouri and Ohio rivers feed into the Mississippi, adding to its strength and volume. The Mississippi River ends at the point where it empties into the Gulf of Mexico.

Since early settlers arrived in America, the Mississippi River has affected the settlement patterns, the economy, and the lifestyles of countless Americans. People have used the river for transportation for hundreds of years.

The Welland Canal (left) begins on Lake Ontario. The St. Lawrence Seaway connects Chicago's harbors to the Atlantic Ocean. (right)

▶ **CRITICAL THINKING**

Describing How do canals aid ship travel?

The Mississippi River is the largest river system in North America.

▶ **CRITICAL THINKING**

Analyzing How do its many tributaries change the Mississippi River?

Ships and steamboats filled with passengers and cargo can follow the wide river and its tributaries for thousands of miles. This vast stretch makes the Mississippi one of the world's busiest commercial waterways.

In the past, the Mississippi would often flood its banks, dumping millions of tons of water and sediment onto the land. The sediment enriched the soil in farm fields, but the floods also destroyed homes and washed away entire fields of crops. The government built **levees**—embankments to control the flooding and reduce the damage to homes and crops. Unfortunately, levees also block the sediment that used to replenish farm fields.

The powerful Mississippi River has influenced the nation's history more than any other river. The river's importance, and the respect humans have for it, is shown in its name: *Mississippi* is a Choctaw word meaning "Great Water" or "Father of Waters."

Rivers as Boundaries

Rivers make natural boundaries. The Mississippi River forms much of the western border of the states of Wisconsin, Illinois, Kentucky, Tennessee, and Mississippi.

Rivers are boundaries for counties and cities, too. For example, the Tennessee River in northwest Alabama forms the border between Lauderdale County and Colbert County.

Rivers are examples of physical systems that form political boundaries. Mountains and lakes are other physical features that may be used as political boundaries. Structures that humans make, such as streets and roads, can also set boundaries.

Scenics of America/PhotoLink/Getty Images

The Ohio River

The Ohio River carries more water to the Mississippi River than any of the other tributaries. The Ohio River begins where the Allegheny and Monongahela rivers combine in western Pennsylvania. From Pennsylvania, the Ohio River flows westward for 981 miles (1,579 km), forming a wide, watery border that separates the states of Ohio, Indiana, and Illinois that lie along the north side of the river from West Virginia and Kentucky on its south side. Like the Mississippi, the Ohio River has long been an important shipping and transportation route. The Ohio River connects much of the Midwest to the Mississippi River. Both of these river systems have affected our nation's land, its people, and its history.

☑ **READING PROGRESS CHECK**

Determining Central Ideas How could a logging company in Kentucky send logs to a buyer on the Gulf of Mexico using an all-water route?

Physical Landscape

GUIDING QUESTION *What characteristics make the physical landscape east of the Mississippi unique?*

The Atlantic Coastal Plain

The East Coast of the United States sits at the edge of a huge continental platform. Most of this platform is underwater, forming a shelf around the Atlantic coastline. But over time, a large area of the platform rose above sea level. Ocean waves washed over the platform for millions of years, leaving behind layers of sandy sediment. As the sediment built up, a flat lowland called the **coastal plain** formed. The coastal plain stretches from the northeastern U.S. to Mexico. In places, the coastal plain was crushed under the weight of glaciers, pushing the land below sea level. These areas often become flooded by fierce storms and heavy rains.

The Appalachian Mountains

The Appalachian Mountain system is the oldest, longest chain of mountains in the United States east of the Mississippi River. It begins in Alabama and continues 1,500 miles (2,414 km) northeast to the Canadian border. Dense forests cover much of the Appalachian Mountains, which are known for their rugged beauty.

The mountains of the Appalachian system stand side by side in **parallel** ranges. Two of the most well-known Appalachian Mountain ranges are the Blue Ridge Mountains in Virginia and the Great Smoky Mountains in Tennessee.

Even though central New York is more than 200 miles (322 km) from the Atlantic Ocean, scientists have found fossils of marine organisms there.

▶ **CRITICAL THINKING**
Analyzing What is an explanation for finding the marine fossils so far from the ocean?

Academic Vocabulary

parallel extending side by side in the same direction, always the same distance apart

The Pisgah National Forest in western North Carolina is a land of heavy forests and many waterfalls.

▶ **CRITICAL THINKING**

Explaining What is a fall line?

These mountain ranges were formed from sedimentary rock by powerful upheavals within Earth's crust. The mountains have worn down over time because of natural erosion. Compared to younger mountain ranges such as the Rockies in the western United States, the Appalachian Mountains show their age in their worn, rounded appearance.

The Appalachian Mountains are home to many natural wonders. Old-growth forests are filled with diverse plant and animal life. Some of the most spectacular features of the Appalachians are the thousands of waterfalls that decorate the landscape. The many waterfalls are evidence that a fall line runs through the region. A **fall line** is an area where waterfalls flow from higher to lower ground. In this region, a fall line stretches for hundreds of miles between New Jersey and South Carolina. This fall line is a long, low cliff that runs parallel to the Atlantic coast. Throughout New England, the Mid-Atlantic, and the Southeast, waterfalls spill over this fall line. The fall line forms a boundary between higher, upland areas and the Atlantic coastal plain. Many cities originally located along the fall line because waterfalls provide water power, a renewable resource.

Climate in the Eastern United States

The climate of the eastern United States is as varied as the landscape. The changing seasons in most places east of the Mississippi are quite noticeable. New England and the Midwest see the most dramatic seasonal changes. These regions have cold winters and hot, humid summers. Autumn is cool and colorful as the leaves change color. Springtime brings rainy and snowy weather and strong storms.

Coastal areas tend to have mild climates. States along the East Coast still experience seasons, but temperatures are less extreme than they are inland. States located farther south experience milder changes in seasons.

Much of the Southeast has a humid subtropical climate. Summers are rainy and hot, and winters are cooler and drier. In general, climates of the eastern United States are more humid and rainy than climates of the West. In late summer and early autumn, **hurricanes**— ocean storms that span hundreds of miles with winds of at least 74 miles per hour (119 km per hour)—can pound the coastline. One of the most damaging hurricanes in history, Hurricane Katrina, struck the Gulf Coast in August 2005. More than 1,800 people died, and hundreds of thousands lost their homes.

Kennan Harvey/Getty Images

Minerals and Energy Resources

A wealth of resources is hidden below the surface of the region, and two of the most valuable materials are minerals and energy resources. Minerals are natural substances such as iron ore, gold, and zinc. These minerals can be processed into metals. Metals and other forms of minerals are used in manufacturing and construction.

Energy resources, such as coal, oil, and natural gas, are called fossil fuels. Burning coal can produce electricity. Oil is processed into fuel for cars and other vehicles. Natural gas is used to heat our homes and to generate electricity. The demand for mineral and energy resources is huge. Mining them is a major industry in the U.S. east of the Mississippi River. Minerals and energy resources are mined from inside mountains and from deep under the ground. Some mining methods harm the environment by damaging the land and polluting the water, soil, and air.

Farming and Industry

One of the most valuable resources east of the Mississippi River is farmland. The rich soil is excellent for growing crops such as grains, fruits, and vegetables. Sandy soils in the Southeast are good for growing cotton. Excellent growing conditions have helped the region become a major producer of meat, dairy foods, wood, cotton, sugar, corn, wheat, soybeans, and other food crops.

Industries such as logging, mining, and fishing are a way of life throughout the region. Products such as automobiles, electronics, and clothing are made in factories in these cities. Information technology (IT) and tourism are also important. Plentiful resources and hard-working people make the eastern United States one of the most productive regions in the world.

Include this lesson's information in your Foldable®.

☑ **READING PROGRESS CHECK**

Explaining Why is farmland considered a natural resource?

LESSON 1 REVIEW

Reviewing Vocabulary

1. How does a *tributary* affect the amount of water flowing through a river?

Answering the Guiding Questions

2. *Determining Central Ideas* Why do you think geographers divide the eastern U.S. into four subregions?

3. *Identifying* Which of the four subregions of the Eastern United States does not border the Atlantic Ocean?

4. *Analyzing* In what ways are the East Coast and the Eastern Gulf Coast alike? In what ways are they different?

5. *Describing* How do locks in the Great Lakes and the St. Lawrence Seaway affect transportation in the Midwest?

6. *Analyzing* Think about how mining minerals and energy resources can damage the environment. Brainstorm a creative solution to this problem. Describe your solution.

7. *Identifying* Crops of citrus fruits, such as oranges and lemons, will die in freezing temperatures. Which subregion of the Eastern U.S. has the best climate for growing citrus fruits?

8. *Argument Writing* Write a persuasive letter encouraging a friend or family member to visit one of the four subregions of the eastern U.S. Include details about the region's physical features, resources, and climate in your letter.

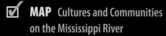

Minneapolis-St. Paul
Beauharnois
Mississippi R.

There's More Online!

☑ **IMAGE** Early Settlements

☑ **MAP** Cultures and Communities on the Mississippi River

☑ **VIDEO**

Reading **HELP**DESK

Academic Vocabulary

- **isolate**

Content Vocabulary

- **indigenous**
- **colonists**
- **agriculture**
- **industry**

TAKING NOTES: *Key Ideas and Details*

Organize Use the graphic organizer below to take notes about the effects of migration on the people and the land of the region.

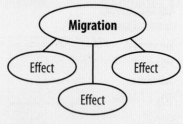

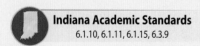

Indiana Academic Standards
6.1.10, 6.1.11, 6.1.15, 6.3.9

Lesson 2
History of the Region

ESSENTIAL QUESTION • *Why is history important?*

IT MATTERS BECAUSE
Learning about our nation's past helps us understand and appreciate its diversity and complexity.

Early America

GUIDING QUESTION *Who were the first peoples to live in the eastern United States?*

Looking back into history to learn about America's past is like examining a colorful quilt. A quilt is made up of bits and pieces of fabrics of many different colors and patterns. Each piece is made up of thousands of threads woven together. These threads were joined together into fabric pieces, and the pieces were joined together to make a quilt.

Like the quilt, our nation is a whole made up of many smaller parts. These parts are peoples and their cultures, interacting with places and environments. Over time, cultures, beliefs, and ways of living have become woven together like the threads of a quilt. The result is a unique and extraordinary nation.

Earlier, you read about the physical geography of the United States east of the Mississippi River and its many different kinds of landforms and bodies of water. This region has a variety of climates and resources. In this lesson, you will learn about the variety of peoples, cultures, and ways of life in the eastern United States, as well as how people have changed the land. You will discover how natural resources have influenced human settlement and how the land, the people, and the cultures of the eastern United States are connected.

Native Americans

Native American peoples were the first humans to settle in North America. Historians believe these peoples came to North America by crossing a land bridge from Asia around 14,000 to 19,000 years ago. Over time, these groups migrated in all directions. Hundreds of different groups settled in locations throughout North and South America. Each group developed a unique culture with its own language, religion, and lifestyle. Some of the groups that settled on lands east of the Mississippi River were the Cherokee, the Iroquois, the Miami, and the Shawnee. These groups are considered **indigenous** to North America. *Indigenous* means "living or occurring naturally in a particular place."

Native American peoples satisfied their needs by using the plants, animals, stones, water, and soil around them. Their way of life was shaped by their environment. For example, peoples who lived in northern woodland areas made their homes out of bark and wood. They burned wood to heat their homes during cold winters. They hunted woodland animals for food and used the animals' skins to make clothing.

Native peoples of the Americas built shelters suitable for the climates where they lived. People who lived in hot climates used grasses, vines, and reeds to build open-air homes. Other groups used stones, caves, and earth to build solid structures and mound cities. Their homes reflected their environments.

Some native groups were **isolated** from other groups, while others had contact with neighboring peoples. Sometimes groups interacted peacefully, such as when trading. Other times, wars over land and resources would develop.

For most of their history, Native Americans had little or no contact with people from other parts of the world. Native peoples lived off the land for thousands of years, resulting in only a minor impact on the natural environment. Then, suddenly, the land that they had relied on for everything was taken from them. When the first Europeans arrived in the Americas in the 1400s, native peoples' ways of life were changed forever.

European Colonization

Across the Atlantic Ocean, Europeans grew interested in the Americas. They heard tales about these wild, bountiful lands. Explorers who had been to the Americas told of endless forests, rivers overflowing with fish, and mountains filled with gold and silver. Kings and queens from England, France, Italy, and Spain wanted to claim land in North America. They wanted to control America's gold and natural resources.

Academic Vocabulary

isolated being alone or separated from others

Sequoyah spent many years developing 86 symbols to represent all the syllables of the Cherokee language.
▶ **CRITICAL THINKING**
Drawing Conclusions Why is having a written language essential for a culture?

In the 1500s, Spanish priests, soldiers, and settlers built military and religious outposts in the Americas. Included was St. Augustine in Florida. Originally founded as a settlement in 1565, settlers soon realized the need for protection after a series of pirate attacks. Settlers were also concerned by the arrival of English settlers. Construction on a stone fort began around 1672. This was the first permanent European settlement in what would become the United States. In the early 1600s, the English began to send **colonists** to the Americas. Colonists are people who are sent to live in a new place and claim land for their home country.

The first English colonists settled along the Atlantic coast of North America. They started early settlements in Jamestown, Virginia, in 1607 and Plymouth, Massachusetts, in 1620. Other settlements soon followed, built on lands that had been home to native peoples. The colonists turned native peoples' hunting grounds into farmland. They used many resources that were important to the Native Americans' survival.

Over the years, more and more Europeans journeyed across the Atlantic Ocean to America. Their colonies grew quickly. By 1650, about 52,000 colonists lived in America. Over the next 50 years, the

Visitors tour the Castillo de San Marcos National monument in St. Augustine, Florida. The fort is over 300 years old and took 23 years to complete.

▶ **CRITICAL THINKING**

Identifying Why do you think settlers chose a site along the water for the fort?

©Jose Fusta Raga/Corbis

number of colonists grew to about 250,000. By 1760, an estimated 1.7 million colonists lived in America. People built towns along the Atlantic coast. Large cities such as Boston and New York City started out as tiny settlements. By the 1750s, thirteen English colonies had been established in North America.

Life was hard for the first colonists. The food supplies they brought from England soon ran out. Many people got sick or died from starvation. In time, however, the colonists learned how to plant crops and hunt for food in this new land. They also adapted the natural resources they found there to make things they needed, such as candles, soap, pots, clothing, tools, and medicine.

The colonies built by English settlers were controlled by English rulers thousands of miles away. In 1707 England and Scotland united to form Great Britain. The colonists did not like the laws and taxes forced upon them by the British government. They made plans to break away and become free from British rule. In 1776, American colonists declared their independence from British rule, which led to the Revolutionary War. The war ended in 1781 when the British surrendered. The thirteen colonies became an independent nation called the United States of America, and the colonists called themselves *Americans.*

✓ READING PROGRESS CHECK

Identifying Point of View Why did European nations want to control land in North America?

European goods important to the survival of the colonists were received in Boston Harbor.
▶ **CRITICAL THINKING**
Determining Central Ideas Why were early colonists dependent on supplies from England?

©Corbis

Settling the Land

GUIDING QUESTION *How has the movement of people shaped the culture of the eastern United States?*

The new nation officially stretched from the Atlantic Ocean to the Mississippi River. However, the land west of the Appalachian Mountains was a mystery to most Americans. The few explorers and settlers who traveled west brought back stories of a wild and dangerous land. Many people were afraid to venture into the West. Others were willing to take their chances to seek new opportunities.

As more European settlers arrived, towns along the Atlantic coast became crowded. People began moving inland, away from the crowded coastal areas, to build new lives. They wanted to claim land for themselves. Many settlers packed everything they owned into wagons and headed west. Some looked for gold or silver. Some hunted animals and sold or traded the animals' skins. Most of these settlers stopped traveling, however, when they found land that looked suitable for farming.

These settlers quickly built homes and planted crops on the land. They hoped to grow enough food to feed their families. Many families of settlers were isolated from other people. They had no neighbors, and they were hundreds of miles from the closest town. These people had to make or grow everything they needed. Like the Native Americans, the early settlers lived off the land.

Daniel Boone helps lead settlers through the Cumberland Gap. This well-traveled path for settlers moving west later became known as the Wilderness Road.
▶ **CRITICAL THINKING**
Analyzing Primary Sources Why would the journey west have been difficult for early settlers?

©Bettmann/Corbis

New Territory

Between 1700 and 1800, thousands of settlers built homes along the Mississippi River. The Mississippi formed a natural boundary because it was wide, deep, and difficult to cross. When they arrived at the banks of the Mississippi, travelers either settled on the eastern side of the river or turned back. Those who did try to drive their horse-drawn wagons through the powerful river risked being swept away by its swift waters.

The young U.S. government wanted to claim as much of America's land as possible. The Land Ordinance of 1785 gave the United States legal claim to lands known as the Ohio Country. These lands were located north of the Ohio River and east of the Mississippi River. The Land Ordinance of 1785 allowed American settlers to buy sections of this land for one dollar per acre. The government divided up some land among settlers and also set aside land to be used for schools. Native Americans living on this land were forced to leave. The United States was slowly settling all territory east of the Mississippi River.

New Technology Changes Farming

Americans have been farming the land east of the Mississippi River for centuries. Much of our country's fruits, vegetables, grains, and cotton are raised on farms in this region. Growing crops and raising livestock to sell is called commercial **agriculture**. Good soil and frequent rains make the region one of the best places in the world for agriculture.

Until recently, planting and harvesting crops was hard work, performed mainly by hand and with the help of animals, such as horses and oxen. In the late 1700s, people designed and built new kinds of machines to do farmwork. Some machines planted large amounts of seeds quickly or harvested crops faster and more thoroughly than was possible by hand. For example, a Massachusetts man named Eli Whitney invented a machine called the cotton gin. The cotton gin made processing cotton faster and easier. This increased profits for cotton farmers.

Over the years, machines began to replace human workers on farms. With the help of machines, farmers planted and harvested more crops. This made farms more productive. As more and more farmwork was done with machines, however, fewer people were

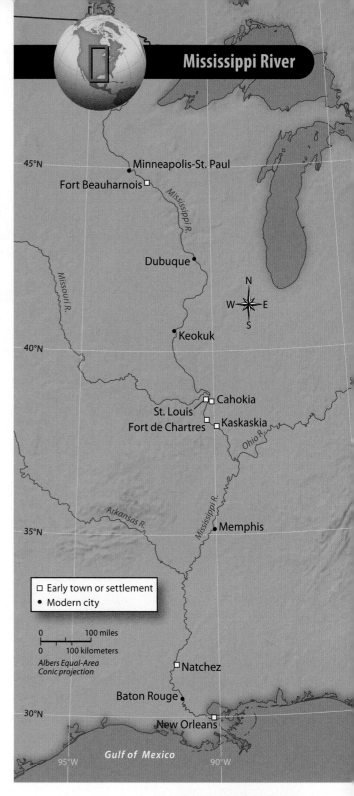

Mississippi River

- □ Early town or settlement
- ● Modern city

0 — 100 miles
0 — 100 kilometers
Albers Equal-Area Conic projection

MAP SKILLS

Throughout history, many cities and towns were built and grew along the Mississippi River.

▶ **CRITICAL THINKING**

Describing How does a river contribute to the growth of a city or town?

Immigrants arrive in New York Harbor in the late 1800s.

▶ **CRITICAL THINKING**

Identifying State three reasons immigrants came to live in the United States.

needed to work on farms. This left thousands of people without work, and many moved to the cities. Today, most farmwork is done with the help of machines, such as tractors and combines.

Industrial Growth

New technology led to jobs in factories for millions of people. Manufacturing, making products to sell, is called **industry**.

Industry is an important part of the economy of the United States east of the Mississippi. This region has been one of the world's leading industrial regions for more than two centuries. Thousands of factories have been built all over the region during this time. Today, factories produce an endless variety of products including clothing, computer parts, shoes, baby formula, and medicines. Some factories process foods and bottled drinks. Cars are also built at automobile assembly plants. Products made east of the Mississippi River are shipped and sold all over the world.

Influence of Immigration

The history of the United States east of the Mississippi River is the history of many cultures. The cultures have become woven together over time, blending and changing into something new. Since the colonial days, people have been moving to the eastern United States from other countries. People immigrate to America because they want to be free. They also want jobs, education, and other opportunities. As a result, an amazing variety of languages, religions, cultures, and customs can be found east of the Mississippi. The cultural traditions brought by people from all over the world have made America a unique and diverse nation.

Forced Migration

In the 1830s, gold was discovered in the Southeast. Word quickly spread, and settlers poured into the southern states. Most did not find gold, but many stayed to start cotton farms in Alabama, Georgia, Mississippi, and the Carolinas. Some of these lands were home to the Cherokee, a large group of Native Americans. U.S. citizens wanted these lands—and their valuable resources.

Ingram Publishing

The government forced the Cherokee to leave their lands. In 1838 thousands of Cherokee men, women, and children were rounded up by U.S. soldiers. These people were forced to make the long and difficult journey to Indian Territory in Oklahoma, far west of the Mississippi River and 1,000 miles (1,609) from their homeland. The land they were given to live on was dry and difficult to farm, which was very different from the land they were forced to leave. Thousands of Cherokee died during and after the journey. This terrible event became known as the Trail of Tears.

The Trail of Tears was just one of many forced migrations in America's history. Native American groups all over the continent were forced to leave the lands that had shaped their ways of life for thousands of years. This movement destroyed some of their cultural traditions. As America gained territory, Native Americans lost their lands and their ways of life.

The Great Migration

The years after the Civil War were a time of change in the Southeast. Slavery became illegal in all states. Angry that slavery was outlawed, many states passed laws that took away the rights the freed Americans had recently gained. During the late 1800s, thousands of African Americans moved to states in the Mid-Atlantic, New England, and the Midwest. The relocation of people from the South to the North was called the Great Migration.

The Great Migration was part of the larger rural-to-urban migration occurring in the United States. This migration increased during the 1900s, when millions more people left rural areas and moved to cities to work in factories. This rural-to-urban movement of people is one of the largest migrations in America's history.

☑ **READING PROGRESS CHECK**

Analyzing How did the invention of farm machinery lead to unemployment in the eastern United States?

Include this lesson's information in your Foldable®.

LESSON 2 REVIEW

Reviewing Vocabulary
1. What makes the United States east of the Mississippi River a good region for *agriculture*?

Answering the Guiding Questions
2. ***Determining Central Ideas*** How did the early Native American peoples utilize resources from their environment?

3. ***Identifying Point of View*** What are reasons why English rulers sent colonists to America in the 1600s?

4. ***Analyzing*** What factors helped the U.S. economy change and grow?

5. ***Identifying Point of View*** How did laws ending slavery affect different populations of people in the Southeast after the Civil War?

6. ***Narrative Writing*** Imagine you are a young Native American living in the 1800s. Your family is being forced to leave its home as colonists take over the land. Write a narrative telling how you feel about what is happening to you and your people. Include details from the lesson in your narrative.

Reading **HELP**DESK

Academic Vocabulary

- **revenue**

Content Vocabulary

- **metropolitan area**
- **tourism**
- **civil rights**
- **Rust Belt**
- **service industry**

TAKING NOTES: *Key Ideas and Details*

Organize As you read about life in the eastern United States, write a one-sentence summary for each of the listed topics.

Topic	Summary Sentence
Metropolitan areas	
U.S. government	
Economy	

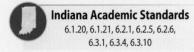

Indiana Academic Standards
6.1.20, 6.1.21, 6.2.1, 6.2.5, 6.2.6, 6.3.1, 6.3.4, 6.3.10

Lesson 3
Life in the Region

ESSENTIAL QUESTION · *What makes a culture unique?*

IT MATTERS BECAUSE
Learning about the human geography of the United States—its people, government, economy, and culture—can help you better understand and appreciate the nation as a whole.

Major Metropolitan Areas

GUIDING QUESTION *What is it like to live in a large metropolitan area in the eastern United States?*

People in the United States live in many different environments. Farmers and ranchers live in rural areas with open land for farming and livestock. Suburbs are popular places for people who want larger homes and their own pieces of property. People who enjoy urban environments often live in large cities called **metropolitan areas**. Metropolitan areas are centers of culture, education, business, and recreation. Large cities such as New York, Boston, Philadelphia, and Miami are home to millions. The populations of metropolitan areas in the eastern United States are large and diverse. The Boston-Washington corridor is a metropolitan area, home to about 50 million people. It is often referred to as a megalopolis.

Some metropolitan areas serve as hubs for international cooperation. The member countries of the United Nations (UN), located in New York City, work together to find and share solutions to problems related to education, science, and culture. For example, the UN's World Heritage program promotes and protects natural and cultural sites around the world. World Heritage sites in the eastern United States include the Everglades National Park in Florida and the Statue of Liberty in New York Harbor.

Tourism is one industry that provides jobs to people in and around metropolitan areas. The **tourism** business provides services to people who are traveling for enjoyment. Businesses such as restaurants, hotels, resorts, travel agencies, and tour companies are part of the tourism industry. The money tourists spend brings **revenue** to the state and local economies.

Some metropolitan areas on the East Coast began as port cities along the shores of the Atlantic Ocean. Port cities are large, busy towns where ships dock and depart. Trade between the United States and the rest of the world began in port cities such as Baltimore; Boston; Charleston, South Carolina; New Haven, Connecticut; and New York.

New York and Chicago

New York and Chicago are two of the largest metropolitan areas east of the Mississippi River. New York City, located on the Atlantic coast at the mouth of the Hudson River, is the most populous city in the United States. This famously diverse city is home to more than 9 million people, making it one of the most heavily populated cities in the world. New York began as a Dutch colonial port city called New Amsterdam. Although it is better known today as a hub of culture and commerce, New York is still home to one of the busiest ports in North America.

New York City is a dense cluster of urban areas, called boroughs, connected by streets, bridges, trains, and water passages. The five boroughs that make up the city are Manhattan, Brooklyn, Queens, the Bronx, and Staten Island. Though it is the smallest of the five boroughs, covering an area of only 22.6 square miles (58.5 sq. km), the island of Manhattan is the cultural, political, and economic center of New York City.

Academic Vocabulary

revenue income generated by a business

New York City is the most populous city in the United States.
▶ **CRITICAL THINKING**
Determining Central Ideas What are some of the advantages of living in a large city? What are some disadvantages?

Population Centers

Geographers study why certain cities became places where many people live. Many population centers in the United States lie in coastal areas where healthy economies support large populations. Some cities are important world trade centers because of their coastal or near-coastal locations. Some population centers are located inland, yet many are situated near rivers and lakes. Other inland cities such as Atlanta grew from agricultural and trading centers.

Manhattan is home to people of every racial, ethnic, and religious background in the world. It is also known worldwide as a center of finance, advertising, and entertainment. An endless variety of visual and performing arts—such as music, dance, drama, painting, sculpture, fashion, and architecture—bring the city to life.

More than 30 million tourists visit New York City each year. The city has an extensive public transportation system that includes subways and buses to help eliminate dependency on cars.

Chicago began as a small settlement between Lake Michigan and the Mississippi River in the early 1800s. The settlement developed into a thriving city after it was connected to the rest of the country by railroads and canals. By the mid-1800s, Chicago had become the center of all railroad travel in the United States. Chicago's importance as a transportation center increased when the St. Lawrence Seaway opened in the mid-1900s. Today, an elevated train system in the center of Chicago helps move its many tourists and residents.

Chicago remains one of the nation's most important centers of shipping, transportation, and industry. The city is one of America's leading producers of steel, machinery, and manufactured products. Several large printing and publishing companies are located in Chicago, as well as major financial institutions such as the Chicago Stock Exchange.

Atlanta and New Orleans

Atlanta and New Orleans are two of the most vibrant cities in the Southeast. Serving as Georgia's capital city, Atlanta is a historic city and a modern metropolis. Its location at the southern edge of the Appalachian Mountain range made it a popular passageway for settlers and other travelers. As railroads brought people and cargoes through the area, Atlanta grew into a thriving economic, cultural, and political center of the South.

During the Civil War, Atlanta served as a supply depot for the Confederate army. Most of the city's buildings were burned to the ground during a devastating invasion by the Union army in 1864. When Atlanta was rebuilt after the war, it became a symbol of strength and rebirth in the South.

Today, Atlanta is a strong center of transportation, industry, trade, education, and culture. It has been called the commercial center of the modern South. A wide array of industries, including publishing, telecommunications, banking, insurance, military supply, and manufacturing, have headquarters in Atlanta. The city's major factories produce electrical equipment, chemicals, packaged foods, paper products, and aircraft.

New Orleans began as a shipping town along the Mississippi River. During times of peace, New Orleans was an important center of transportation and trade. In times of war, such as during the Revolutionary War, the War of 1812, and the Civil War, New Orleans

became a major stop along supply lines that served the military and civilians. Located only 110 miles (177 km) from the Gulf of Mexico, New Orleans eventually was connected to the Gulf by river channels, making it easier for ships to enter and leave the city.

In 2005, Hurricane Katrina struck the Gulf of Mexico's coast. The storm raged from Louisiana to Florida. In New Orleans, levees failed when the storm surged, flooding low-lying areas and trapping many people. Thousands of people were left homeless and many people died. Hurricane Katrina was one of the worst natural disasters in U.S. history. Some neighborhoods and areas of New Orleans have been rebuilt. Other areas have not recovered.

New Orleans remains an important commercial trade center in the eastern United States. Manufacturing and transportation still contribute to the city's economy. New Orleans is renowned for its rich cultural traditions including its spicy Cajun and Creole foods, original musical styles, and colorful celebrations such as Mardi Gras. New Orleans was one of the first centers of jazz music. The city and its people, art, music, language, and architecture are a bold and unique mixture of French, Spanish, Caribbean, and African cultures. Tourism has long been vital to the economy of New Orleans.

New York, Chicago, Atlanta, and New Orleans each have a unique character. These remarkable cities are just a few of the places that make the eastern United States such a fascinating region.

✓ **READING PROGRESS CHECK**

Citing Text Evidence How are New York and New Orleans alike? How are they different?

Millions visit the city of New Orleans every year to take part in the festivities of Mardi Gras.

▶ **CRITICAL THINKING**

Identifying Besides Mardi Gras, what attractions draw tourists to New Orleans?

The Nation's Capital

The planned city of Washington, D.C., is the capital of the United States. Washington, D.C., is located along the banks of the Potomac River between the states of Virginia and Maryland. These two states gave land to the government to form the federal District of Columbia. Thus, the District was free of any single state's influence. *Why was the capital located in the eastern United States near the Atlantic Coast?*

The U.S. Government

GUIDING QUESTION *How have the government's actions affected the land and people of the United States?*

The United States declared its independence from Great Britain on July 4, 1776. On that day, Americans became citizens of a free and independent nation. Like all nations, the new country needed a system of government. It was important to Americans that their government protect the rights and freedoms of the people. The U.S. government was designed as a representative democracy, a system in which the people elect representatives to operate the government.

In 1787 representatives from each of the 13 states gathered to write a plan of government called a constitution. The U.S. Constitution is still the law of our country. The United States is a federal republic. The national government shares power with the states. Government leaders must promise to obey the Constitution. Amendments, or changes to the Constitution, have been made to meet the nation's changing needs. The first 10 amendments—the Bill of Rights—guarantee the basic rights of citizens.

The Three Branches of Government

One of the main functions of the U.S. Constitution is to make sure government power is shared. The men who wrote the Constitution did not want a single person or group to have all the power to make laws and decisions for the country. They divided the government into three separate but equal branches. Each branch has important functions, and each branch must work with the other two branches to govern the country. The three parts of the U.S. government are the executive, legislative, and judicial branches. The U.S. government's system of shared power was created to ensure that all parts of the government work together in a balance of power.

The legislative branch, called Congress, makes laws for the nation. Congress has two parts: the House of Representatives and the Senate. Members of Congress are elected by the people and come from all 50 states. To pass a law, the House of Representatives and the Senate must agree on what the law states.

The executive branch is the office of the president of the United States. The president's main duties are to carry out laws, to lead the military, to appoint judges to the Supreme Court, to plan the national budget, to meet with foreign leaders, and to appoint advisors to help make decisions for the nation.

The judicial branch is made up of state and federal courts. The role of the government's judicial branch is to decide if laws are fair and if they follow the Constitution. The Supreme Court, with nine judges called justices, is the most powerful court in America.

Government Actions Affect the Land

The three branches of the U.S. government make laws and decisions that affect our nation's land. Some laws and decisions protect the environment. For example, the president has the authority to set aside land for use as national parks. This protects plant and animal habitats and creates recreational areas. Local governments enforce laws that reduce water pollution and littering.

Another environmental program is Superfund. The purpose of the program is to clean up abandoned hazardous waste sites. Superfund was implemented after the discovery of toxic waste sites such as Love Canal and Times Beach in the 1970s. The federal government works in conjunction with the state government and the communities to implement cleanup plans.

Government Actions Affect People

Some actions of the United States government that affected Native American peoples had severe consequences. Other government actions—such as building roads and bridges, providing aid to people in need, and establishing national parks—help people and enrich our lives.

One example of how government actions have affected Americans positively is the civil rights movement. **Civil rights** are basic rights that belong to all citizens, such as the right to be treated equally under the law. To answer the demand for civil rights for African Americans, the federal government made laws ending segregation. These laws were meant to help people. Some local governments, however, did not agree with these laws and refused to enforce them.

National parks, such as Maine's Acadia National Park, are pieces of land that are protected by the federal government.

▶ CRITICAL THINKING

Describing What are the purposes of national parks?

Another example of positive action is how the government helps people during emergencies. Government agencies such as the Federal Emergency Management Agency (FEMA) provide food, water, medical care, and transportation to people affected by tornadoes, hurricanes, earthquakes, floods, terrorist attacks, and other disasters. FEMA has saved many lives, but some people claim the agency does too little to help and is slow to respond.

When thinking about how government actions affect people and the land, it is important to remember that no government is perfect. The U.S. government is made up of many different people doing many different jobs.

✓ READING PROGRESS CHECK

Determining Central Ideas Why is the U.S. government divided into three branches?

Richard Freeda/Aurora/Getty Images

Everyday Life

GUIDING QUESTION *How has diversity shaped the culture of the United States?*

Religion and Ethnicity

In the United States, the religion with the largest number of followers is Christianity. The influence of the Christian faith in the United States has continued since colonial times. However, many Americans practice the Jewish, Muslim, Buddhist, or Hindu faith. Research shows that 16 percent of Americans do not participate in organized religion. Nearly half of all American adults say they have changed their faiths or beliefs at least once in their lives.

In gathering population statistics, the Census Bureau classifies the people of the United States into several different categories. The most recent data collected from the 2010 Census show that the four most populous ethnic groups are Caucasian, Latino, African American, and Asian American. In the last two censuses, people of Hispanic or Latin-American origin comprised the second largest ethnic group after Caucasians. In the Census of 1990, African Americans made up the second most-populous group.

Economy

America has one of the largest and strongest economies in the world. Advancements in technology help keep industries such as farming and manufacturing productive. The U.S. economy has slowed in the past decade, however. Many Americans remain unemployed. It could take years for the economy and employment rate to rebound.

Agriculture has long been an important industry. Today, the businesses of farming and raising animals for food are changing. Farms owned and operated by families are being replaced by corporate farms. They are managed by people who do not own or live on the land. Some Americans believe the growth of corporate farms is bad for the economy and the nation's people. They believe corporate farms agree to sell crops to large grocery-store chains at

---

GRAPH SKILLS ›

ETHNIC ORIGIN
The graph shows census statistics from 1980 to 2010.

▶ CRITICAL THINKING

1. Analyzing Which groups more than doubled in percentage of population from 1980 to 2010?

2. Analyzing Which group from 1980 to 2010 shows the biggest decrease in percentage of population?

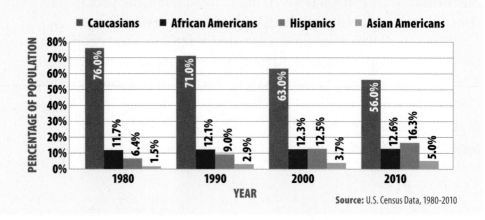

Source: U.S. Census Data, 1980-2010

such low prices that family farms can not compete. A growing number of Americans now support organic farming. Organic farms use only natural pesticides and fertilizers. Some people believe that organically-grown foods are safer and healthier than foods that have been treated with chemicals.

In the late 1800s and early 1900s, the U.S. shifted away from an agricultural-based economy to a more industrial economy. Factories producing automobiles, appliances, machinery, electronics, and other items have employed millions of Americans. During the 1980s, however, the manufacturing industry began to weaken. Businesses closed, and factories were abandoned. Workers lost their jobs. So many factories closed across the Midwest, the Mid-Atlantic, and New England that these areas earned the nickname "the **Rust Belt**."

In recent years, more Americans have found jobs in service industries. **Service industries** are businesses that provide services rather than products. Child care centers, restaurants, grocery stores, hair salons, electricians, moving companies, and auto repair shops are some examples of service industries. Other industries such as finance, insurance, and education still provide many jobs and revenue today. Many businesses, including banks and insurance companies, are located in cities east of the Mississippi River.

Different types of industries are important to the economy of the eastern U. S. Today, another shift is taking place as the nation moves from a service-based economy to an economy that is connected to the computer-information age.

✔ **READING PROGRESS CHECK**

Citing Text Evidence What is the main difference between service industries and manufacturing industries?

Every March, Cuban Americans in Miami celebrate Calle Ocho, the single largest Latino celebration in the United States.

▶ **CRITICAL THINKING**
Analyzing Why are ethnic celebrations important for the community as well as for the people?

FOLDABLES
Study Organizer

Include this lesson's information in your Foldable®.

Diversity
Geographic Barriers
East and West
The US East of the Mississippi

©Nik Wheeler/Corbis

LESSON 3 REVIEW

Reviewing Vocabulary

1. What are the names of the major *metropolitan* areas located in your state?

Answering the Guiding Questions

2. *Integrating Visual Information* What are some of the largest metropolitan areas located in the eastern United States?

3. *Determining Central Ideas* How have the government's actions affected the land and people of the United States?

4. *Determining Central Ideas* How have cultures from other parts of the world shaped the culture and character of the United States?

5. *Citing Text Evidence* What types of businesses and industries are important to the economy of the eastern United States today?

6. *Informative/Explanatory Writing* Write a paragraph explaining how tourism might affect the economy of a major metropolitan area.

What Do You Think?

Is Fracking a Safe Method for Acquiring Energy Resources?

Fracking is a process for obtaining natural gas and oil through high-pressure blasting of underground rock. Fracking involves pumping a specially blended liquid—a mix of water, sand, and chemicals—into wells drilled deep below Earth's surface. The fluids pour in with such force that the rock formations fracture, or crack, releasing precious oil and natural gas. Oil and gas are key resources for heating our homes, creating electricity, and fueling our cars. But critics of fracking worry that the process is harming the environment and people's health. Is fracking a safe way to meet our energy needs?

No !

PRIMARY SOURCE

" The form of natural gas drilling called fracking has caused livestock and crops to die from tainted water, people in small towns to black out and develop headaches from foul air, and flames to explode from kitchen taps. . . . [I]n recent years, we have learned that extracting gas through fracking poses unacceptable risks to the public. Fracking uses large quantities of water and a cocktail of toxic chemicals that have been shown to poison water resources. To date, thousands of cases of water contamination have been reported near drilling sites around the country. In many cases, residents can no longer drink from their taps, and in one instance, a home near a fracking site exploded after a gas well leaked methane into its tap water. . . . [S]tudies . . . found that 25 percent of fracking chemicals can cause cancer and 40 to 50 percent can affect the nervous, immune and cardiovascular [heart and blood vessel] systems. "

—Sam Schabacker, senior organizer for Food & Water Watch

Environmental activists protest fracking in New York.

Water is unloaded from trucks at a treatment plant. The plant will separate water, oil, and sediment that is mixed during the fracking process.

Yes !

PRIMARY SOURCE

" "Typically, steel pipe known as surface casing is cemented into place at the uppermost portion of a well for the explicit (specific) purpose of protecting the groundwater. . . . As the well is drilled deeper, additional casing (large pipe) is installed . . . which further protects groundwater. . . .

Casing and cementing are critical parts of the well construction that not only protect any water zones, but are also important to successful oil or natural gas production. . . . Industry well design practices protect sources of drinking water from . . . oil and natural gas well with multiple layers of impervious (hard to pass through) rock.

"While 99.5 percent of the fluids used consist of water and sand, some chemicals are added to improve the flow." "

—American Petroleum Institute

What Do You Think? DBQ

1. *Identifying Point of View* According to Sam Schabacker, how does fracking put people's health at risk?

2. *Identifying Point of View* How does each side support its position?

Critical Thinking

3. *Analyzing* Some people believe that fracking should be halted until experts have studied the risks more thoroughly. Do you think a temporary ban on fracking is reasonable? What would be the advantages and disadvantages?

Chapter 4 ACTIVITIES

Directions: Write your answers on a separate piece of paper.

❶ Exploring the Essential Question

INFORMATIVE/EXPLANATORY WRITING Choose one of the subregions located in the United States east of the Mississippi. Write an essay explaining how the physical geography of the subregion influenced its settlement and economic development.

❷ 21st Century Skills

INTEGRATING VISUAL INFORMATION Use a map to trace the Ohio River from its birthplace at the mouths of the Allegheny and Monongahela Rivers in Pennsylvania and the Mississippi River from its birthplace in Minnesota to the point where they meet. Then follow the path of the Mississippi to the Gulf of Mexico. List the states that border these two rivers and five cities along the rivers' banks.

❸ Thinking Like a Geographer

INTEGRATING VISUAL INFORMATION Create a graph showing how quickly the population of the colonies grew between 1650 and 1700 and from 1700 to 1750.

❹ GEOGRAPHY ACTIVITY

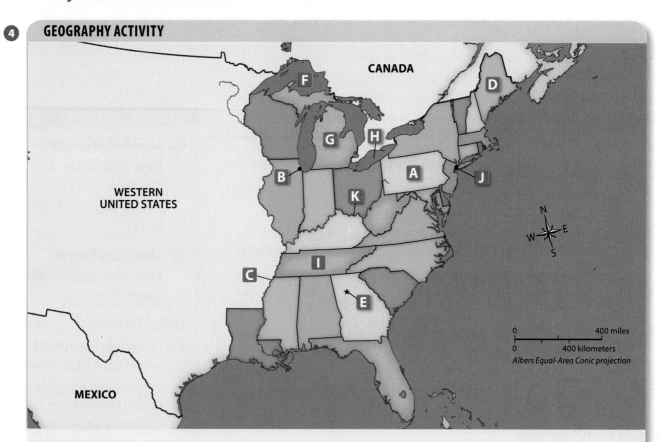

Locating Places
Match the letters on the map with the numbered places below.

1. New York City
2. Chicago
3. Atlanta
4. Maine
5. Mississippi River
6. Ohio River
7. Tennessee
8. Lake Erie
9. Michigan
10. Pennsylvania
11. Lake Superior

REVIEW THE GUIDING QUESTIONS

Directions: Choose the best answer for each question.

1 In which subregion of the United States is the nation's capital located?

A. Southeast

B. Midwest

C. Mid-Atlantic

D. New England

2 What is the name of the river system that connects the Great Lakes to the Atlantic Ocean?

F. the Hudson River

G. the St. Lawrence Seaway

H. the Ohio and Mississippi rivers

I. the Monongahela and Allegheny rivers

3 Why did Europe's kings and queens claim lands in North America?

A. They wanted to spread Catholicism to the Western Hemisphere.

B. They wanted control over America's gold and natural resources.

C. They wanted to reduce the population in overcrowded cities.

D. They wanted to protect the native population from exploitation.

4 Why were the Cherokee people forced to leave their land in the Southeast and walk 1,000 miles (1,609 km) to Oklahoma?

F. A hurricane destroyed their homes.

G. Their land was needed for forts to protect America from the Spanish.

H. White settlers wanted their land and the resources on it.

I. Locusts destroyed their crops.

5 New York City is made up of five unique urban areas called

A. counties.

B. boroughs.

C. townships.

D. districts.

6 Whose job is it to carry out the laws passed by Congress, appoint federal judges, lead the military, meet with foreign leaders, and plan the budget?

F. the Senate

G. the Speaker of the House of Representatives

H. the president of the United States

I. the secretary of state

DBQ **ANALYZING DOCUMENTS**

❼ DETERMINING CENTRAL IDEAS The government reports on the future need for workers who create software:

"*Employment of software developers is projected to grow . . . much faster than the average for all occupations. . . . The main reason . . . is a large increase in the demand for computer software. Mobile technology requires new applications. Also, the healthcare industry is greatly increasing its use of computer systems.*"

—from the *Occupational Outlook Handbook*

Which generalization can you make from the information in this quote?

A. Computer use is expected to go down in the future.

B. More software developers will be needed than most other jobs.

C. Software developers are not likely to work for health care companies.

D. The government will be the biggest employer of software developers.

❽ ANALYZING What will happen to the number of software developer jobs if smartphone sales go down in the future?

F. increase at the expected rate

G. increase more rapidly than expected

H. stay the same instead of increase

I. decrease or increase at a slower rate

SHORT RESPONSE

"*We the People of the United States, in Order to form a more perfect Union, establish Justice, insure domestic Tranquility [calm], provide for the common defence, promote the general Welfare, and secure the Blessings of Liberty . . ., do . . . establish this Constitution.*"

—from the United States Constitution

❾ IDENTIFYING What is the purpose of this section of the United States Constitution?

❿ DETERMINING WORD MEANINGS What do you think the writers of the Constitution meant by "the Blessings of Liberty"?

EXTENDED RESPONSE

⓫ INFORMATIVE/EXPLANATORY WRITING If you could choose to live in any state east of the Mississippi River, which one would it be and why? Have you ever lived in or visited that state? Do you have friends or family who live there? Write a report explaining your choice. Be sure to give details about the state's features that appeal to you. Consider things like climate, job opportunities, education, and recreational opportunities when making your choice.

Need Extra Help?

If You've Missed Question	❶	❷	❸	❹	❺	❻	❼	❽	❾	❿	⓫
Review Lesson	1	1	2	2	3	3	3	3	3	3	1

THE UNITED STATES WEST OF THE MISSISSIPPI RIVER

networks

There's More Online about The United States West of the Mississippi River.

CHAPTER 5

Lesson 1
Physical Features

Lesson 2
History of the Region

Lesson 3
Life in the United States West of the Mississippi

ESSENTIAL QUESTIONS · *How does geography influence the way people live?* · *How do people make economic choices?* · *How does technology change the way people live?*

Modern-day cowhands still ride in the cattle country of the West.

Just One Film/The Image Bank/Getty Images

The Story Matters...

Within the region are several mountain ranges, including the Rocky Mountains, the longest mountain range in North America. Its many mountains, plateaus, basins, and valleys mean that this region contains a range of elevations. The region is rich in land, mineral, and energy resources, all of which contributed to westward expansion to the Pacific Ocean in the 1800s, and continues to influence the way of life of its residents.

FOLDABLES®
Study Organizer

Go to the Foldables® library in the back of your book to make a Foldable® that will help you take notes while reading this chapter.

THE UNITED STATES WEST OF THE MISSISSIPPI RIVER

The United States west of the Mississippi River is one of the two regions that make up the United States. As you study the map, identify the states and cities of the region.

Step Into the Place

MAP FOCUS Use the map to answer the following questions.

1 THE GEOGRAPHER'S WORLD What is the name of the state just north of Missouri?

2 THE GEOGRAPHER'S WORLD Which four states meet at one point?

3 PLACES AND REGIONS What natural feature separates Texas from Mexico?

4 CRITICAL THINKING
Identifying The contiguous United States consists of the states between Canada and Mexico. Which two states are not contiguous?

MONUMENT VALLEY Buttes are a common landform in Arizona's Monument Valley Navajo Tribal Park.

GOLDEN GATE BRIDGE Named after the Golden Gate Strait, the bridge stands where water from the Pacific Ocean enters San Francisco Bay.

Step Into the Time

DESCRIBING Select one event on the time line and write a paragraph describing how social, political, ecological, and/or economic factors of the time period led to the occurrence of that event.

1598
Spain settles Santa Fe

1846
The Mexican-American War begins

1800

1803 The U.S. purchases the Louisiana Territory from France

Cities and States of the Region

CANADA

National boundary
★ State capital
• City

Washington
Seattle
Olympia ★ • Tacoma
• Portland
Salem ★
Oregon

Idaho
★ Boise

Helena ★ **Montana**

Minnesota
St. Paul ★

North Dakota
Bismark ★

South Dakota
Pierre ★

Missouri R.

Wyoming

A

Nevada
Sacramento ★ ★ Carson City
San Francisco •

PACIFIC OCEAN

California

Great Salt Lake
Salt Lake City ★
• Provo

Utah

Cheyenne ★

Colorado R.

B

★ Denver
Colorado

Arkansas R.

Nebraska
Omaha •
Lincoln ★

Iowa
★ Des Moines

Kansas City •
Topeka ★
Kansas

Jefferson City ★
Missouri

Mississippi R.

• Las Vegas

Los Angeles •
• Long Beach

0   400 miles
0   400 kilometers
Albers Equal-Area Conic projection

Arizona
★ Phoenix

★ Santa Fe

New Mexico

Rio Grande

Oklahoma City ★
Oklahoma

Arkansas
Little ★
Rock

120°W

30°N

Ft. Worth • • Dallas

Louisiana

Texas

★ Austin

Baton ★
Rouge

• Houston

70° N

Arctic Circle

Alaska

Valdez •
60° N
Juneau •

0   300 miles
0   300 kilometers

160° W   150° W   140° W

110°W

MEXICO

0   200 miles
0   200 kilometers

Honolulu •

Hawaii

20° N

160° W

100°W

Gulf of Mexico

TROPIC OF CANCER

90°W

40°N

Rio Grande

1869
Transcontinental Railroad
completed

1959
Alaska becomes a state

1900

2000

1930s Dust storms destroy
farmland in Great Plains

1989 *Exxon Valdez* oil spill
damages environment

Reading HELPDESK

Academic Vocabulary

- significant
- create

Content Vocabulary

- cordillera
- timberline
- contiguous
- Continental Divide
- irrigation
- chinook
- ethanol
- national park

TAKING NOTES: *Key Ideas and Details*

Organize As you read, take notes on different characteristics of the Great Plains.

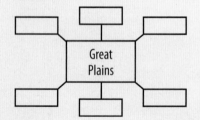

Great Plains

Indiana Academic Standards
6.1.18, 6.1.19, 6.3.5, 6.3.7

Lesson 1
Physical Features

ESSENTIAL QUESTION • *How does geography influence the way people live?*

IT MATTERS BECAUSE
The region includes many resources, but its rapidly growing population is causing overuse of some of them.

Physical Landscape

GUIDING QUESTION *How do the physical features of the western United States make the region unique?*

The Mississippi River divides the United States into two parts. These two regions are not equal in size. The area west of the Mississippi River is larger than the area to the east.

The states to the west differ from each other in some ways. At the same time, they have more in common with each other than with the states east of the Mississippi River. The western states are typically larger than the eastern states. Their human populations are generally more spread out across these vast distances. Their landforms are steeper and rockier than in the eastern states, and their climates overall are much drier. Many of the western states are rich in natural resources.

The Great Plains

Just west of the Mississippi River lie the Great Plains. In many places, the Plains appear flat. In other places, the Plains are gently rolling land. In spite of their flat appearance, however, the Plains are tilted downward toward the east. In eastern Nebraska, for example, the elevation of the land is less than 1,500 feet (457 m) above sea level. But on western edge of Nebraska, the land rises to about 6,000 feet (1,829 m).

The Great Plains were once covered by wild grasses. Vast herds of bison and pronghorn—an American antelope—grazed there. Today, the Great Plains are covered by farms and ranches.

Mountains and Hills

Toward the north, the Great Plains are interrupted by the Black Hills. These hills were once mountains, but over time they eroded. Evergreen trees appear to darken the hills, giving them their name.

West of the Great Plains tower the Rocky Mountains. The Rockies are not a single mountain chain, but a cordillera. A **cordillera** is a region of parallel mountain chains. The Rockies include dozens of different mountain systems. They extend from the Canadian border to the Mexican border. Peaks soar up to 14,000 feet (4,267 m), and valleys plunge thousands of feet below. Many of the mountains are snow capped. Trees cover the slopes, but not above the **timberline**. At that elevation, the climate is too cold for trees to grow.

Several different mountain ranges tower over the Pacific coast. Many of these mountains formed because plate tectonics exert pressure on Earth's lithosphere. This causes the lithosphere to crack and the broken land to rise into steep, rugged mountains. Among them are the Olympic Mountains of Washington. Heavy rainfall and cold temperatures form glaciers on these mountains.

About 150 miles (241 km) east of the Pacific coast are two higher ranges. The Cascades run from north to south through western Oregon and Washington State. The Cascades are volcanic, and some of the volcanoes are still active. The Sierra Nevada range runs along the California-Nevada border. The name *Sierra Nevada* comes from the Spanish for "snowy mountains." These mountains include Mount Whitney, which at about 14,500 feet (4,419 m) is the highest point in the 48 contiguous states. **Contiguous** means "connected to." The contiguous states are those that stretch from the Atlantic to the Pacific oceans. Alaska and Hawaii are not among them.

Think Again?

The bison and buffalo are the same animal.

Not true. The bison and buffalo are animals that belong to the *Bovidae* biological family, but they differ in their physical appearance and habitat. The buffalo is native to Asia and Africa. The bison is native to North and South America. *Bison* is the correct scientific name for the American animal, but the term *buffalo* is widely used.

Today, bison live mainly in parks and reserves.

Identifying In what area do most bison live?

Hikers stroll toward the peaks called the Maroon Bells. The peaks are part of the Elk Mountains near the ski town of Aspen, Colorado.

Identifying What three sections of landforms are located between the Rocky Mountains and the mountain ranges along the Pacific Coast?

Between the coastal mountains and the line of the Sierras and the Cascades are long, low valleys. This lowland area is called Central Valley in California. In Oregon, it is called the Willamette Valley.

Basins and Plateaus

Between the Rockies and the Sierra Nevada and Cascade Range is a mix of landforms. They can be grouped into three sections.

To the south and east is the Colorado Plateau. A plateau is a large area of generally flat land. This highland area is marked by smaller, flat-topped features called mesas that are sometimes separated by canyons. In addition, many canyons cut deep into the Colorado Plateau. Among these canyons is the Grand Canyon. Winding along the canyon floor, more than a mile (1.6 km) below the rim, is the Colorado River. Rising up to the plateau are rocks of many colors and shapes. The spectacular sight attracts more than 4 million visitors every year.

West of the Colorado Plateau and extending to the north is the Basin and Range region. This name refers to a pattern on the land in which clusters of steep, high mountains are separated by low-lying basins.

To the north is the Columbia Basin. This large area was formed mainly by vast amounts of lava that flowed from volcanoes and then cooled and hardened. Much of the area is flat, but rivers cut deep valleys and canyons.

Landforms of Alaska

Alaska—the largest U.S. state in land area—lies to the west of Canada. Mountains run along its southern and northern edges. The Alaska Range, also in the south, is the home of the highest point in the United States, Mount McKinley. Also called Denali, the mountain soars 20,320 feet (6,194 m) high. Lowland plains cover the area between the Alaska Range and the northern mountains.

Landforms of Hawaii

Nearly 2,400 miles (3,862 km) southwest of California is Hawaii. An archipelago, Hawaii includes more than 130 islands. The eight largest ones are in the eastern part of this chain of islands. Volcanoes formed these islands. Two volcanoes—Mauna Loa and Kilauea—are still active. Wind and the sea have eroded some mountains to make steep cliffs. Along the shore, some islands have sandy beaches that draw many tourists.

☑ **READING PROGRESS CHECK**

Citing Text Evidence What are two ways in which the states west of the Mississippi River are similar to each other?

John Kieffer/Peter Arnold/Getty Images

Bodies of Water

GUIDING QUESTION *How do the bodies of water in the region affect people's lives?*

The United States west of the Mississippi River is much drier than the eastern part of the country. As a result, it has fewer rivers and lakes. Those that do exist play **significant** roles in the area's economy.

Ocean and Gulf

The chief body of water for this region is the Pacific Ocean. This vast ocean meets the western shores of the continental United States and of Alaska. Hawaii sits in its midst. Inlets from the ocean **create** many excellent harbors along the Pacific coast. As a result, the coast has many major ports. They include San Diego, Long Beach, and Los Angeles, California; Portland, Oregon; and Seattle and Tacoma, Washington. Valdez, Alaska, and Honolulu are also important Pacific ports.

Louisiana and Texas border the Gulf of Mexico. They both have several major ports, including those of New Orleans and Houston. The coastlines of the Pacific Ocean and the Gulf of Mexico are very different. Plate tectonics are active along the Pacific coast. This not only causes earthquakes but also makes for a steep coastline with rocky cliffs. The Gulf coast is much more stable and very flat. It has many swamps and shallow water.

Lakes of the Region

Because the western United States has a fairly dry climate, lakes are not as common in this region as east of the Mississippi. But thousands of years ago, the climate was wetter. A huge, freshwater lake covered many of the basins that are located in what are now Utah, Idaho, and Nevada. Today, only a few isolated lakes remain.

Academic Vocabulary

significant important

Academic Vocabulary

create to cause to form

Tall cliffs surround volcanic Crater Lake in Oregon. A volcanic hill rises out of the middle of the lake.
▶ **CRITICAL THINKING**
Analyzing Are lakes more plentiful in the western United States than in the eastern region? Explain.

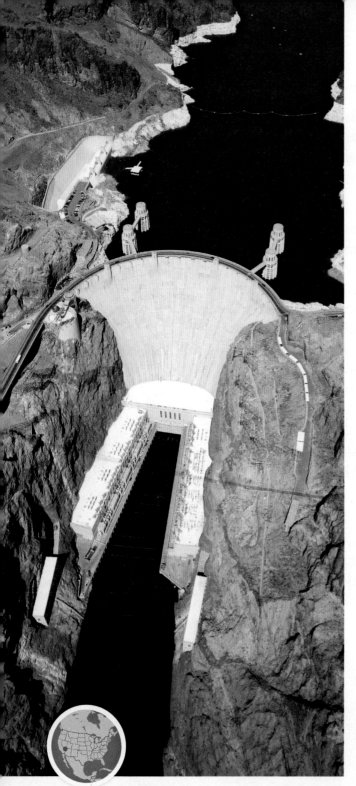

One of them is Utah's Great Salt Lake. It is the largest salt lake in the Americas. The lake changes in size gradually because a varying amount of water evaporates from it. The lake is salty because it does not have an outlet. Tributary rivers bring salt to the lake. Evaporation takes the water away but leaves the salt behind.

Two other important lakes in the region are Lake Tahoe and Lake Mead. Lake Tahoe sits high in the Sierra Nevada. Lake Mead is a human-made lake formed when Hoover Dam was built on the Colorado River. Both lakes are used for boating and other water recreation.

Rivers

Rivers are important in the western United States. The Colorado is one of the major rivers of the region. It begins along the western slope of the Rocky Mountains and twists its way south and west to the Gulf of California. At many places along its course, the Colorado has been dammed. So much of the river's water is used for farming and in cities that no water at all reaches the Gulf of California. To the north, the Columbia River flows from the Rocky Mountains to the Pacific Ocean. It has been dammed in several places to provide hydroelectric power. The Snake and Willamette rivers feed into this river.

All these rivers flow west. But some of the rivers in the region flow east toward the Gulf of Mexico. The **Continental Divide**, an imaginary line through the Rocky Mountains, separates these two sets of rivers. The eastward-flowing rivers include the Missouri, the Platte, the Kansas, the Arkansas, and the Rio Grande. The first four of these rivers feed into the Mississippi.

The dams on all the rivers bring much benefit to the people in the region. They control floods, generate hydroelectric power, and provide water for urban and rural areas. The dams also greatly affect the hydrosphere and biosphere. Without dams, the rivers flow fast and cold during springtime when snows are melting. Then in summer, they flow slowly and are warmer.

Hoover Dam is located on the Colorado River along the border between Arizona and Nevada.
▶ **CRITICAL THINKING**
Describing How do dams benefit the people of the region?

✓ READING PROGRESS CHECK

Integrating Visual Information What landform is crossed by the rivers that flow from the Rocky Mountains to the Mississippi River?

Climates of the Region

GUIDING QUESTION *What factors influence the climates of the region?*

The United States west of the Mississippi River has many different climates. Tropical rain forests cover parts of Washington and Oregon because of the many storms that come from the North Pacific Ocean. Dry, hot deserts cover large parts of the Southwest.

Coastal and Highland Climates

High mountains play a role in forming climates in the region. The western mountains cause what is called a rain shadow. West-facing slopes of the Pacific Coast Ranges, the Cascades, and the Sierra Nevada receive plentiful rain and snow from Pacific Ocean storms. Heavy rains give northern California, Oregon, and Washington a marine west coast climate. Vast forests grow there.

The valleys to the east of the coastal ranges lie in the rain shadow, so they are dry. Although California's Central Valley lies in the rain shadow, it is a major farming region. It is hot and dry, but it is located near the western slopes of the Sierra Nevada, which receive abundant rain. Mountain rainwater is used for irrigation in the valley. **Irrigation** is the process by which water is supplied to dry land.

Climates in the Interior

The high mountains of the Cascades and the Sierra Nevada produce a rain shadow effect that keeps the interior of the region dry. This dry climate is what causes so much evaporation from Great Salt Lake. Some areas are covered by large deserts.

Advances in technology have produced new methods of watering farmland.

Determining Word Meaning What is irrigation?

Don Farrall/Stockbyte/Getty Images

The mountain regions in Colorado can have large variations in climate. Snow-covered mountains have cold nighttime temperatures in winter. Bright sunshine can make summer days comfortably warm.

Identifying What area of the region has a humid subtropical climate?

The western Great Plains have a semiarid climate. A semiarid area receives more rain than a desert but not enough for trees to grow. Instead, semiarid areas have bushes and grasslands. Temperatures in the Great Plains get hot in the summer and cold in the winter. This is what is known as a humid continental climate. Sometimes in the winter, a dry wind called the **chinook** blows over the region. It originates in the mountains where it is cold. But the air heats up as it blows down the eastern slopes of the mountain.

The eastern half of the Plains has two climate types. The northern part has a humid continental climate. It is influenced by cold air masses moving down from the Arctic. The climate of the southern part is humid subtropical. It is shaped by warm, moist air from the Gulf of Mexico. It has higher temperatures than the northern Plains. When cold, dry air from the north collides with warm, moist air from the south, thunderstorms and even tornadoes can result.

Climates of Alaska and Hawaii

Climates in Alaska are generally moderate but cool to cold. More moderate temperatures occur toward the south and colder ones to the north. Winters are cold in the far north. Snow can be heavy in the south and southeast. Valdez can receive as much as 200 inches (508 cm) of snow per year.

Hawaii has a tropical rain forest climate with high temperatures and high levels of rainfall. Rain tends to be heaviest in the winter. More rain falls on the northeastern side of mountains because the moist wind comes from that direction. Steady ocean breezes keep the air comfortable even when temperatures are high.

☑ **READING PROGRESS CHECK**

Citing Text Evidence Why does so much water evaporate from Great Salt Lake?

Resources of the Region

GUIDING QUESTION *What resources does the region have?*

The United States west of the Mississippi River has a great variety of natural resources. In addition to land that supports raising livestock, rich reserves of petroleum, minerals, and a variety of energy sources are found here.

Energy Resources

The United States west of the Mississippi River has large reserves of energy resources. Petroleum is found in the Gulf of Mexico, near Louisiana and Texas; in the southern Great Plains; in California; and in Alaska. Natural gas is found in the same areas. Coal is abundant in Wyoming.

A growing source of energy coming from the region is ethanol. **Ethanol** is a liquid fuel made from plants. In the United States, ethanol is made from corn and blended with gasoline. The United States is one of the world's leading producers of ethanol.

Hydroelectric power is an important source of energy in this region. Dams along the Columbia and Colorado rivers supply this power. Wind power is a growing source of energy here. South Dakota gets nearly a quarter of its electricity from wind power—more than any other state. Solar power is also becoming more important.

Minerals and Other Resources

The Rocky Mountains are important sources of gold, silver, copper, zinc, and lead. Timber is an important resource, too. Fertile soil makes the Plains, California's Central Valley, and parts of Oregon and Washington major farming regions.

Another important resource in the region is its natural beauty. Large areas of great natural beauty have been set aside in **national parks**. These parks attract millions of visitors every year.

☑ **READING PROGRESS CHECK**

Analyzing Does this region of the United States rely too much on one energy resource? Explain?

Think Again

Nothing grows in a desert.

Not true. Deserts can support plant life—and some desert plants can reach large sizes. A fully grown saguaro cactus can be as much as 50 feet (15 m) high. Desert plants have to be well-suited to the dry conditions, though. They need large root systems and thick leaves that trap and retain moisture.

Include this lesson's information in your Foldable®.

LESSON 1 REVIEW

Reviewing Vocabulary

1. Why is *irrigation* needed in California's Central Valley?

Answering the Guiding Questions

2. *Determining Central Ideas* How would the landforms and climate of the region affect where people live?

3. *Identifying* What are two characteristics that Washington and Hawaii share?

4. *Describing* What can result in the Great Plains when cold, dry air collides with warm, moist air?

5. *Analyzing* Are you more likely to find hydroelectric power in Washington and Oregon or on the Great Plains? Why?

6. *Informative/Explanatory Writing* In a paragraph, describe the scenery you would see on a drive from the Mississippi River to the Pacific coast.

Reading **HELP**DESK

Academic Vocabulary

• **establish**
• **data**

Content Vocabulary

• **nomadic**
• **pueblo**
• **mission**
• **frontier**
• **Manifest Destiny**
• **annex**
• **extinct**
• **reservation**

TAKING NOTES: *Key Ideas and Details*

Organize As you read about westward expansion, use a graphic organizer like the one below to identify two land acquisitions and how the land was acquired.

Land Acquired	How Acquired

Indiana Academic Standards
6.1.10, 6.1.11

Lesson 2
History of the Region

ESSENTIAL QUESTION • *How do people make economic choices?*

IT MATTERS BECAUSE
Westward expansion is an important story in U.S. history.

Early Settlements

GUIDING QUESTION *How did life in the region change for Native Americans?*

The first people to live in the western states of what is now the United States were Native Americans. Native Americans belonged to dozens of different groups. Each group had its own language and culture and followed a lifestyle well-suited to the area where they lived.

Native American Ways of Life

The tribes of the Great Plains adopted different ways to live on these grasslands. Some farmed and hunted. They settled along rivers, where they tended fields that grew corn, squash, and other foods.

Other Native Americans of the Plains hunted the herds of bison. Along with obtaining meat, the people used other parts of the animals for clothing and homes. These peoples were **nomadic**, always on the move. Few trees grew there, so the nomadic peoples of the Plains built homes called teepees, using animal hides stretched over long poles. Teepees could be folded up and moved fairly easily, allowing Plains peoples to take their homes with them as they traveled.

The Pueblo people of the Southwest lived in villages that the Spanish called **pueblos** ("towns" or "villages" in Spanish). These villages' multistoried homes were made of dried mud. The Pueblo practiced dry farming, conserving scarce water to grow corn, beans, and squash.

(l to r) ©George H.H. Huey/Corbis; ©Bettmann/Corbis; ©Corbis; ©Corbis

In the Northwest, Native Americans fished for salmon and hunted sea mammals. On land, they hunted small game. Taking advantage of the thick forests that grew in the region's climate, they built large homes of wood.

When Europeans came to North America, the lives of Native Americans changed dramatically. For example, Europeans introduced new animals such as horses and sheep. Horses made hunting bison easier for the Plains peoples. The Navajo of the Southwest began herding sheep. But Europeans also brought diseases that killed large numbers of Native Americans.

Colonial Times

The Spanish were the first Europeans to come to this region. In 1598 they **established** the first European settlement in the region near what is now El Paso, Texas. By the early 1600s, they had founded Santa Fe, New Mexico. Soon they spread out along the upper reaches of the Rio Grande.

In the 1700s, the Spanish settled parts of California and Texas. Central to some settlements were **missions**. These church-based communities led by Catholic priests were meant to house native peoples. The priests hoped the Native Americans would adopt Christianity and the Spanish way of life. They relied on the work of the native peoples to grow food.

The land and climate across much of the region were similar to what they had left in Spain. The Spanish settlers introduced numerous types of crops that grew in Spain but did not exist in the Western Hemisphere. These included oranges, grapes, apples, peaches, pears, and olives. The Spanish also adopted crops that grew in the Western Hemisphere but not in Spain. These crops included corn, tomatoes, and avocados.

(l to r) ©George H.H. Huey/Corbis; Peter Pearson/Getty Images

Academic Vocabulary

establish to start

This ancient pueblo (left) is near Taos, New Mexico. Located near Tucson, Arizona, San Xavier del Bac (right) was founded as a Catholic mission by Father Eusebio Kino.

▶ **CRITICAL THINKING**

Describing In what ways was life for Native Americans in pueblos similar to life in the mission? In what ways was it different?

In the 1680s, France claimed the land drained by the Mississippi River, which included much of the land east of the Rocky Mountains. It called this vast area Louisiana. Over the years, the French placed a few settlements along the Mississippi River. The most important was New Orleans. It was founded in the early 1700s as a port for shipping goods from the river's valley.

✓ **READING PROGRESS CHECK**

Determining Central Ideas What was the Native American lifestyle like before Europeans came to the region?

Westward Expansion

GUIDING QUESTION *Why and how did Americans move into this region?*

When the American Revolution ended in 1783, the territory of the United States was entirely east of the Mississippi River. Much of this land was still unsettled by white Americans. But after the Revolution, they quickly began moving to the frontier in large numbers. A **frontier** is a region just beyond or at the edge of a settled area. Soon, Americans turned their eyes west of the river, eager for more land.

Exploring the West

In 1803 President Thomas Jefferson purchased the vast Louisiana Territory from France. The Louisiana Purchase gave the United States most of the land between the Mississippi River and the Rocky Mountains.

Soon after, Meriwether Lewis and William Clark led nearly 50 men to explore parts of the area. The Lewis and Clark expedition lasted more than two years, as they traveled from St. Louis to what is now the coast of Oregon and back. They traveled along the Missouri River as far as they could, and then they proceeded overland by horseback. They mapped the land and rivers they saw. They recorded **data** about the plants and animals living there. They also made peaceful contact with Native American peoples. By reaching the Pacific Ocean, they helped set an American claim to Oregon and Washington.

Over the next decades, other explorers helped open new areas. Meanwhile, hardy adventurers called mountain men began to trap beavers in the Rocky Mountains. Their travels added more knowledge about the geography of the American West. They also discovered ways through the mountains that settlers would use later.

Settling the West

By the 1830s, some Americans had come to believe in the idea of **Manifest Destiny**. According to this concept, the United States had a right to extend its boundaries to the Pacific Ocean. This belief helped promote the nation's westward movement.

Academic Vocabulary

data information

Settlers began moving to the rich farmlands in what is now Oregon. Traveling in wagons, they took a long route called the Oregon Trail. It carried them across the Great Plains and through passes in the Rocky Mountains. Over the years, thousands of people moved to Oregon. What is now Oregon and Washington, however, were claimed by both the United States and Great Britain. In 1846 the two countries reached an agreement. Under the deal, the United States gained control of those two future states. Britain kept control of lands to the north, which became part of Canada.

Meanwhile, some American settlers had moved to what is now Texas, which belonged to Mexico at the time. In the next decade, they declared independence. They set up an independent country, though many Texans wanted to join the United States. In 1845 the United States **annexed**, or took control of, Texas.

Some Americans hoped to gain California and other lands that were part of Mexico. This desire led to a war with Mexico, which the United States won. In the Treaty of Guadalupe Hidalgo (1848), Mexico gave the United States a vast area that later formed all of California, Utah, and Nevada and parts of Colorado, Arizona, Wyoming, and New Mexico. This territory added a sizable Spanish-speaking population to the United States.

MAP SKILLS

1 **THE GEOGRAPHER'S WORLD** Where did the Spanish found most of their western settlements?

2 **ENVIRONMENT AND SOCIETY** How were settlers able to reach Salt Lake City from the central part of the United States?

Westward Expansion

PURCHASED FROM GREAT BRITAIN, 1818

1851 - Seattle

1811 - Astoria, Oregon

OREGON TERRITORY, 1846

Oregon Trail

LOUISIANA PURCHASE, 1803

40°N

1839 - Sacramento

1847 - Salt Lake City

Mormon Trail

1776 - San Francisco (founded by Spanish)

1777 - San Jose (founded by Spanish)

MEXICAN CESSION, 1848

1858 - Denver

PACIFIC OCEAN

1781 - Los Angeles (founded by Spanish)

Santa Fe Trail

30°N

1769 - Mission San Diego (founded by Spanish)

1868 - Phoenix

1841 - Dallas

130°W

• Settlement/City
1845 Year acquired

GADSDEN PURCHASE, 1853

TEXAS ANNEXATION, 1845

1718 - San Antonio (founded by Spanish)

0   400 miles
0   400 kilometers
Albers Equal-Area Conic projection

120°W

110°W

MEXICO

100°W

Gulf of Mexico

90°W

When gold was discovered in California, thousands of people streamed there in hopes of making their fortunes. This mass migration is called the California Gold Rush. San Francisco and other cities grew rapidly as a result.

Western Lands in the Late 1800s

Later in the 1800s came more discoveries of mineral wealth in other areas in the region. Each new discovery brought more people to the region in hopes of becoming wealthy. A huge reserve of silver lured them to Nevada in the 1870s. Also during that decade, gold attracted people to South Dakota's Black Hills and to Colorado.

At the same time, the nation was building railroads to join the eastern and western areas. The first line from the Mississippi River to the Pacific Ocean was completed in 1869. Others followed. Trains carried settlers to the western states. Some started farms or ranches. Others settled in towns that sprang up along rail lines or near mines.

As a result of these changes, the population of the West grew rapidly. In 1900 more than twice as many people lived in the West as in 1880, just 20 years earlier.

The Great Plains changed dramatically during this time. The vast grasslands were turned into farms and ranches. Settlers hunted the huge herds of bison and other animals. Some of the animals became **extinct**, or disappeared from Earth. These were huge changes to the biosphere of the Plains.

Passengers board stagecoaches in Virginia City, Nevada. The discovery of silver drew settlers to the town and made the area wealthy almost overnight.

Identifying What other western areas drew settlers as a result of the discovery of minerals?

©Bettmann/Corbis

The spread of white settlements came at the expense of Native Americans. Native Americans suffered from the changes to the environment and the growth of the population. Farms, ranches, railroads, and mines took away land that Native Americans had farmed or hunted on. Some Native American groups resisted the changes, but they were outnumbered. Finally, Native Americans were forced to live on **reservations**. These are lands that were set aside for them. Reservations were often located in areas with poor soil that made farming difficult.

Gaining New Lands

During the late 1800s, the United States made its last land acquisitions. The first new territorial gain was Alaska. In 1867, the United States purchased Alaska from Russia for just over $7 million. The future state was so large that the cost was only about two cents per acre. The purchase of Alaska was not entirely popular. Some newspaper editors criticized Secretary of State William Seward for agreeing to the sale. They called the area "Seward's Icebox" or "Seward's Folly." In 1898, however, gold and copper were found in Alaska. These discoveries awakened new interest in the land.

Americans also took an interest in Hawaii in the late 1800s. Businesspeople began to grow sugar there. By the late 1880s, American sugar planters feared Hawaii's royal family would take away the power and land they had acquired. Instead, they seized the government and requested that the United States annex Hawaii. In 1900 the government agreed to do so.

☑ **READING PROGRESS CHECK**

Analyzing How were the acquisitions of Texas and Hawaii similar?

Shows as He Goes, a Native American chief (left), fought U.S. pioneers and soldiers on the Great Plains. Native American boys in uniform (right) attended a white-run school opened in Pennsylvania during the late 1800s.

▶ **CRITICAL THINKING**
Determining Central Ideas How did Native Americans in the West live before the arrival of white settlers? How did they live after whites settled the area?

Agriculture and Industry

GUIDING QUESTION *How did people in the states west of the Mississippi live?*

As people moved into the states west of the Mississippi, they developed various ways of earning a living. For many decades, their choices depended on the resources of the area where they settled. Most made their living in primary economic activities that extract resources directly from the earth. These include farming, ranching, mining, lumbering, and fishing.

Farming and Ranching

In the 1800s, many Americans were farmers. Many dreamed of starting farms in the West. In 1862 Congress made that easier by passing the Homestead Act. This law made public land in the western states free to anyone who claimed the land, built a farm, and stayed on it for five years. Hundreds of thousands of people settled on the Great Plains to start farms.

Life on these farms was not easy. The lack of trees made it difficult to find wood to build homes. People covered homes with sod—chunks of soil held together by the roots of grasses.

Another important activity in the western states was raising cattle. Cowboys in places such as Texas herded the cattle and drove them north to towns in Colorado and Kansas that had railroad

Cowboys on horseback round up cattle near Colorado's Cimarron River.

▶ **CRITICAL THINKING**
Describing How did railroads help the cattle industry grow?

©Corbis

stations. There, the animals were shipped east to cities like Chicago. Meatpacking companies butchered the animals into meat that could be sold in growing eastern cities. Cities like Denver; Kansas City, Missouri; and Omaha, Nebraska, became major centers for processing crops and meat.

Industry

The first industries in the western states were also primary economic activities. The industries concentrated on using the resources of the region. Companies set up silver and copper mines. Others cut trees in the Northwest to provide lumber for building homes and ships. Fish canneries were important along the Pacific coast and the Gulf coast.

In the 1900s, new primary industries developed. The growing popularity of cars created rising demand for oil. The oil industry boomed in Texas, Oklahoma, and California. Oil in northern Alaska became usable in the 1970s with completion of a major construction project: the Trans-Alaska Pipeline System (TAPS). That project built a pipeline from the northern oil fields to the port of Valdez in the south so that oil could be shipped to the continental states.

Although primary economic activities remained dominant as the region developed, more secondary economic activities began to emerge. These are industries that turn raw materials into manufactured goods. Many cities across the region became major manufacturing centers. Los Angeles specialized in machine tools and automobiles. San Francisco became a major shipbuilding center.

Recreation and Entertainment

Other new industries focused on recreation and entertainment. Southern California became home to the movie industry. Las Vegas turned into a major resort city known for its casino-hotels, shops, and restaurants. Areas in the Rockies with great natural beauty—or excellent slopes for skiing—became favored vacation spots.

☑ **READING PROGRESS CHECK**

Determining Central Ideas What resources attracted Americans to the western region in the 1800s?

Include this lesson's information in your Foldable®.

LESSON 2 REVIEW

Reviewing Vocabulary
1. How did *pueblos* and *missions* differ?

Answering the Guiding Questions
2. *Describing* What was one change in the way of life for Native Americans after the Spanish and French came to this region?

3. *Determining Central Ideas* Why was the Lewis and Clark expedition important?

4. *Identifying* Why did the movement of white settlements into the West cause problems for Native Americans?

5. *Analyzing* In what way was the recreation industry in the region similar to farming, ranching, and mining in earlier times?

6. *Informative/Explanatory Writing* In a paragraph, explain how the industries that developed in the region late in the 1900s were different from those of earlier times.

(l to r) Somos/Veer/Jupiterimages; Kevork Djansezian/Getty Images News/Getty Images; Roy Delgado/www.Cartoonstock.com; Stephen Brashear/Getty Images News/Getty Images

Reading **HELP**DESK

Academic Vocabulary

- **annual**
- **decline**

Content Vocabulary

- **Mormon**
- **Dust Bowl**
- **topsoil**
- **agribusiness**
- **aerospace**

TAKING NOTES: *Key Ideas and Details*

Organize As you study the lesson, take notes on the topics shown below.

Urban and Rural Life
Challenges
The Economy

 Indiana Academic Standards
6.1.17, 6.3.14

Lesson 3
Life in the United States West of the Mississippi

ESSENTIAL QUESTION • *How does technology change the way people live?*

IT MATTERS BECAUSE
The states west of the Mississippi are a source of technological change.

The Region's Cities and Rural Areas

GUIDING QUESTION *Where do the people of the region live?*

Modern cities are in many ways similar. Glass, steel, and concrete skyscrapers rise into the sky. Networks of highways carry heavy traffic. Cities in this region have distinct characters, though. The French flavor of New Orleans differs from the Spanish style of Santa Fe, New Mexico. Denver, near towering mountains, is unlike Omaha, Nebraska, on the relatively flat Plains. What could be more different than tropical Honolulu and cold Anchorage?

One characteristic common to almost all cities in the western United States is dependence on the automobile for transportation. Most western cities have limited or no subway or light-rail systems. In addition, most large cities are spread out, and the distances between cities are often great.

Major Port Cities

The region has many major ports. Los Angeles and Long Beach in California are essentially one port. They handle more than one-half the value of all imports into the United States that come through Pacific ports. San Diego and San Francisco in California, and Seattle and Tacoma in Washington State, are also vital to U.S. trade with Asia.

Although only a small part of the region is located along the Gulf of Mexico, Gulf ports are important. Three of the nation's top 10 ports in terms of the **annual**, or yearly, value of goods they handle are the Texas port cities of Houston, Beaumont, and Corpus Christi.

Many of these cities are diverse. Los Angeles—the nation's second-largest city—is home to people who collectively speak about 90 languages other than English at home. Los Angeles County has more Latinos and Native Americans than any other county in the United States. The city of Los Angeles has more people from South Korea and Nicaragua than any other city outside those nations.

Interior Cities

Denver has an unusual location for a major city. It is not a seaport or on the navigable part of a river. Denver owes its vibrance to the mountains nearby. It originally grew as a mining town. In the late 1900s, the city attracted people who wanted to enjoy the mountains. Its economy is based on software, finance, and communications.

Some of the nation's most rapidly growing cities are in the interior of the United States west of the Mississippi. They include Austin, Texas; Boise, Idaho; Las Vegas; Phoenix; Provo, Utah; and Riverside, California. In Texas, the location of San Antonio, Dallas, and Fort Worth near Mexico makes them important to trade with that country.

Somos/Veer/Jupiterimages

Many urban schools today reflect the growing ethnic diversity of America's cities.

Life in Rural Areas

Small towns and villages remain home to millions of people in the states west of the Mississippi. Many rural Americans rely on the rich resources of the land and sea. They might be farmers of the Plains or fishers in rural Alaska. They may drill for oil in Texas or run a ranch in Montana. Although these occupations have existed for centuries, modern technology often makes these jobs easier. For example, farmers and ranchers use GPS devices to map regions, manage the land, and track cattle.

☑ **READING PROGRESS CHECK**

Citing Text Evidence In what ways are Las Vegas and Phoenix different from Los Angeles and Seattle?

Challenges Facing the Region

GUIDING QUESTION *What issues will face the region in the coming years?*

Americans celebrate their ethnic and religious diversity. Diversity has long been a strength of the nation. While diversity enriches American life, other population changes pose challenges to the region's future.

Population Changes

In recent decades, much of the growth in U.S. population has taken place in the states west of the Mississippi River. This region attracts new residents because of the mild climate and growing businesses. But more people means a strain on natural resources such as water, which is already scarce in much of the region.

Ethnic and racial diversity are common in the states in this region. In Hawaii, Asian Americans and other distinct ethnic or racial groups form the majority of the population. In California, New Mexico, and Texas, Latinos constitute a large part of the population. In these four states, non-Hispanic whites are in the minority.

A man from East Asia takes the oath to become a U.S. citizen. He was part of a group of 7,000 candidates who became U.S. citizens in a ceremony held in Los Angeles, California.

▶ **CRITICAL THINKING**

Identifyng What are some of the ethnic groups that make up the diverse population of the western United States? Why do you think many people from East Asia have settled in this region?

Kevork Djansezian/Getty Images News/Getty Images

The United States west of the Mississippi is also marked by great religious diversity. A few Christian groups are particularly important in this region. Lutherans are numerous in the northern Great Plains states. Catholics are prominent in the states from New Mexico to California. Utah and some neighboring states have many Mormons. **Mormons** are members of the Church of Jesus Christ of Latter Day Saints. Large numbers of Mormons settled in Utah in the mid-1800s.

Another population trend raises important economic challenges. The share of the population over age 65 has been growing. Many older people require extensive health care. That increases the costs for Medicare, a government-run program that pays the health care costs of the elderly. Most older people collect monthly retirement checks from the Social Security system. Meeting the costs of Social Security and Medicare will be a challenge in the coming years.

Relations With Neighbors

Relations between the United States and neighboring Canada and Mexico are strong. In the early 1990s, the three nations signed a trade agreement called the North American Free Trade Agreement (NAFTA). In that treaty, they pledged to remove all barriers to trade among themselves. This created the world's largest free trade area.

Today, Canada and Mexico are the largest markets for exports from the United States. They are also the second- and third-largest sources of U.S. imports, behind China. States west of the Mississippi form a vital part of this trade. Their food products and manufactured goods form a share of the exports to these nations. In addition, much of the trade that takes place flows into and out of the United States through ports in this region.

Open borders help trade. They also make it possible for people to enter the country illegally. In the late 1900s, the problem of illegal immigrants drew a great deal of attention. The U.S. government has taken steps to reduce the flow of illegal immigrants. These efforts have had some success. Illegal immigration dropped from about 550,000 people in 2005 to around 300,000 in 2008. Changes in Mexico also help explain this reduction. A better economy, smaller families, and better education have meant better chances of landing good jobs in Mexico.

The Water Problem

Most of the United States west of the Mississippi usually receives little rain. Years of low rainfall can easily lead to drought. Between 1930 and 1940, a severe drought dried the southern Great Plains so thoroughly that crops died. Strong winds carried dry soil away, covering other areas with dust. The area came to be called the **Dust Bowl**. Many farmers lost their homes and left the area looking for work.

Thinking Like a Geographer

Making the Desert Bloom

Why did Mormon pioneers settle in Utah? Utah in the mid-1800s was a harsh desert land. The Mormons, however, needed a safe, isolated place where they could practice their religion free of persecution. In 1847 the first Mormon settlers reached Utah after traveling 1,000 difficult miles (1,609 km) from the Midwest. The land was dry and wild. Nevertheless, the Mormons stayed in Utah. They built irrigation canals to support farms and towns. Life at first was difficult, but the Mormons made their Utah communities prosper because of their hard work and determination to succeed. By 1860, many other Mormons had arrived, and numerous Mormon settlements dotted the Utah region.

" Bad news . . we've run out of unlimited resources. "

An editorial cartoon makes a point about a political issue or event.

▶ **CRITICAL THINKING**

Analyzing What issue do you think this cartoon is about? What do you think the cartoonist's opinion on this issue is?

The Dust Bowl resulted in new practices in land use and farming. These practices were intended to reduce the chance of another dust bowl. Nevertheless, unusually low rainfall for several years could cause similar difficulties in the future.

The lack of water is more of a challenge because of population growth. As more and more people settle in the area, more water is needed. When farmers expand their operations, they also need more water. This strains the limited amount of water present in the region, which could limit economic growth.

Limited water resources have prompted scientists to develop ways to remove salt from seawater to make it usable for drinking and farming. This process, called desalination, is expensive but might provide a long-term solution to the problem in the future.

Human Actions and the Environment

Oil—another precious resource—is also a potential cause of environmental damage. In 1989 the oil tanker *Exxon Valdez* ran aground in Alaska's Prince William Sound. Its shattered hull released more than 250,000 barrels of oil into the sea. The oil killed plants and animals and severely hurt the local economy. Even worse was the Deepwater Horizon disaster of 2010. An explosion destroyed an oil drilling platform deep in the Gulf of Mexico. By the time the underwater leak was stopped, 5 million barrels of oil had gushed into the sea. It devastated the coastal economy.

Erosion and the Environment

Erosion is another environmental problem in the region. Harvesting trees for lumber has left some mountain slopes bare. With no tree roots to hold the soil in place, soil runs off with the rain. This runoff affects the surface of the mountain. Pieces of the mountain become smaller pieces and go down the side. Erosion of rich **topsoil**, the fertile soil that crops depend on to grow, is a problem in some farming areas.

The region west of the Mississippi River also experiences a variety of natural disasters. Washington, Alaska, and Hawaii have active volcanoes that can cause damage if they erupt. Wildfires and

mud slides can also strike the region. The wildfires occur when wooded areas become too dry. Mud slides can result from heavy rains or severe shaking from an earthquake. During a mud slide, the soil moves like a liquid, flowing downhill in a sea of mud that dislodges trees and buries houses in its path.

Some parts of the region are prone to earthquakes. A major fault system called the San Andreas Fault, cuts through western California. Movement along this fault has caused several major earthquakes over the years. Scientists think more are likely to occur. As a result, severe damage is possible—as well as the loss of many lives—in major cities such as Los Angeles and San Francisco.

☑ READING PROGRESS CHECK

Determining Central Ideas Why is the high likelihood of a major earthquake along the San Andreas Fault so worrisome?

In 1994 the deadly Northridge earthquake struck near Los Angeles. It resulted in 57 deaths and was one of the costliest natural disasters in U.S. history.
Citing Text Evidence Why is California so prone to earthquakes?

The Economy

GUIDING QUESTION *How do the people of the region make their living?*

Land and resources have attracted many Americans to the western states. Some people of the region still rely on these advantages to earn their living. Most people, however, work in modern settings that are far removed from the land.

Modern Agriculture

Agriculture remains an important part of life west of the Mississippi River. The Great Plains are the center of the nation's wheat industry with eight of the nation's top wheat-producing states. Spring wheat— planted in spring and harvested in autumn—is grown on the northern Plains. Winter wheat—planted in the fall and harvested in the spring—grows in the southern Plains. Wheat plants would wither in this southern region if they faced the hot summer temperatures. Other important crops in the Plains include cotton, corn, hay, and sorghum, another grain. Cattle and sheep ranching are major activities from Montana to New Mexico.

Other areas in the region are known for other crops. Washington and Oregon produce dairy products and fruit. Idaho is famous for potatoes. California, with its warm, year-round Mediterranean climate, provides fruits and vegetables throughout the year.

Today, the number of small family farms in the region is **declining**. They are being replaced by large **agribusinesses**, firms that rely on machines, advanced technology, and mass-production methods to farm large areas.

The Mining Industry

Mining the region's vast resources remains a vital part of its economy. Oil and gas fields in the southern Plains, Wyoming, and southern California provide energy. Mines in Montana furnish copper, and those in the Colorado Plateau provide uranium. The Northwest is still a major producer of timber.

The Aerospace Industry

In the early 1900s, the airplane industry was born. Decades later, the states of Washington, California, and Texas became important to the aerospace industry. The **aerospace** industry makes vehicles that

Stephen Brashear/Getty Images News/Getty Images

Academic Vocabulary

decline to reduce in number

Workers install an engine cover on a passenger plane at an aerospace factory in Washington state. *Identifying* What other states became important centers of the aerospace industry?

travel in the air and in outer space. This industry brought many engineers and other highly skilled workers to the United States west of the Mississippi.

The aerospace industry remains important in southern California and around Seattle. Employment in this industry has declined in recent years, however. Taking its place has been the computer industry. The first area to experience rapid growth from this field was California's Silicon Valley. Today, centers of computer research and manufacturing are also found in Texas, Washington, and New Mexico.

Service Industries in the Region

Advanced technologies, such as robotics and computerized automation, have transformed manufacturing in the region. As with farming, the region's factories produce greater quantities of goods with fewer workers than in the past.

The growth of computer technology has spawned other industries. Software and information science companies are based in the same areas that are home to computer manufacturing. Utah and Colorado have also become important in these fields. Los Angeles, San Francisco, Denver, Dallas, and Seattle have become major financial centers. Telecommunications—telephone and related services—is an important industry in Denver and Dallas.

Over time, new industries have focused on tertiary economic activities such as retail sales, entertainment, and tourism. Visitors flock to the area to see the soaring mountains, stunning rock formations, and dense forests. They enjoy unusual features like Alaska's glaciers and Hawaii's tropical beaches. The western region includes some of the most-visited national parks in the country.

Include this lesson's information in your Foldable®.

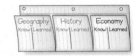

✅ **READING PROGRESS CHECK**

Citing Text Evidence Why did tourism and the tourist industry develop in the United States west of the Mississippi?

LESSON 3 REVIEW

Reviewing Vocabulary

1. Why would the loss of *topsoil* threaten *agribusiness*?

Answering the Guiding Questions

2. *Determining Central Ideas* Why is it unusual that Denver developed into a major city?

3. *Identifying* What are the advantages and disadvantages of open borders?

4. *Analyzing* What steps could be taken to prevent the damage caused by mud slides? Explain your answer.

5. *Describing* Would you describe the economy of these states as diversified? Explain.

6. *Informative/Explanatory Writing* Write a short essay explaining why water conservation is important in the region.

Directions: Write your answers on a separate piece of paper.

❶ Use your FOLDABLES to explore the Essential Question.

INFORMATIVE/EXPLANATORY WRITING Write an essay explaining why the physical geography of the area west of the Mississippi River made settlement of the region difficult.

❷ 21st Century Skills

ANALYZING Working in small groups, research and prepare a presentation explaining how an alternative resource can be used to reduce America's dependence on foreign oil.

❸ Thinking Like a Geographer

CITING TEXT EVIDENCE Create a time line like the one shown and place these six events in the correct order on it.

- Lewis and Clark expedition
- The American Revolution
- Louisiana Purchase
- Spanish establish settlements in the West
- Alaska is purchased from Russia
- The Pueblo flourish in the Southwest

❹ GEOGRAPHY ACTIVITY

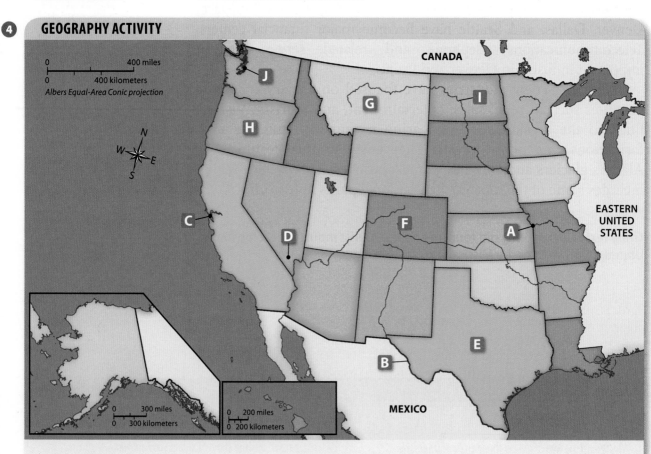

Locating Places
Match the letters on the map with the numbered places listed below.

1. Oregon   **3.** San Francisco   **5.** Las Vegas   **7.** Montana   **9.** Kansas City

2. Missouri River   **4.** Texas   **6.** Colorado   **8.** Rio Grande   **10.** Seattle

REVIEW THE GUIDING QUESTIONS

Directions: Choose the best answer for each question.

1 The Rocky Mountains are
   A. the shortest mountain range in the United States.
   B. a cordillera that extends from Canada to Mexico.
   C. an archipelago.
   D. worn down from erosion.

2 Which of the following is a contiguous state?
   F. Puerto Rico
   G. Alaska
   H. Colorado
   I. Hawaii

3 How did western settlement affect Native Americans?
   A. They were able to sell their lands for a great deal of money.
   B. Those who did not die of disease were forced onto reservations.
   C. They moved east to work in coal mines and factories.
   D   Many became cowboys or joined the army.

4 During the Civil War, Congress passed the Homestead Act, which
   F. guaranteed low-interest loans to home buyers.
   G. established the border with Canada.
   H. set aside land to be used for schools.
   I. gave away western lands to people who were willing to move there and build farms.

5 Population in the states west of the Mississippi River is
   A. growing slowly.
   B. declining.
   C. growing rapidly.
   D. showing little change.

6 Which resource is the most critical factor to further development in the western states?
   F. oil
   G. natural gas
   H. solar power
   I. freshwater

DBQ ANALYZING DOCUMENTS

7 **DETERMINING WORD MEANING** Read the following quotation about the American population.

> "*With an overall 20 percent growth rate, the [population of the] West grew more rapidly than any other region. The South was the second fastest growing region, increasing 17 percent. The Midwest and the Northeast grew almost 8 percent and 6 percent, respectively.*"

—from *National Atlas of the United States*

What does the term *growth rate* in the passage refer to?

A. the number of states in one year compared to another

B. the amount of land in one year compared to another

C. the number of people in one year compared to another

D. the economic output in one year compared to another

8 **CITING TEXT EVIDENCE** How did population growth in the West compare to that in the Northeast?

F. much lower

G. more than three times higher

H. about the same

I. nearly twice as high

SHORT RESPONSE

> "*In the Western United States, the availability of water has become a serious concern. . . . The climate . . . in the West . . . is best known for its low precipitation, aridity [dryness], and drought. . . . The potential for departures from average climatic conditions threatens to disrupt society and local to regional economies.*"

—from Mark T. Anderson and Lloyd H. Woosley, Jr.,
Water Availability in the Western United States

9 **IDENTIFYING** Why might a change in climate threaten to disrupt societies and economies?

10 **ANALYZING** What do you think the people of the western states should do in light of this problem? Why would it work?

EXTENDED RESPONSE

11 **INFORMATIVE/EXPLANATORY WRITING** Imagine that you are the governor of any one of the states west of the Mississippi River and you want to bring more tourism dollars into your state. Use what you have learned from the chapter and do additional Internet research on that state. Write a press release to newspaper travel editors that promotes all the reasons people should vacation in your state.

Need Extra Help?

If You've Missed Question	❶	❷	❸	❹	❺	❻	❼	❽	❾	❿	⓫
Review Lesson	1	1	2	2	3	3	3	3	3	3	1

From "Water Availability for the Western United States—Key Scientific Challenges" by Mark T. Anderson and Lloyd H. Woosley, Jr., U.S. Geological Survey Circular 1261. Department of the Interior/USGS. The USGS home page is http://www.usgs.gov.

CANADA

Michelle Gilders Canada West/Alamy

ESSENTIAL QUESTIONS • *How do people adapt to their environment?*
• *What makes a culture unique?*

Blackfoot girl takes part in First Nations
Pow Wow held in Alberta, Canada.

Lesson 1
The Physical Geography of Canada

Lesson 2
The History of Canada

Lesson 3
Life in Canada

The Story Matters...

Nearly 4 million square miles (10.4 million sq. km) in total area, Canada is the second-largest country in the world. The landscape of this immense country is known for its beauty and bountiful natural resources. Even though Canada is immense in size, it is sparsely populated. Canada's population of First Nations and Inuit people, French, English, and immigrants from around the world reflects its history and diversity.

FOLDABLES®
Study Organizer

Go to the Foldables® library in the back of your book to make a Foldable® that will help you take notes while reading this chapter.

North | South

Past and Present | World Relations

In land area, Canada is the second-largest country in the world. It occupies most of the northern part of North America and is bordered by three major oceans. As you study the map, look for the geographic features that make Canada unique.

Step Into the Place

MAP FOCUS Use the map to answer the following questions.

1 **THE GEOGRAPHER'S WORLD** Which Canadian province borders the Great Lakes?

2 **THE GEOGRAPHER'S WORLD** What body of water divides Canada nearly in half?

3 **PLACES AND REGIONS** What is the national capital of Canada?

4 **CRITICAL THINKING Analyzing** What direction would you travel when flying from Winnipeg to Yellowknife?

A

CANADIAN ARCTIC Canada's polar bears live in icy Arctic terrain surrounded by open water.

B

VICTORIA, BRITISH COLUMBIA British Columbia's Legislative Assembly assembles in the Parliament Buildings to pass laws for the province.

Step Into the Time

DESCRIBING Select one event from the time line and write a paragraph describing how that event changed Canada.

1000 Vikings arrive in Canada

1700

1534 Cartier explores Gulf of St. Lawrence

1608 Quebec founded

1670 Hudson's Bay Company formed

Canada

ARCTIC OCEAN

Greenland (Denmark)

Alaska (U.S.)

Yukon

Whitehorse

Great Bear Lake

Mackenzie R.

Northwest Territories

Nunavut

Yellowknife

Great Slave Lake

A

Iqaluit

ATLANTIC OCEAN

Hudson Bay

British Columbia

Alberta

Lake Athabasca

Saskatchewan

Manitoba

Edmonton

Calgary

Victoria Vancouver

B

PACIFIC OCEAN

Regina

Winnipeg

Lake Winnipeg

Seven R.

Ontario

Lake Superior

Newfoundland and Labrador

St. John's

Québec

Gulf of St. Lawrence

Prince Edward Island

Fredericton Charlottetown

Québec Halifax

Nova Scotia

New Brunswick

Montréal

Ottawa

Toronto

Lake Huron

Lake Ontario

Lake Erie

UNITED STATES

Lake Michigan

St. Lawrence R.

National boundary
⊗ National capital
★ Provincial/territorial capital
• City

| 0 | 500 miles |
| 0 | 500 kilometers |

Lambert Azimuthal Equal-Area projection

120°W   110°W   100°W   90°W   80°W   60°W

70°N   60°N   50°N   40°N

ARCTIC CIRCLE

1896
Gold discovered in Yukon Territory

2010
Vancouver hosts Winter Olympics

1763 France cedes Canada to Britain

1800

1900

2000

1774 Quebec Act becomes law

1993 Canada ratifies NAFTA

1999 Inuit win rights to their own territory

Reading **HELP**DESK

Academic Vocabulary

- comprise
- access

Content Vocabulary

- province
- territory
- shield
- coniferous
- deciduous
- archipelago
- tundra
- fishery

TAKING NOTES: *Key Ideas and Details*

Summarizing As you read about Canada's regions, take notes about them using the graphic organizer below.

Region	Provinces or Territories	Key Facts
Atlantic Provinces		

Indiana Academic Standards
6.1.18, 6.1.19, 6.3.5, 6.3.8, 6.3.12, 6.4.7

Lesson 1
The Physical Geography of Canada

ESSENTIAL QUESTION • *How do people adapt to their environment?*

IT MATTERS BECAUSE
Canada shares many physical features with the United States.

Canada's Physical Landscape

GUIDING QUESTION *How is Canada's physical geography similar to and different from that of the United States?*

For millions of U.S. residents, another nation, Canada, is a short drive away. Canada is a vast and sprawling land of some 3.86 million square miles (10 million sq. km). That size makes it six times larger than the state of Alaska. Canada is larger than every nation in the world except Russia.

Canada and the United States share the longest undefended border in the world. Several of the crossings along this 5,523-mile (8,888-km) border are busy. About 400,000 people and $1.4 billion worth of goods cross it every day.

Overview of Canada

Although huge, Canada is home to relatively few people. Its population ranks thirty-sixth in the world. The great majority of those people live in the southern part of the country. Almost 90 percent of them live within 150 miles (241 km) of the Canada-U.S. border.

Canada is divided into smaller units. The main divisions are Canada's 10 provinces. **Provinces** are administrative units similar to states. Each province has its own government. Canada also has three territories. The **territories** are lands administered by the national government. Canada's provinces and territories can be grouped into five regions.

The Atlantic Provinces

Nova Scotia, New Brunswick, Prince Edward Island, and Newfoundland and Labrador are called the Atlantic Provinces. These four relatively small provinces border the Atlantic Ocean. Except for Newfoundland and Labrador, they are distinguished mainly by lowlands and plateaus. Highlands cover western Newfoundland and Labrador.

Much of the Atlantic Provinces has a humid continental climate. Winters are cold, and summers are warm. Toward the north, winters can last as long as six months. To the south, winters last only four months. More rain falls here than in Canada's interior.

Quebec and Ontario

The provinces of Quebec and Ontario reach from the St. Lawrence River and Great Lakes to northern Canada. They are the heart of Canada and home to more than 6 out of every 10 Canadians.

The land along the St. Lawrence River and the Great Lakes tends to be lowland plains with fertile soil. A massive plateau called the Canadian Shield covers the northern area of these two provinces. A **shield** is a large area of relatively flat land **comprised** of ancient, hard rock. This plateau extends south, east, and west of Hudson Bay. The Shield holds many valuable minerals, such as iron ore, uranium, gold, and copper.

Like the Atlantic Provinces, the southern parts of Ontario and Quebec have a humid continental climate. Temperatures are colder to the north than to the south, and winters are longer. In the northern part of these two provinces, a subarctic climate prevails. These areas have long, cold winters and mild summers. Much of the northern part of these provinces is covered by forests that include coniferous and deciduous trees. **Coniferous** evergreen trees produce cones that hold seeds, and they have needles instead of leaves. **Deciduous** trees shed their leaves in the autumn.

Rainfall and relatively mild temperatures give the southern edge of these two provinces a longer growing season than areas in northern Canada. That, combined with fertile soil, makes it a productive farming area.

MAP SKILLS

1 **PLACES AND REGIONS** What subregion lies directly north of the Great Lakes?

2 **THE GEOGRAPHER'S WORLD** What subregions border the Pacific Ocean?

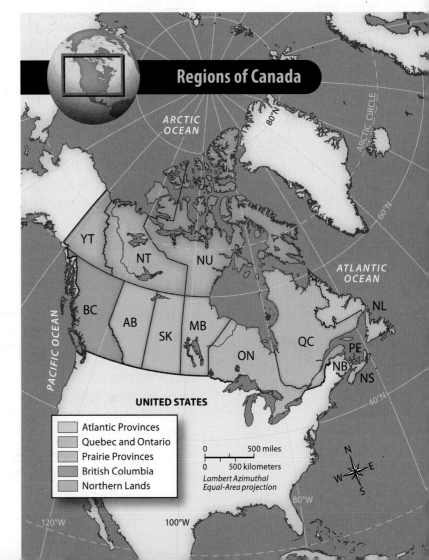

Regions of Canada

- Atlantic Provinces
- Quebec and Ontario
- Prairie Provinces
- British Columbia
- Northern Lands

0   500 miles
0   500 kilometers
Lambert Azimuthal Equal-Area projection

Cape Breton Highlands National Park (left) is located in Nova Scotia. Alberta's Jasper National Park (right) is one of Canada's oldest and largest national parks.

▶ **CRITICAL THINKING**

Describing What is the climate of Alberta and the other Prairie Provinces?

The Prairie Provinces

The Prairie Provinces comprise Manitoba, Saskatchewan, and Alberta. They are covered chiefly by plains extending north from the United States. As in the United States, the plains are tilted from west to east. Their elevation is higher in Alberta than in Manitoba.

Highland areas rim the plains to the northeast and the west. The northeastern highlands, in Manitoba, are part of the Canadian Shield. The western highlands, in Alberta, are the eastern edge of the Canadian Rockies.

The flat land of the Prairie Provinces receives cold blasts of Arctic air from the north. The average daily temperature in January for Regina, Saskatchewan, is –1°F (–18°C). Summers are much warmer, with the average temperature in July being 67°F (19°C). The prairies have reached the highest temperature ever recorded in Canada. On one hot day in 1937, two communities in Saskatchewan had temperatures of 113°F (45°C).

Precipitation in the Prairie Provinces is light. The area receives only about 15 inches (38 cm) of rain or snow each year.

British Columbia

The landscape of British Columbia is like that of the northwestern United States. The Rocky Mountains tower over the eastern edge of the province. A deep valley and plateaus separate the mountains from another high range farther west, the Coast Mountains. The Canadian Rockies include 30 mountain peaks that reach more than 10,000 feet (3,048 m). They are also home to five of Canada's national parks. The Coast Mountains rise even higher—several of them soar more than 15,000 feet (4,572 m).

 (l to r) Design Pics/Bilderbuch; Design Pics/Richard Wear

Inlets of the Pacific Ocean cut into the mountains along British Columbia's coast. Steep cliffs rise directly from the water to heights of more than 7,000 feet (2,134 m).

British Columbia has a marine west coast climate. Temperatures are mild because warm air blows west from the ocean. Rainfall is heavy in some areas—as much as 100 inches (254 cm) per year.

Northern Lands

Canada's three territories—Nunavut, Northwest Territories, and Yukon Territory—lie to the north. The Arctic Ocean laps their northern shores. Within that ocean lies an archipelago. An **archipelago** is a group of islands. Some of the roughly 1,000 islands are tiny. Baffin Island, though, is nearly 200,000 square miles (518,000 sq. km)—larger than the state of California.

The vast center of these territories is covered by lowland plains. To the east, the rim of the Canadian Shield rises. The Shield continues onto the islands north of Hudson Bay.

The western part of the far north has high mountains. Among them is Mount Logan, Canada's highest peak, at 19,524 feet (5,951 m).

Much of the land of the territories, called the far north, is covered by a subarctic climate zone marked by cold winters and mild summers. The areas farthest north, however, have a tundra climate. The name of this climate zone comes from the landscape. A **tundra** is a flat, treeless plain with permanently frozen ground.

✓ **READING PROGRESS CHECK**

Determining Central Ideas Why do most of Canada's people live in southern Canada?

Cold and ice do not stop Yellowknife residents in the Northwest Territories from enjoying outdoor activities (left). Canada's wettest area, in terms of rain and snow, is the Pacific coast.

▶ **CRITICAL THINKING**
Describing What is the climate of the far north?

Think **Again**

Northern Canada receives a huge amount of snow.

Not true. The cold Arctic air that controls the climate of northern Canada holds little moisture. The dry air brings little snowfall to northern Canada. The Rocky Mountains and the Gulf of St. Lawrence region receive more snow than the far north.

(l to r) Design Pics/Carson Ganci/Getty Images; Anita Erdmann/Flickr/Getty Images

Bodies of Water

GUIDING QUESTION *What bodies of water are important to Canada?*

Some bodies of water shape Canada's climate. Others have economic importance.

Oceans, Bays, and Gulfs

Three oceans border Canada—the Atlantic, the Pacific, and the Arctic. The Pacific Ocean brings rain and mild temperatures to western Canada. Cold air blows over the Arctic Ocean to chill northern and central Canada. The Atlantic Ocean moderates the temperatures of eastern Canada.

The Atlantic and Pacific oceans are also important economically. Ships cross the oceans to bring goods to and from Canada. They are important for another reason, too. East of the Atlantic Provinces is an area called the Grand Banks. This part of the Atlantic Ocean is one of the world's great fisheries. A **fishery** is an area where fish come to feed in huge numbers. The Grand Banks is visited by fishing fleets from all over the world. However, overfishing has severely hurt the populations of some kinds of fish in recent years.

Tugboats tow logs along the Fraser River to sawmills. Forest industries have always been an important part of the country's economy.

North Light Images/age fotostock

The Pacific coast is another important fishery. Many transitions are occurring in the Arctic Ocean because of climate change. Less ice is forming, which allows more ship traffic through the region. It is also altering how native peoples live.

The Gulf of St. Lawrence is another important body of water. It is an extension of the Atlantic Ocean and serves as the mouth of the St. Lawrence River. It provides **access**, or a way in, to the interior of Canada.

An inland sea called Hudson Bay covers much of east central Canada. Native peoples live around its shores. They catch fish and hunt sea mammals for food.

Lakes and Waterways

The Gulf of St. Lawrence connects to the St. Lawrence River, which flows into the Great Lakes. For centuries, rapids and steep drops in elevation blocked ships from moving along the river west of Montreal. In the 1950s, though, the United States and Canada worked together to build the St. Lawrence Seaway. They made canals and a system of locks that can raise and lower ships from water at two different levels. As a result, oceangoing ships can reach as deep into the interior as western Lake Superior.

Canada shares four of the five Great Lakes with the United States. The one it does not share is Lake Michigan. Three other major lakes are found in lowland areas west of the Canadian Shield. They are Great Slave Lake and Great Bear Lake in the Northwest Territories, and Lake Winnipeg in Manitoba.

The Mackenzie River and its tributaries dominate much of the lowlands of the far north. Beginning at Great Slave Lake, the Mackenzie flows north and west to empty into the Arctic Ocean. It is a wide river—from 1 mile to 4 miles (1.6 km to 6.4 km) across— and flows for more than 1,000 miles (1,609 km). The Fraser River flows through the mountains of British Columbia. The river basin is important for many activities such as farming, fishing, and mining.

✔ **READING PROGRESS CHECK**

Analyzing Why is the St. Lawrence River economically more important to Canada than the Mackenzie River?

FOLDABLES®
Study Organizer

Include this lesson's information in your Foldable®.

North | South
Past and Present | World Relations

LESSON 1 REVIEW

Reviewing Vocabulary

1. What is the difference between a *province* and a *territory* in Canada?

Answering the Guiding Questions

2. *Identifying* What landforms in Canada are similar to those in the United States?

3. *Analyzing* Why do most of Canada's people live in southern Ontario and Quebec?

4. *Identifying Point of View* Which body of water do you think is most important to Canada? Why?

5. *Analyzing* Which of Canada's regions do you think has benefited most from the fisheries of the Grand Banks? Why?

6. *Informative/Explanatory Writing* Write a paragraph describing the physical geography of one of Canada's regions and the impact of its landforms and climate on the people in that region.

Reading **HELP**DESK

Academic Vocabulary

- **occupy**
- **eventually**
- **migrate**

Content Vocabulary

- **aboriginal**
- **Métis**
- **transcontinental**
- **granary**

TAKING NOTES: *Key Ideas and Details*

Sequencing Events As you read about Canada's history, use the graphic organizer below to record important dates and events.

1530s	Cartier claims St. Lawrence River and lands around it for France
1763	
1867	
1931	

Indiana Academic Standards
6.1.11

Lesson 2
The History of Canada

ESSENTIAL QUESTION • *What makes a culture unique?*

IT MATTERS BECAUSE
Canada has one of the world's largest economies.

The First Nations of Canada

GUIDING QUESTION *How did native peoples of Canada live before Europeans came to the area?*

In the United States, the **aboriginal**, or native, peoples who lived in North America before Europeans are called Native Americans. Canadians call them the First Nations.

Coming to Canada

The first people to arrive in Canada came from Asia. They came during a long period of intense cold called the Ice Age. The first groups to arrive moved south because Canada was covered with ice. Over thousands of years, though, Earth's climate warmed. The ice sheets over most of Canada melted. As Canada warmed, people **occupied** the land there. Their way of life depended on the resources where they lived.

Different Ways of Life

The peoples of the eastern woodlands lived by farming, hunting, and fishing. They built villages where they lived for most of the year. They traded with one another. Two important eastern groups were the Huron and the Iroquois. The two nations were rivals and often fought.

The peoples of the Pacific coast lived in a region of plenty. Rivers and the waters of the Pacific Ocean provided fish and sea mammals. They hunted game animals in the forests. They used wood from the region's trees to make houses. They also made oceangoing canoes, which they used to hunt and fish.

Life was more difficult in the far north. There, peoples like the Inuit had to find food in a land where few plants grow. Shelters had to be built without using trees. They had to protect themselves from fierce cold. They hunted caribou, a large animal related to deer, on land and seals and whales on the water. Caribou skins were used to make clothes and shoes to keep them warm in winter.

☑ **READING PROGRESS CHECK**

Determining Cause and Effect How did the presence and absence of ice affect the early settlement of Canada?

Exploration and Settlement

GUIDING QUESTION *How did migration and settlement change Canada?*

The first Europeans to reach what is now Canada were Vikings. They began their travels around A.D. 1000 and settled in southern Newfoundland. They soon abandoned their settlements and left, however. More than 500 years later, other Europeans came to the Americas.

Europeans in Canada

The next Europeans to explore Canada were the French. An explorer named Jacques Cartier sailed up the St. Lawrence River in the 1530s. He claimed the St. Lawrence and the lands around it for France. The whole area of French control **eventually** was called New France.

In the 1600s, the French made the first serious effort to settle the region. Explorer Samuel de Champlain founded the first French settlement, Quebec, in 1608.

Over time, more French settlers **migrated** to Canada. Some became fur traders. These traders exchanged European goods with the Huron, a First Nations people. The French received beaver furs that could be shipped back to Europe. Some settlers were priests. They came to Canada to minister to the French settlers, who were Roman Catholic. They also hoped to convert native peoples to Christianity. Some settlers farmed. Their crops fed the other settlers.

©Bettmann/Corbis

Academic Vocabulary

occupy to settle in a place

eventually at some later time

migrate to move to an area to settle

Inuit and other native North Americans lived in Canada long before European settlers arrived.

▶ **CRITICAL THINKING**

Describing How did the lives of the First Nations people of the far north differ from the lives of the Pacific coast people?

British Canada

France was a powerful nation in the 1600s and 1700s. It had a rival for power, however. Britain competed with France in the Americas. In the late 1600s, some British merchants formed a company called the Hudson's Bay Company. They set up trading posts around the bay in the hope of gaining some of the profitable fur trade.

The two nations fought wars in the 1700s. As a result of a British victory in 1763, France was forced to give up much of its land in North America. However, the Quebec Act, passed by the British in 1774, gave French settlers in Canada the right to keep their language, religion, and system of laws.

The next big change in Canada came in the 1770s and 1780s. When the American Revolution broke out, thousands of Americans remained loyal to Britain. During and after the war, many of them moved to Canada. Some settled in what are now the Atlantic Provinces and others in modern Ontario.

During the early 1800s, English and French communities disagreed over colonial government policies. Fears of a U.S. takeover, however, forced them to work together. In 1867 the British colonies of Quebec, Ontario, Nova Scotia, and New Brunswick united as provinces of the Dominion of Canada. This new nation was partly self-governing within the British Empire. Other territories would join Canada over the next 100 years.

Known as outstanding commanders, the British General Wolfe (below) and the French General Montcalm lost their lives in the Battle of Quebec. The war ended in 1763.

▶ **CRITICAL THINKING**

Analyzing What was the result of the war?

World History Archive/Alamy

Canada Expands Westward

Canada's leaders began looking westward. They hoped to expand the nation all the way to the Pacific. In 1869 Canada gained the vast territory held by the Hudson's Bay Company. Many Métis lived on some of this land. **Métis** are the children of French and native peoples. They wanted more say in governing themselves. The province of Manitoba was created to give them that chance.

By this time, settlers had already arrived in British Columbia on the Pacific coast. There, they traded furs and searched for gold. The people of British Columbia agreed to join Canada in 1871. In the 1880s, Canadians built a **transcontinental**, or continent-crossing, railroad that united the eastern and western parts of their nation.

New Provinces

Canada's leaders made agreements with some native peoples of the west. According to the government, the native peoples agreed to give up their lands in exchange for aid. Native people believed they gave settlers only the right to use the land for farming. Canada's westward expansion came at a price. New settlers pushed native peoples off their lands. In 1905 Saskatchewan and Alberta entered Canada as provinces.

Meanwhile, gold had been discovered in the Yukon Territory. This discovery led to a gold rush. Helping create order in the west was a police force that formed in 1873. Called the North West Mounted Police, the group is known today as the Royal Canadian Mounted Police.

✓ READING PROGRESS CHECK

Determining Central Ideas How did European rivalries affect the development of Canada?

Canada honored two African Canadian heroes in 2012. John Ware (left) was important in starting the ranching industry in western Canada. Viola Desmond (right) worked to repeal unjust laws.

▶ CRITICAL THINKING
Describing Who were the Métis? What right did they obtain?

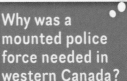

Thinking Like a Geographer

Why was a mounted police force needed in western Canada?

The vast size and sparse population of the region explain it. Police needed to be able to ride horses to travel quickly from one place to another. Because settlements were far-flung, they had to travel often. The Mounties filled an important need in a time before car travel.

Canada Grows and Unites

GUIDING QUESTION *How did Canada change in the 1900s?*

During the 1900s, Canada grew into a prosperous, independent country. Its population grew and became more diverse. Canada also reached the territorial size it is today. In 1949 Newfoundland and Labrador became the last territory to join Canada as a province.

Economic Growth and Immigration

Early in the 1900s, however, Canada had many economic problems. The country's economy was based on agriculture and mining. When prices for those resources fell, Canada suffered. In the late 1900s, Canada became an industrial nation. Canadians built factories and took advantage of their mineral resources. They developed hydroelectric projects and transportation systems. Agriculture also grew as farmers in the west increased production. Granaries stored wheat to feed the growing population. A **granary** is a building used to store harvested grain.

To help increase the nation's industrial power, Canada needed more workers. Canada's leaders made it easier for people to enter the country. Canada's population began to grow, particularly after World War II, which helped Canada meet its need for more workers. Canada's population jumped from 12 million people in 1945 to nearly 35 million in 2012.

The Royal Canadian Mounted Police take part in Canada Day celebrations in Ottawa. The annual celebration commemorates when Canada became a self-governing dominion within the British Empire.

▶ **CRITICAL THINKING**

Identifying What major change in Canada occurred in 1931?

©Zou Zheng/Xinhua Press/Corbis

In recent decades, Canada has welcomed immigrants from all over the world. Today it is a multicultural society. People descended from settlers who came from the British Isles or France account for only about half the population. Another 15 percent trace their background to some other European country. About 6 percent of the people have an African or an Asian background. About 2 percent of Canada's people are from the First Nations. More than a quarter of Canadians have mixed backgrounds.

Independence

A major change came to Canada in 1931. That year, Britain granted almost complete independence to Canada. Canadians were now able to make their own laws without interference from Britain. The British government, however, maintained the right to approve changes to Canada's constitution. This link to Britain finally ended in 1982.

During the later 1900s, Canada became less connected to Britain. Instead, Canada developed closer ties to the United States. It also became active in international bodies, such as the United Nations.

The Growth of Industry

Canada's industries boomed after World War II. Part of the boom was the result of increased demand for Canada's mineral resources. Part was the result of industrial growth. Part, too, came from population growth and movement. More people meant more demand for more goods. Also, people were moving from rural areas to cities and suburbs. That created a need for construction of homes, stores, offices, and roads.

Construction of the St. Lawrence Seaway helped the economy grow, too. Canada could more easily—and cheaply—ship its products around the world. It also could bring in needed goods at lower costs. At the same time, agriculture expanded. Farmers in the Prairie Provinces boosted production. Granaries were built. By 2013, Canada's economy was strong in all sectors: agriculture, industry, services, and information technology (IT).

Include this lesson's information in your Foldable®.

☑ **READING PROGRESS CHECK**

Identifying What are two ways that Canada changed in the 1900s?

LESSON 2 REVIEW

Reviewing Vocabulary
1. Why do the Prairie Provinces have *granaries*?

Answering the Guiding Questions
2. *Describing* Why did people of the First Nations have different ways of life in different parts of Canada?

3. *Analyzing* Why was the province of Manitoba created?

4. *Analyzing* Why do you think people settled what is now British Columbia before settling the Prairie Provinces?

5. *Determining Central Ideas* Why has Canada developed closer ties to the United States?

6. *Informative/Explanatory* Write a summary that highlights the key points of Canada's development as a nation.

Reading **HELP**DESK

Academic Vocabulary

- **via**
- **vary**

Content Vocabulary

- **metropolitan area**
- **bilingual**
- **peacekeeping**
- **separatist**
- **autonomy**
- **acid rain**

TAKING NOTES: *Key Ideas and Details*

Summarizing As you read about life in Canada, take notes using the graphic organizer below.

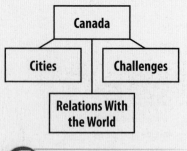

Indiana Academic Standards
6.2.6, 6.3.1, 6.3.14

Lesson 3
Life in Canada

ESSENTIAL QUESTION • *What makes a culture unique?*

IT MATTERS BECAUSE
Canada plays a major role in world affairs.

City and Country Life

GUIDING QUESTION *Where and how do Canadians live?*

Where do you think most Canadians live? Do you think of a fishing village or a farm? Do you picture a remote settlement in the far north? Actually, four out of every five Canadians live in cities or suburbs. Canada is a modern urban society.

Ottawa and Toronto

Ottawa is located in the province of Ontario. It is Canada's capital city and home to the national government. Canada's government is similar to those of the United States and the United Kingdom. Like the United States, Canada has national and regional governments. In Canada, those regions are the provinces. Like the United Kingdom, Canada has a parliamentary system. Voters elect members of the legislature, or Parliament. The party with the most members chooses the prime minister, who carries out the laws.

Another Ontario city, Toronto, is Canada's largest metropolitan area. A **metropolitan area** is a city and its surrounding suburbs. Metropolitan Toronto has more than 5 million people. Toronto has access to the Atlantic Ocean **via** the St. Lawrence Seaway. As a result, it is a major port. Rail lines and highways make it possible to ship imported goods from the seaway across Canada. About half of Canada's manufactured goods are made in Ontario. Toronto is an ideal location for shipping Canadian products around the world.

Each summer, Toronto is home to the Canadian National Exhibition, a combination fair and business meeting. The city also has major sports teams and a wide array of cultural opportunities. Its restaurants offer food from many different ethnic groups.

Montreal and Quebec

Montreal and Quebec are the major cities of Quebec Province. Canada's French heritage remains strong throughout this province. Canada is a **bilingual** nation, meaning it has two official languages— English and French. Most people in Quebec speak French.

Montreal is Canada's second-largest city. It is the economic hub, or center of activity, of Quebec Province. Like Toronto, it is a major port because of the St. Lawrence Seaway. Montreal is an important center of manufacturing, banking, and insurance. It offers a special combination of historic European charm and modern life.

Quebec, the capital of Quebec Province, attracts tourists who are eager to see buildings that reflect its 400-year history. Costumed performers reenact life in the early days of New France to make that history come alive.

Western Cities

Canada's third-largest metropolitan area is Vancouver, in British Columbia. It is Canada's busiest port. Vancouver's port ships food products from the nearby Prairie Provinces. Because of its Pacific Ocean location, it is a vital center of trade with Asia. Many people have moved to the city from Asian nations in recent years.

Canada's other major western cities are Calgary and Edmonton in Alberta. Both benefit from Alberta's large oil and natural gas reserves, which have fueled an economic boom in the province. The cities have grown rapidly in recent years. Both are centers for processing the grain and meat produced in the province.

Life in Rural Areas

Life in rural Canada **varies** from place to place. People of the First Nations who live in the far north live in a harsh landscape. Many follow traditional ways, although modern aspects of life can be found as well. For example, they may travel on snowmobiles instead of dogsleds. Fishing villages in the Atlantic Provinces have suffered in recent years. Overfishing has reduced fish stocks—and thus income from fishing.

✓ **READING PROGRESS CHECK**

Analyzing Why are Toronto and Vancouver more important to trade than Calgary and Edmonton?

Brand X Pictures/PunchStock

Canada's National Tower in Toronto is a symbol of the country, and it is the tallest, free-standing structure in the Western Hemisphere.
▶ **CRITICAL THINKING**
Identifying What are Canada's two most populous cities?

Outdoor winter sports are a popular pastime in both Canada and the northern United States.

▶ **CRITICAL THINKING**

Analyzing Why is Canada's government making a strong effort to promote the nation's culture?

Economic and Political Relationships

GUIDING QUESTION *What is Canada's relationship with other nations?*

Canada has one of the world's largest economies. It also plays a leading role in many world issues.

Canada and the United States

Canada has close economic, defense, and cultural ties to the United States. In the early 1990s, Canada joined the United States and Mexico in signing the North American Free Trade Agreement (NAFTA). This agreement eliminated trade barriers in North America. In 2010 three-quarters of all of Canada's exports went to the United States. The same year, three-quarters of its imports came from the United States.

Canada and the United States also cooperate in defense. They work together to defend the air space against possible attack from planes or missiles. They also work together to combat terrorism.

Canada and the United States share many cultural features. Canadians watch American movies and television shows. Many Canadian singers and actors enjoy success in the United States. Canadians worry, though, about cultural dominance from the United States. Canada's government promotes the production of Canadian movies and television shows.

Canada and Britain

Canada's government is modeled on Britain's. So are its laws, except in Quebec Province. Canadian culture draws on British culture. Many Canadians trace their ancestry to Britain. For these reasons, Canada maintains close ties with that nation. The British king or queen is officially Canada's king or queen, as well. A Canadian official called the governor-general acts in his or her place, though.

Canada and the World

Canada is active in many world organizations. It is a member of the United Nations and the North Atlantic Treaty Organization (NATO). NATO links Canada, the United States, and many nations of Europe in defense matters. Canada has worked in recent years to connect more closely with Asian nations.

Canada plays a major role in efforts to aid poorer nations. It has often taken part in peacekeeping efforts. **Peacekeeping** is sending trained members of the military to crisis spots to maintain peace and order.

☑ READING PROGRESS CHECK

Analyzing Why is Canada more similar to the United States and the United Kingdom than to other nations?

An efficient and comfortable train system is an important part of Canada's economic development.
▶ CRITICAL THINKING
Identifying Point of View The nation's transcontinental train makes only eight planned stops en route from Toronto to Vancouver. How does this help travelers?

Canada's Challenges

GUIDING QUESTION *What challenges do Canadians face?*

Canada's biggest challenge might be staying together as a nation. Some people in Quebec want to create their own nation.

Unity and Diversity

Canada's constitution guarantees the rights of French-speaking people in Quebec and elsewhere. Still, tension is evident. English speakers have dominated the nation. They controlled the economy of Quebec for many years. As a result, some French speakers felt they were treated as second-class citizens.

In the late 1900s, some Quebec leaders launched a separatist movement. **Separatists** are those who want to break away from control by a dominant group. Voters in Quebec have twice defeated attempts to make it independent. Still, the issue remains unsettled.

The government did succeed in giving more power to people of the First Nations. In 1999 the government created the new territory called Nunavut. Most of its people are from the First Nations. The government gave them greater **autonomy**, or self-government, than they had in the past.

A Montreal rally calls for a "No" vote on independence for Quebec.
▶ **CRITICAL THINKING**
Identifying Point of View Why do some Canadians support independence for Quebec?

Environmental Challenges

Some scientists worry about the effects of climate change. Milder weather threatens plants and animals that are adapted to the cold of the far north. Experts also believe fisheries will suffer further. In addition, they fear water shortages and more extreme weather. Extreme weather includes long periods of drought and sudden damaging storms.

Canada's government is taking some steps to reduce its use of fossil fuels such as oil, coal, and natural gas. Burning these fuels is thought to contribute to climate change. The government is also encouraging research into clean energy.

Nevertheless, Canada depends greatly on fossil fuels to power its industries, transportation systems, and homes. Primary economic activities such as fossil fuel production are a large part of Canada's economy. Fossil fuel extraction has become an environmental concern in Athabasca Tar Sands in northwestern Alberta. This area has sand located near Earth's surface that contains a form of crude oil that is as thick as tar. The tar can be refined into oil, but the process is difficult and requires a great deal of energy and water.

Acid Rain

Another environmental problem for Canada is acid rain. **Acid rain** is produced when chemicals from air pollution combine with precipitation. When the rain falls to Earth, the chemicals may kill fish, land animals, and trees. Even when they do not kill living things, the chemicals weaken them. That makes them more vulnerable to damage from pests, disease, or severe weather.

Damage from acid rain has been particularly bad in eastern Canada. The government has made efforts to reduce acid rain. But many of the chemicals that cause acid rain in Canada enter the air in the United States. Canada cannot solve its problem without U.S. help. Canada's government wants to reach an agreement that includes tough steps to prevent acid rain.

☑ **READING PROGRESS CHECK**

Determining Central Ideas Why can Canada not meet its environmental challenges by itself?

Include this lesson's information in your Foldable®.

LESSON 3 REVIEW

Reviewing Vocabulary

1. What does it mean to say that Canada is a *bilingual* nation?

Answering the Guiding Questions

2. *Determining Central Ideas* How do the cultures of Quebec and Montreal reflect Canada's history?

3. *Describing* How does life in Calgary and Edmonton relate to Alberta's resources?

4. *Describing* In what ways has Canada tried to build a better world?

5. *Identifying Point of View* Why is separatism in Quebec an ongoing problem for Canada?

6. *Argument Writing* Take the role of Canada's prime minister. Write a letter to the president of the United States urging stronger American action to reduce air pollution that causes acid rain.

Directions: Write your answers on a separate piece of paper.

1 Use your **FOLDABLES** to explore the Essential Questions.

ANALYZING Look at the population map of Canada at the beginning of the unit. In a short essay, explain why Canada's major cities and population centers developed where they did.

2 **21st Century Skills**

ANALYZING In a small group, research to find a historical painting about an important event in Canada's history. Find a secondary source about the same event. Compare the two sources and develop an answer to the question: Did one of the sources help you better understand the event? Why?

Benefits	Disadvantages

3 **Thinking Like a Geographer**

INTEGRATING VISUAL INFORMATION Reseach the benefits and disadvantages of separatism for Quebec and list them in a chart like the one shown.

4 **GEOGRAPHY ACTIVITY**

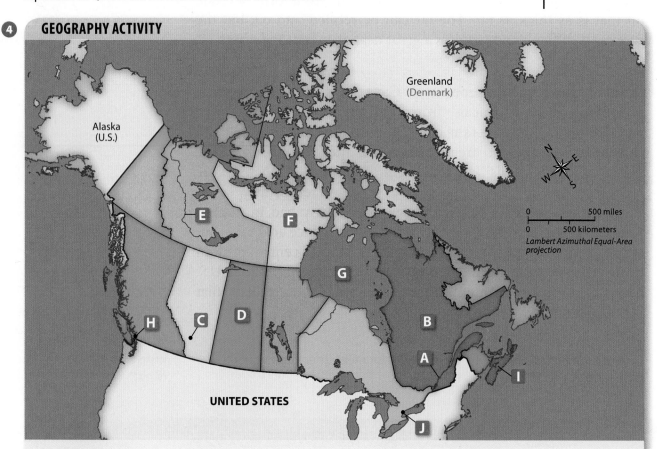

Locating Places

Match the letters on the map with the provinces, territories, or locations listed below.

1. Vancouver
2. Quebec
3. Calgary
4. Toronto
5. Mackenzie River
6. Nunavut
7. Hudson Bay
8. St. Lawrence River
9. Nova Scotia
10. Saskatchewan

REVIEW THE GUIDING QUESTIONS

Directions: Choose the best answer for each question.

1 Manitoba, Saskatchewan, and Alberta are called the
- A. archipelago.
- B. Northwest Territories.
- C. Prairie Provinces.
- D. Atlantic Provinces.

2 Canada's highest mountain is
- F. Mount McKinley.
- G. Mount Logan.
- H. Hudson Mountain.
- I. Pikes Peak.

3 The man who founded Quebec and led the French effort to settle Canada was
- A. Henry Hudson.
- B. Father Junipero Serra.
- C. Samuel de Champlain.
- D. the Métis.

4 Which of Canada's provinces is predominately French in language and culture?
- F. Nova Scotia
- G. Prince Edward Island
- H. Quebec
- I. Yukon Territory

5 Canada's largest metropolitan area, which includes the city and its surrounding suburbs, is
- A. Ottawa.
- B. Vancouver.
- C. Montreal.
- D. Toronto.

6 The territory created in 1999 to give the First Nations greater self-government is
- F. Nunavut.
- G. Saskatchewan.
- H. Northwest Territory.
- I. Yukon Territory.

DBQ DOCUMENT-BASED QUESTIONS

7 **DETERMINING CENTRAL IDEAS** Read the following statement about Canada's people.

"*Immigration . . . was responsible for two-thirds of [Canada's] population growth in the period 2001 to 2006. . . . The effect of immigration is mostly felt in Canada's largest urban [centers] and their surrounding municipalities.*"

—from *The Atlas of Canada*

Which generalization can be made from this quotation?

A.   Without immigration, Canada's population would have declined.

B.   The immigration rate is higher in Canada today than ever before.

C.   The immigration rate to Canada is higher than to the United States.

D.   Immigrants were the major reason for Canada's population growth.

8 **DETERMINING CENTRAL IDEAS** What does the second sentence in the quotation tell you about Canada's immigrants?

F.   their occupations

G.   their origin in urban or rural areas

H.   their settlement patterns

I.   their economic status

SHORT RESPONSE

"*[Samuel de Champlain] thought that the conquering spirit in which the Spanish usually approached the new Indian groups was itself mistaken. . . . A new colony should seek friends and allies amidst the indigenous peoples.*"

—from Arthur Quinn, *A New World*

9 **ANALYZING** How did Champlain benefit from France not being the first European nation to place colonies in the Americas?

10 **IDENTIFYING** In what ways could French colonists benefit from peaceful relations with Native Americans?

EXTENDED RESPONSE

11 **INFORMATIVE/EXPLANATORY WRITING** Choose the Canadian province or territory where you think you would be most comfortable living. Write an essay describing the province or territory, and explain why you chose it over the others. Your essay should show your understanding of the similarities and parallels between life in Canada and the United States.

Need Extra Help?

If You've Missed Question	❶	❷	❸	❹	❺	❻	❼	❽	❾	❿	⓫
Review Lesson	1	1	2	3	3	3	2	2	2	2	1

FROM *The Atlas of Canada* Web site, http://atlas.nrcan.gc.ca/auth/english/maps/peopleandsociety/immigration, Natural Resources Canada.

MEXICO, CENTRAL AMERICA, AND THE CARIBBEAN ISLANDS

networks

There's More Online about Mexico, Central America, and the Caribbean Islands.

CHAPTER 7

ESSENTIAL QUESTIONS · *How does geography influence the way people live?*
· *Why does conflict develop?* · *Why do people trade?*

Lesson 1
Physical Geography

Lesson 2
History of the Region

Lesson 3
Life in the Region

Girl from the highlands of Guatemala

©Sergio Pitamitz/Robert Harding World Imagery/Corbis

The Story Matters...

Early advanced civilizations developed in this region of the Americas. Their people developed economies based on farming and trade. They built planned cities and developed highly organized societies and governments. The arrival of the Spanish and other Europeans had a dramatic impact on the region and its indigenous peoples. The influence of European colonial rule and the struggles for independence can still be seen in the economies, politics, and cultures of the region today.

FOLDABLES®
Study Organizer

Go to the Foldables® library in the back of your book to make a Foldable® that will help you take notes while reading this chapter.

○	Trade and Commerce
○	Civilizations
○	Gulf of Mexico
○	Mexico, Central America, and the Caribbean Islands

MEXICO, CENTRAL AMERICA, AND THE CARIBBEAN ISLANDS

Mexico, Central America, and the Caribbean islands sit between North America and South America. The region is surrounded by oceans and seas and is located close to the Equator. As you study the map, look for the geographic features that make this area unique.

Step Into the Place

MAP FOCUS Use the map to answer the following questions.

1 THE GEOGRAPHER'S WORLD What is the largest country in this region?

2 ENVIRONMENT AND SOCIETY Why was the Panama Canal built where it is?

3 THE GEOGRAPHER'S WORLD Which of the Caribbean islands is part of the United States?

4 CRITICAL THINKING
Analyzing Given their location, what might be a key economic industry of the Caribbean islands?

A

HISTORIC CITY Willemstad, capital of the Caribbean island of Curacao, was founded in 1634 by Dutch settlers.

B

MAYAN RUINS Early Americans known as the Maya built cities in the rain forests of southern Mexico.

Step Into the Time

DESCRIBING Select one location on the time line and describe the impact of European colonization on the lives and environment of the people who lived there.

1325
Aztec found Tenochtitlán

1492 Christopher Columbus arrives in Americas

1300

1500

Map of the Region

UNITED STATES

Bermuda *(U.K.)*

ATLANTIC OCEAN

40°N

30°N

TROPIC OF CANCER

Gulf of California

Rio Grande

Gulf of Mexico

MEXICO

Mexico City

PACIFIC OCEAN

B

Havana

Nassau

BAHAMAS

CUBA

DOMINICAN REPUBLIC

20°N

Cayman Islands *(U.K.)*

Port-au-Prince

Kingston

Santo Domingo

HAITI

JAMAICA

Puerto Rico *(U.S.)*

see inset to left for detail

BELIZE
Belmopan

GUATEMALA

Guatemala

San Salvador

EL SALVADOR

HONDURAS
Tegucigalpa

NICARAGUA

Managua

San José

Panama City

Caribbean Sea

A

10°N

110°W

100°W

20°N

British Virgin Islands *(U.K.)*

Anguilla *(U.K.)*

St. Martin *(Fr.)*

St. Maarten *(Neth.)*

St. Barthélemy *(Fr.)*

San Juan

Puerto Rico *(U.S.)*

Virgin Islands *(U.S.)*

ANTIGUA AND BARBUDA

Montserrat *(U.K.)*

Guadeloupe *(Fr.)*

ST. KITTS AND NEVIS

DOMINICA

Martinique *(Fr.)*

ST. LUCIA

Caribbean Sea

ST. VINCENT AND THE GRENADINES

BARBADOS

GRENADA

TRINIDAD AND TOBAGO

0   200 mi

0   200 km

60°W

10°N

COSTA RICA

PANAMA

Panama Canal

SOUTH AMERICA

EQUATOR   0°

— National boundary

✪ National capital

○ Territorial capital

90°W

80°W

0   500 miles

0   500 kilometers

Lambert Azimuthal Equal-Area projection

60°W

1804
Led by Toussaint-Louverture, Haiti achieves independence from France

1810
Father Hidalgo leads Mexico rebellion

1600   1700   1800   1900   2000

1914 Panama Canal links Atlantic and Pacific Oceans

2010 Deadly earthquake strikes Haiti

1848 Mexico cedes large areas of territory to U.S.

1959 Fidel Castro takes power in Cuba

netw⊙rks

There's More Online!

☑ **GRAPHIC ORGANIZER**
Landforms and Waterways

☑ **IMAGE** 360° View: Mexico City

☑ **ANIMATION** Waterways as
Political Boundaries

☑ **VIDEO**

Reading **HELP**DESK

Academic Vocabulary

- **similar**
- **benefit**

Content Vocabulary

- **isthmus**
- ***tierra caliente***
- ***tierra templada***
- ***tierra fría***
- **bauxite**
- **extinct**
- **dormant**

TAKING NOTES: *Key Ideas*
and Details

Organize As you read about the
region, take notes on the physical
geography using a graphic organizer
like the one below.

Area	Landforms
Mexico	
Central America	
Caribbean islands	

Indiana Academic Standards
6.1.18, 6.1.19, 6.3.5, 6.3.7, 6.3.12, 6.4.7

Lesson 1
Physical Geography

ESSENTIAL QUESTION • *How does geography influence the way*
people live?

IT MATTERS BECAUSE
Mexico and Central America are southern neighbors to the
United States.

Physical Geography of Mexico and Central America

GUIDING QUESTION *What landforms and waterways do Mexico and*
Central America have?

Mexico and the seven nations of Central America act like a
bridge between two worlds. Geographically, they form an
isthmus that connects North and South America. An **isthmus**
is a narrow piece of land that connects two larger landmasses.
Culturally, they join with South America and some Caribbean
islands to make up Latin America. Latin America is a region
of the Americas where the Spanish and Portuguese languages,
based on the Latin language of ancient Rome, are spoken.
Economically, the nations have close ties to the United States.
They also trade with their Latin American neighbors.

Shaped like a funnel, the region is wider in the north than
in the south. To the north, Mexico has a 1,951-mile (3,140-
km) border with the United States. At the southern end, the
Central American country of Panama is only about 40 miles
(64 km) wide.

Land Features

Mexico is the largest nation of the region, occupying about
two-thirds of the land. Imagine a backwards *y* along the
western and eastern coasts of Mexico, with the tail to the
south. That backwards *y* neatly traces the mountain systems

on Mexico's two coasts and south central region. The coastal ranges are called the Sierra Madre Occidental (Spanish for "western") and the Sierra Madre Oriental (Spanish for "eastern"). They join in the southern highlands. Coastal plains flank the western and eastern mountains. The eastern plain is wider.

Between the two arms of the *y* is a vast highland region called the Central Plateau. It is the heartland of Mexico. This plateau is home to Mexico City, which is the capital, and a large share of the nation's people.

Mexico has two peninsulas. The Yucatán Peninsula bulges northeast into the Gulf of Mexico. Baja California (*baja* means "lower" in Spanish) extends to the south in western Mexico.

Central America has landforms **similar** to those of south central Mexico. Mountains run down the center of these countries. Narrow coastal lowlands flank them on the east and west.

Mexico and Central America lie along the Ring of Fire that rims the Pacific Ocean. Earthquakes and volcanoes are common in the Ring of Fire. The Sierra Madre Occidental are made of volcanic rocks, but they have no active volcanoes. The mountains in the southern part of the central plateau and in Central America, however, do have numerous active volcanoes. These volcanoes bring a **benefit**. Volcanic materials weather into fertile, productive soils.

Academic Vocabulary

similar much like
benefit advantage

Mountains appear on the hazy horizon of Mexico City. Mexico's capital lies about 7,800 feet (2,377 m) above sea level.

Popocatépetl volcano stands in the background as a farmer plows the land. Popocatépetl, also known as "smoking mountain," has experienced eruptions since ancient times.

▶ **CRITICAL THINKING**

Explaining Why are earthquakes common in some parts of Mexico and Central America?

Earthquakes are common in the area, too. A magnitude 8.0 earthquake that hit Mexico City in 1985 killed thousands of people. One that struck El Salvador in 2001 produced another kind of disaster. A hill weakened by the earth's movement collapsed onto the town of Las Colinas. It crushed homes and killed hundreds of people.

Bodies of Water

Mexico and Central America are bordered by the Pacific Ocean to the west. The Gulf of California, an inlet of that ocean, separates Baja California from the rest of Mexico. To the east, the region is surrounded by the waters of two arms of the Atlantic Ocean. They are the Gulf of Mexico and the Caribbean Sea.

The region has few major rivers. In the northern half of Mexico, the climate is dry. This means that few rivers flow across the rocky landscape. Southern Mexico and Central America receive more rain, but the landscape is steep and mountainous, and the rivers are short. An important river is the Río Bravo. In the United States, this river is called the Rio Grande. The largest lake in the region is Lake Nicaragua, in Nicaragua.

©Francisco Guasco/epa/Corbis

An important waterway in the region is not a river, but a feature built by people. It is the Panama Canal, built in the early 1900s. The Panama Canal makes it possible for ships to pass between the Atlantic and Pacific oceans without journeying around South America. It saves thousands of miles of travel, which in turn saves time and money. It is one of the world's most important waterways.

Climates

Most of Mexico and Central America lie in the Tropics. Because of their location near the Equator, it might seem that the climate would be hot. Although the coastal lowlands are hot, areas with higher elevation are not. The highlands are much cooler.

Nearly the entire region can be divided into three vertical climate zones. Soil, crops, animals, and climate change from zone to zone. The **tierra caliente**, or "hot land," is the warmest zone. It reaches from sea level to about 2,500 feet (762 m) above sea level. Major crops grown here are bananas, sugarcane, and rice.

Next highest is the **tierra templada**, or "temperate land." This climate zone has cooler temperatures. Here farmers grow such crops as coffee, corn, and wheat. Most of the region's people live in this climate zone.

Higher in elevation is the **tierra fría**, or "cold land." This region has chilly nights. It can be used only for dairy farming and to grow hearty crops such as potatoes, barley, and wheat.

Think Again?

Because the Panama Canal connects the Atlantic and Pacific oceans, it must go east to west.

Not really! Central America twists to the east where Panama is located. As a result, the Panama Canal is cut from the north to the south.

The Rio Bravo, or Rio Grande, carves its way through rugged countryside. It forms part of the border between Mexico and the United States.
▶ **CRITICAL THINKING**
Describing Why are there so few rivers in the northern part of Mexico?

©iStockphoto.com/yuhirao

CLIMATE ZONES

Although the region is located in the Tropics, many inland areas of Mexico and Central America have relatively cool climates.

▶ **CRITICAL THINKING**

1. *Identifying* What products are grown in *tierra caliente*?

2. *Analyzing* Why are many inland areas of Mexico and Central America relatively cool?

Tierra Helada
20°F – 55°F
(-7°C – 13°C)

— 10,000 feet
(3,048 m)

Wheat

Barley

Tierra Fría
55°F – 65°F
(13°C – 18°C)

Potatoes   Apples

— 6,000 feet
(1,829 m)

Coffee

Tierra Templada
65°F – 75°F
(18°C – 24°C)

Corn

Citrus

— 2,500 feet
(762 m)

Tierra Caliente
75°F – 80°F
(24°C – 27°C)

Bananas   Sugarcane

Cacao

Rice

— Sea Level

Geographers also designate other vertical climate zones. Few human activities take place on the *tierra helada,* or "frozen land". This vertical climate zone is more common in other regions of the Americas.

Tropical Wet/Dry Climate

Much of Mexico and Central America have a tropical wet/dry climate. The climate is characterized by two distinct seasons. The wet season, during the summer months, is when most of the precipitation falls. The dry season occurs during the winter months. The dry season is longer in the areas farther from the Equator and closer to the polar regions.

The region's tropical location exposes it to another natural hazard. Ferocious hurricanes can strike in the summer and early autumn months. These storms do great damage. For example, a 1998 hurricane killed more than 9,000 people in Honduras and destroyed 150,000 homes.

Natural Resources

Oil and natural gas are Mexico's most important resources. They are found along the coast of the Gulf of Mexico and in the gulf waters. Mexico is an important oil-producing country. It has enough oil and gas to meet its own needs and still export a large amount. The exports help fuel the nation's economy. However, Mexico's oil production has declined since 2004. Many oil fields are old and are starting to run out of oil.

When Spanish explorers first came to Mexico, they were attracted to the area's gold and silver. Mexico still produces silver, which is mined in the central and north central parts of the country. Gold also is still mined in Mexico. Other minerals include copper, iron ore, and **bauxite**. Bauxite is used to make aluminum.

The seven smaller nations of Central America have few mineral resources. Nicaragua is an exception, with gold, silver, iron ore, lead, zinc, and copper. The nation is so poor, however, that it has not been able to take advantage of these deposits. Guatemala also has some oil, and its mountains produce nickel.

☑ **READING PROGRESS CHECK**

Analyzing Why are different climate zones found in this region, even though most of the region is in the Tropics?

Physical Geography of the Caribbean Islands

GUIDING QUESTION *How are the Caribbean islands alike and different from one another?*

Hundreds of islands dot the Caribbean Sea. The islands are home to more than 30 countries or territories belonging to other countries. Some are large, with millions of people living on them. Others are tiny and home to only thousands.

Major Islands

The Caribbean islands can be segmented into three different groups. The first group is the Greater Antilles. The four islands, the largest Caribbean islands, include Cuba, Jamaica, Hispaniola, and Puerto Rico. Cuba and Jamaica are independent countries. Hispaniola is home to two countries: Haiti in the west and the Dominican Republic

This scenic bay in the Caribbean island of Antigua provides an ideal harbor for yachts and other sailing ships. Antigua is part of the Lesser Antilles.

▶ **CRITICAL THINKING**
Describing What islands make up the Greater Antilles?

Pixtal/age fotostock

Why are the Caribbean islands hit so often by earthquakes and volcanoes?

The islands of the Caribbean are not along the edge of the Pacific Ocean. So why do they experience these disasters? The Ring of Fire is not the only place where tectonic plates meet. Many Caribbean islands are located at the boundaries of different plates where volcanoes form and earthquakes occur. Hence, the islands are vulnerable to these disasters. A 2010 earthquake near Haiti's capital of Port-au-Prince might have killed as many as 200,000 people and forced 1 million others out of their homes.

in the east. Puerto Rico is a commonwealth of the United States. Although it is a possession of the United States, it has its own government. The people of Puerto Rico are American citizens. They can travel freely between their island and the United States.

The second group of islands is the Lesser Antilles. Dozens of smaller islands make up this group. They form an arc moving east and south from Puerto Rico to northern South America. Most of the islands are now independent countries. At one time, they were colonies of France, Britain, Spain, or the Netherlands. Each has a culture reflecting its colonial period.

The third island group is the independent nation of the Bahamas. The islands lie north of the Greater Antilles and east of Florida. The Bahamas include more than 3,000 islands, although people live on only about 30 of them.

The Greater Antilles are a mountain chain, much of which is under water. On a map, you can see that this chain extends eastward from Mexico's Yucatán Peninsula. These islands include some mountains, such as the Sierra Maestra in the eastern part of Cuba and the Blue Mountains of Jamaica. The highest point in the Caribbean is Duarte Peak, in the Dominican Republic. The Lesser Antilles are formed by volcanic mountains. Many of the volcanoes are **extinct**, or no longer able to erupt. Some islands have **dormant** volcanoes, or ones that can still erupt but show no signs of activity.

The Caribbean Sea

The Caribbean Sea is a western arm of the Atlantic Ocean. In the past, sailing ships traveling west from Europe followed trade winds blowing east to west to reach the sea. Christopher Columbus used the winds to reach the Bahamas in 1492. There, he first sighted land in the Americas. Columbus explored the Caribbean, too. These voyages sparked European settlement of the Americas.

The warm waters of the Caribbean help feed the Gulf Stream. This current carries warm water up the eastern coast of the United States.

The Climate of the Caribbean Islands

The Caribbean islands have a tropical wet/dry climate. Temperatures are high year-round, though ocean breezes make life comfortable. Humidity is generally high, but rainfall is seasonal and varies significantly. Islands like Bonaire receive only about 10 inches (25 cm) of rain per year. Dominica, on the other hand, receives about 350 inches (899 cm) of rain each year. That is an average of almost an inch of rain every day.

Like Central America and Mexico, the Caribbean islands are prone to hurricanes. These storms are more likely to occur in the northern areas, toward the Gulf of Mexico, than to the south. On average, seven hurricanes strike the Caribbean islands each year.

Natural Resources

The waters of the Caribbean are rich in fish. Some are fished for food and others for sport. The islands have few timber resources today. People have cut down most of the trees already to use for fuel or to make farmland.

Mineral resources are generally lacking too, although some Caribbean islands have important resources. Trinidad and Tobago has reserves of oil and natural gas. The Dominican Republic exports nickel, gold, and silver. Cuba is a major producer of nickel. Jamaica has large amounts of bauxite.

Perhaps the most important resources of the Caribbean are its climate and people. Warm temperatures and gracious hosts attract millions of tourists to the region each year. Some enjoy the white sandy beaches and clear blue water. Some scuba dive to see the colorful fish darting through coral reefs.

☑ **READING PROGRESS CHECK**

Citing Text Evidence How did the islands of the Caribbean form?

Tourists on a boat near the French-ruled island of Guadeloupe learn to dive in Caribbean waters.

Identifying What natural resource of the Caribbean Sea benefits the region's island nations?

Include this lesson's information in your Foldable®.

LESSON 1 REVIEW

Reviewing Vocabulary

1. What is the *tierra templada*? Why do most people in Mexico and Central America live in this vertical climate zone?

Answering the Guiding Questions

2. *Describing* How are the physical geography of Mexico and Central America similar?

3. *Analyzing* What impact does the Panama Canal have on the cost of shipping goods? Why?

4. *Determining Central Ideas* How do the locations of Mexico and Central America increase the possibility of natural hazards striking the region?

5. *Determining Word Meanings* Why are some islands in the Caribbean called the *Greater* Antilles and others called the *Lesser* Antilles?

6. *Narrative Writing* Imagine you are taking a cruise that stops at a Caribbean island, a port in Central America, and a port in Mexico. Write three diary entries describing what you would see in each place.

Reading **HELP**DESK

Academic Vocabulary

- **feature**
- **transform**

Content Vocabulary

- **staple**
- **surplus**
- **conquistador**
- **colonialism**
- **revolution**
- **plantation**
- **cash crop**
- **caudillo**
- **Columbian Exchange**

TAKING NOTES: *Key Ideas and Details*

Summarize As you read about the history of Mexico, take notes using the graphic organizer below.

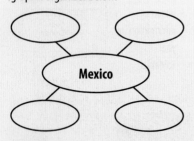

Indiana Academic Standards
6.1.1, 6.1.6, 6.1.10, 6.1.11, 6.2.5, 6.3.11

Lesson 2
History of the Regions

ESSENTIAL QUESTION • *Why does conflict develop?*

IT MATTERS BECAUSE
The region was home to highly developed Native American civilizations.

Mexico's History

GUIDING QUESTION *How did economic and governmental relationships between Spanish and Native Americans in Mexico change over time?*

Mexico was first inhabited by Native American groups. Later, Spanish soldiers conquered the groups and ruled them. Since the early 1800s, Mexico has been independent. Its history is long and rich, and its accomplishments are many.

Early Civilizations
Native peoples first grew corn in Mexico about 7,000 years ago. They also grew other foods that have become **staples**, or foods that are eaten regularly, such as corn, squash, chilies, and avocados. Farming allowed people to produce food **surpluses**, or more than they needed to survive. Surpluses helped people specialize in jobs other than getting food.

About 3,000 years ago, the Maya formed the major civilization in the region. They lived mainly in the lowland plains of Mexico's Yucatán Peninsula and in what is now Guatemala and Belize. One **feature** of their culture was great cities. The Maya erected pyramids with stepped sides and temples on top. They invented a complex system of writing. By studying astronomy, they were able to make accurate calendars. The height of Maya civilization was from about A.D. 300 to A.D. 900. Then their power suddenly collapsed. Archaeologists do not know exactly why.

(l to r) DEA/G. DAGLI ORTI/DEA PICTURE LIBRARY/Getty Images; Russell Kord/Alamy

The Aztec ruled the region next. They settled in central Mexico in about 1300. Their impressive capital city was Tenochtitlán. Mexico City occupies the site where it once stood. Tenochtitlán was built on an island in the middle of a lake. Causeways connected it to the mainland.

The Aztec had a complex social and religious system. They conquered many of their neighbors and made slaves of captured soldiers. Priests performed rituals to win the favor of their gods. The Aztec were also skilled farmers. They built up land in the lake to form small islands called *chinampas,* which they used to grow crops.

The Spanish Arrive

In the early 1500s, a rival power appeared. Around 1520, Hernán Cortés led a small force of Spanish **conquistadors**, or conquerors, to Mexico. Within two years, these explorers and soldiers had defeated the Aztec and taken control of their empire.

How could the Spanish conquer the Aztec with only a few hundred men? Spanish guns and armor were better weapons than Aztec spears. Another major factor was European diseases. The diseases did not exist in the Americas until Europeans unknowingly brought them. Native Americans had no resistance to them, so the diseases killed many thousands. Cortés also took advantage of the anger of other native peoples who resented Aztec rule. Several groups joined him as allies.

Winning the Aztec Empire brought Spain riches in gold and silver mines. The conquest completely **transformed** life in Mexico. Roman Catholic priests converted native peoples to Catholicism. Conquistadors forced native peoples to work on farms or in mines. Spanish rule in Mexico was an example of colonialism.

DEA/G. DAGLI ORTI/DEA PICTURE LIBRARY/Getty Images

The Aztec city of Tenochtitlán was linked by canals, bridges, and raised streets built across the water.
▶ **CRITICAL THINKING**
Describing How were the Aztec able to build a city and farms in an area covered by a lake?

Under **colonialism**, one nation takes control of an area and dominates its government, economy, and society. The colonial power uses the colony's resources to make itself wealthier. In colonial Mexico, settlers from Spain had the most wealth.

Independence and Conflict

After almost 300 years of Spanish rule, a priest named Miguel Hidalgo led a rebellion in Mexico in 1810. The goal of the rebellion was to win independence from Spain. Some people hoped it would also create a more nearly equal society. The Spanish captured and executed Hidalgo, but by 1821 Mexico had gained its independence. Spanish rulers, though, were replaced by wealthy Mexican landowners. Native peoples remained poor.

Through much of the 1800s, Mexico was troubled by political conflict. Rival groups fought one another for power. Most of Mexico's people remained poor.

Revolution and Stability

By the early 1900s, dissatisfaction was widespread. A revolution erupted in Mexico. A **revolution** is a period of violent social and political change. One change was the land reform plan, which divided large estates into parcels of land that were then given to poor people to farm. National public schools were established, and a new constitution was written detailing the responsibilities of the government toward the people. Only one political party, however, held power until the 1990s.

☑ **READING PROGRESS CHECK**

Determining Central Ideas How were the Spanish able to conquer the Aztec?

A History of Central America

GUIDING QUESTION *How did the nations of Central America develop?*

The nations of Central America developed in similar ways to Mexico. But there were differences, as well.

Early Civilizations and Conquest

The Maya had flourished in Guatemala and Belize, as well as in southern Mexico. Even after their great cities were abandoned, the Maya continued to live in the region. After conquering Mexico, the Spanish moved south. By the 1560s, Spain had seized control of most of Central America. During the early 1800s, Britain claimed the area that is now Belize.

Independence

Central America gained its independence soon after Mexico. In 1823 the territories of Central America united to form one government. By 1840, they had separated into five independent

countries: Guatemala, Honduras, El Salvador, Nicaragua, and Costa Rica. The area that is now Belize was still a British colony. Panama was part of Colombia.

Central American countries were subjected to economic colonialism. This means that foreign interests dominate a people economically. These foreign interests were large companies from other countries. They set up **plantations**, or large farms, where poorly paid workers produced cash crops. **Cash crops** are crops sold for profit. The most important were bananas, coffee, and sugarcane.

Heading the governments for much of this time were military strongmen called **caudillos**. The caudillos helped ensure the foreigners' success. In turn, the foreigners made sure that the caudillos remained in power.

Conflict in Modern Times

Around 1900, Panama gained its independence from Colombia. It was helped by the United States, which wanted to build a canal there. The United States controlled the canal until 2000. Then, by agreement, Panama took control of the canal.

The late 1900s was a time of conflict. New wealth came to the upper classes, but most people remained poor. Various groups demanded reforms. Several countries were ravaged by civil wars. Only Costa Rica and Belize remained peaceful. One of Costa Rica's presidents, Óscar Arias Sánchez, helped bring peace to the region.

☑ **READING PROGRESS CHECK**

Analyzing How did Central America and Mexico's history differ?

Built by the United States in the early 1900s, the Panama Canal is now owned and operated by the Republic of Panama.

Identifying Which country held Panama shortly before the United States built the Panama Canal?

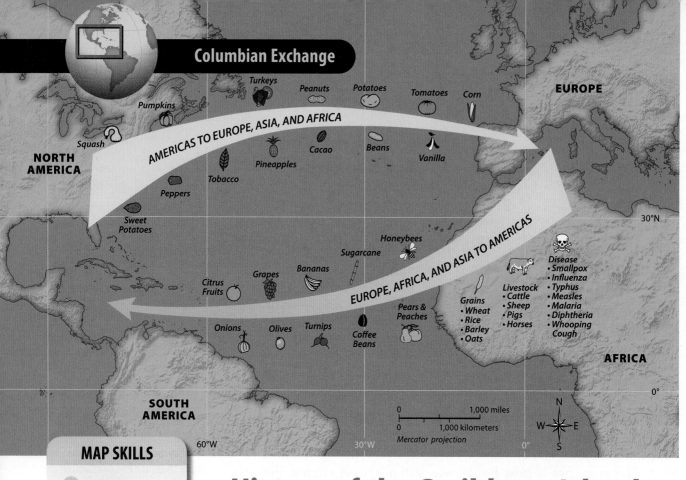

AMERICAS TO EUROPE, ASIA, AND AFRICA

EUROPE, AFRICA, AND ASIA TO AMERICAS

NORTH AMERICA

SOUTH AMERICA

EUROPE

AFRICA

Turkeys
Peanuts
Potatoes
Tomatoes
Corn
Pumpkins
Squash
Cacao
Beans
Vanilla
Pineapples
Tobacco
Peppers
Sweet Potatoes

Honeybees
Sugarcane
Grapes
Bananas
Citrus Fruits
Onions
Olives
Turnips
Coffee Beans
Pears & Peaches
Grains
• Wheat
• Rice
• Barley
• Oats
Livestock
• Cattle
• Sheep
• Pigs
• Horses
Disease
• Smallpox
• Influenza
• Typhus
• Measles
• Malaria
• Diphtheria
• Whooping Cough

30°N
0°
60°W
30°W
0°

0 — 1,000 miles
0 — 1,000 kilometers
Mercator projection

N W E S

MAP SKILLS

1 THE USE OF GEOGRAPHY What happened as a result of the Columbian Exchange?

2 HUMAN GEOGRAPHY How did the Columbian Exchange affect African peoples?

History of the Caribbean Islands

GUIDING QUESTION *How did the Caribbean islands develop?*

The history of the Caribbean islands is similar to that of Mexico and Central America. The islands have greater diversity, though, because several European countries ruled them as colonies.

Indigenous Peoples and European Settlers

Europeans changed the way the native peoples of the Caribbean lived. Like the Native Americans of the mainland, they suffered from diseases carried by the Europeans. This is why their numbers declined sharply soon after the arrival of the Europeans. Overwork and starvation also reduced their numbers. The Spanish set up colonies in what are now Cuba, the Dominican Republic, and Puerto Rico. Later, the French settled in what is now Haiti and on other smaller islands. The British and Dutch had some colonies, too.

Colonialism

During the 1600s, the Caribbean colonies became the center of the growing sugar industry. European landowners hoped to make money by selling the sugar in Europe. Because so many Native American workers had died, Europeans brought in hundreds of thousands of enslaved Africans to work the plantations.

The transport of enslaved Africans was part of the **Columbian Exchange**. This term refers to the transfer of plants, animals, and people between Europe, Asia, and Africa on one side and the Americas on the other. Foods such as wheat, rice, grapes, and apples were introduced to the Americas as were cattle, sheep, pigs, and horses. At the same time, products from the Americas were introduced into Europe, Africa, and Asia. They included corn, chocolate, and the potato. The Columbian Exchange also resulted in the introduction of new diseases into different parts of the world.

Independence

The first area in the Caribbean to gain independence was Haiti, then called Saint Domingue. Led by Toussaint-Louverture, Haiti gained its independence from France in 1804. The Dominican Republic won its independence in 1844. Cuba and Puerto Rico remained Spanish until 1898. When Spain lost the Spanish-American War, it gave independence to Cuba. Puerto Rico passed into American hands. Other islands of the Caribbean did not win the right to self-government until the middle 1900s.

Turmoil in the Twentieth Century

Independence did not mean freedom or prosperity. Rule by caudillos and widespread poverty have remained a problem in Haiti and the Dominican Republic.

Cuba, too, was often subject to dictatorial rule following its independence. Then in 1959, revolutionaries led by Fidel Castro took over. Castro soon cut all ties with the United States. He said his government would follow the ideas of communism. Communism involves government control of all areas of the economy and society. His rule did not bring economic success to Cuba.

The other islands of the Caribbean have had their own difficulties. Some countries in the region are trying to improve conditions and bring economic benefits to all their citizens. Small and with few resources, they have been unable to develop strong economies. Many of the islands depend on aid from the governments that used to run them as colonies.

Include this lesson's information in your Foldable®.

☑ **READING PROGRESS CHECK**

Analyzing What caused the population of the Caribbean islands to grow in colonial times?

LESSON 2 REVIEW

Reviewing Vocabulary
1. What is the difference between a *conquistador* and a *caudillo*?

Answering the Guiding Questions
2. *Describing* How did the Maya and the Aztec differ?

3. *Analyzing* How were relations between people with European and Native American heritage similar in Mexico during colonial times and the 1800s?

4. *Determining Central Ideas* How did economic colonialism affect the nations of Central America?

5. *Analyzing* How was the development of Cuba and of Haiti similar and different?

6. *Informative/Explanatory Writing* Write a summary of the history of Mexico, Central America, or the Caribbean islands after independence.

Reading **HELP**DESK

Academic Vocabulary

- **circumstance**
- **initiate**

Content Vocabulary

- **maquiladora**
- **mural**
- **dependence**
- **free-trade zone**
- **remittance**
- **reggae**

TAKING NOTES: *Key Ideas and Details*

Summarize As you read about Mexico, use the graphic organizer below to take notes about its economy and culture.

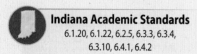

Mexico
Economy   Culture

Indiana Academic Standards
6.1.20, 6.1.22, 6.2.5, 6.3.3, 6.3.4, 6.3.10, 6.4.1, 6.4.2

Lesson 3
Life in the Region

ESSENTIAL QUESTION • *Why do people trade?*

IT MATTERS BECAUSE
Mexico and other countries in the region have close ties to the United States.

Modern Mexico

GUIDING QUESTION *What is life like in Mexico today?*

When you think of Mexico, you might think of Mexican food like tacos. You might think of mariachi musicians playing lively music and wearing large sombreros. But Mexico has a rich and complex culture and is a rising economic power.

The Economy

Mexico has close economic ties to the United States and Canada. These ties are a result of joining with them in the North American Free Trade Agreement (NAFTA). About 80 percent of Mexico's exports go to NAFTA partners. More than 60 percent of Mexico's imports come from members of NAFTA. Most of this trade is with the United States.

In recent decades, Mexico has developed its manufacturing industry. Factories account for about a third of Mexico's output. Some of them are **maquiladoras**. These are factories where parts made elsewhere are assembled into products. Many of the factories are located in northern Mexico. The goods are then exported. Food processing is another major industry in Mexico. The textile and clothing industries are important, too. Mexico also has heavy manufacturing, producing iron, steel, and automobiles.

Farming remains important. Cotton and wheat are grown in the dry north using irrigation. Along the southeastern coast, farms produce coffee, sugarcane, and fruit. On the central plateau, farmers grow corn, wheat, and fruits and vegetables. In the poor south, many farmers engage in subsistence farming—growing just enough food to feed themselves and their families.

Service industries are important in Mexico. Banking helps finance economic growth. A major service industry is tourism. Visitors from around the world come to visit ancient Maya sites or to see the architecture of Spanish colonial cities. Tourists also come to relax in resorts along the warm and scenic tropical coasts.

Culture

Mexicans are proud of their blend of Spanish and native cultures. They have long celebrated the folk arts that reflect native traditions. In the early 1900s, several Mexican painters drew on these traditions to paint impressive murals celebrating Mexico's history and people. **Murals** are large paintings made on walls. The Ballet Folklorico performs Mexican dances.

Sports reflect Mexico's ties to Spain and the United States. Soccer is popular there, as it is in Spain. So is baseball.

Challenges

With nearly 9 million people, Mexico City is one of the largest cities in the world. Including the city's suburbs, it has more than 21 million people—nearly 20 percent of Mexico's population. Overcrowding is a major problem.

Pollution is another problem, particularly air pollution. Because Mexico City is at a high elevation, the air has less oxygen than at sea level. This makes breathing difficult for some people in normal **circumstances**, but conditions in Mexico City are not normal. A great deal of exhaust from cars and factories is released into the air. The polluted air is held in place by the mountains around the city.

Think Again?

All Mexican food is spicy.

Not true! Mexican food varies from region to region. Most dishes are based on traditional native foods —corn, beans, and squash—along with rice, introduced from Spain. Not all Mexican food is spicy, although chilies are used in many dishes.

Academic Vocabulary

circumstance a condition

Farmers sell produce at a village market in Mexico.
▶ **CRITICAL THINKING**
Describing What is farming like in southern Mexico?

Nicaraguan workers make clothing in a factory.

▶ **CRITICAL THINKING**

Identifying What types of products have Central American manufacturers recently begun to make?

Sometimes a layer of cold air high in the atmosphere keeps the pollution from rising. The result can be a serious threat to health.

Another challenge facing Mexico is the power of criminals who sell illegal drugs. Drug lords use violence to fight police and to intimidate people. Mexico has mounted a major effort to battle this problem with some success.

Poverty is yet another major challenge facing Mexico. Anywhere from one-fifth to nearly half of Mexico's people are poor. Continued economic growth would help, and seems to be working. Some economists are predicting that Mexico will overtake Brazil in the 2010s as the leading economy in Latin America.

✅ **READING PROGRESS CHECK**

Analyzing How have close ties with the United States helped Mexico's economy?

Modern Central America

GUIDING QUESTION *What is life like in Central America?*

The nations of Central America have fewer resources than Mexico. The region must also deal with political problems.

Central America's Economies

The countries of Central America long showed **dependence**, or too much reliance, on cash crops. In recent years, some have begun to escape this trap. A good sign is the growth of manufacturing. This consists mostly of food processing and production of clothing and textiles. Tourism has grown as well. Tourists come to Belize and Guatemala to see ancient Maya sites. They travel to Costa Rica to see the varied plants and animals in its rain forests.

Panama benefits economically from the Panama Canal. Working for additional benefit, Panama **initiated** a major building program to expand the canal so it can accept larger cargo ships.

Academic Vocabulary

initiate to begin

High rates of population growth create an economic challenge. The countries need to grow their economies fast to provide enough jobs. One hope for promoting growth is trade agreements between the countries of the region and other countries.

In the 2000s, the United States and the Dominican Republic signed a series of agreements with five Central American countries (Costa Rica, El Salvador, Guatemala, Honduras, and Nicaragua). The agreement, called the Central America Free Trade Agreement (CAFTA-DR), was the first agreement among the United States and smaller developing economies. CAFTA-DR creates a **free-trade zone** that lowers trade barriers between the countries. Often, however, such trade agreements help the United States more than the other countries.

Challenges Facing the Region

Another challenge to the area is natural disasters. Earthquakes and hurricanes can have a devastating effect on the region's fragile economies. Nicaragua was making some economic progress in the 1990s when Hurricane Mitch hit. The destruction set the nation's economy back significantly.

The need to solve long-standing political problems also holds the region back. The civil wars of the 1980s and 1990s are over, but some of the issues that caused them remain unsolved. If these issues again become more severe, conflict may resume.

Culture

The culture of Central America is strongly influenced by European and native traditions. Spanish is the chief language in all countries except Belize, where English is the official language. English is spoken in many cities in the region as well. In rural Guatemala, native languages are common.

The population is mainly of mixed European and native heritage. Some people of African and Asian descent live there as well. Most people of the region are Roman Catholics. In recent years, however, Protestant faiths have gained followers.

☑ **READING PROGRESS CHECK**

Identifying What are the causes of poverty in Central America?

A tourist descends from a tree in a rain forest in Costa Rica.
▶ **CRITICAL THINKING**
Identifying Point of View Why do visitors travel to Costa Rica's rain forests?

James P. Blair/National Geographic/Getty Images

Haiti's brown, barren landscape contrasts sharply with the richly forested terrain of the neighboring Dominican Republic.

▶ **CRITICAL THINKING**

Analyzing Why has economic development been held back in Haiti?

The Caribbean Islands

GUIDING QUESTION *What is life like on the Caribbean islands?*

The Caribbean islands are mostly small countries with small populations and few resources. Although they have a rich and vibrant culture, they face many challenges.

Island Economies

The biggest challenge for the islands is to develop economically. Many people on the islands are poor. Even in Puerto Rico, a large share of the population lives in poverty. One reason for the poverty is high unemployment.

Cuba's economy is in poor condition after decades of communism. The government has been unable to promote economic development. It relied on aid first from the Soviet Union and more recently from Venezuela. Conditions are worse now than in the 1980s. Cubans also have little political freedom. Those who criticize the government are often arrested.

In Haiti, a history of poor political leadership has held back economic development. Haiti ranks among the world's poorest nations. Poverty is not the country's only problem. Widespread disease is another threat. In addition, as many as one in eight Haitians have left the country. Many of those who emigrated were among Haiti's most educated people. This loss hurts efforts to improve the economy. Finally, the country has not yet recovered from a deadly 2010 earthquake. Despite these problems, Haiti's people are determined to succeed.

Trinidad and Tobago has one of the more successful economies in the region. Sales of its oil and natural gas have funded economic development. Its location near Venezuela and Brazil has helped make its ports busy. The smaller Caribbean islands have had more political success than the larger ones. Governments are democratic and stable, but the economies are plagued by few resources and poverty.

Another important economic factor in the region is remittances. A **remittance** is money sent back to the homeland by people who migrated someplace else to find work. Many Dominicans came to the United States for work and send money home to support their families.

Tourism is a major part of the economy of several islands. Resorts in the Bahamas, Jamaica, and other islands invite tourists to come and relax in pleasant surroundings. The resorts often separate tourists from the lifestyle of the islanders, but they provide jobs for island citizens.

Island Cultures

The cultures of the Caribbean islands show a mix of mainly European and African influences. Large numbers of Asians also came to some of the islands in the 1800s and 1900s. Those from China went mainly to Cuba. South Asians settled in Jamaica, Guadalupe, and Trinidad and Tobago.

The languages spoken on the islands reflect their colonial heritage. English is the language of former British colonies such as the Bahamas and Jamaica. Spanish is spoken in Cuba, the Dominican Republic, and Puerto Rico. English is also taught in Puerto Rico's schools. French and Creole, a blend of French and African languages, are spoken in Haiti.

The Caribbean islands have strongly influenced world music. Much of the music blends African and European influences. Cuba is famous for its salsa, and Jamaica for reggae. Both forms of music rely on complex drum rhythms. **Reggae** has become popular around the world not only for its musical qualities but also for lyrics that protest poverty and lack of equal rights.

☑ **READING PROGRESS CHECK**

Citing Text Evidence How do economic conditions in Jamaica relate to the development of reggae?

Include this lesson's information in your Foldable®.

LESSON 3 REVIEW

Reviewing Vocabulary

1. What is a *free-trade zone*, and why do the nations of the region want to be in one?

Answering the Guiding Questions

2. *Determining Central Ideas* Do you think Mexico has a strong economy? Why or why not?

3. *Identifying* What challenges does Mexico face?

4. *Describing* How have the economies of the Central American countries changed in recent years?

5. *Analyzing* How do the languages of the Caribbean islands reflect their colonial history?

6. *Argument Writing* Take the role of a government official in one of these countries. Write a brief report to the nation's president explaining whether you think promoting tourism is good or bad for the nation's economy. Give reasons.

NAFTA
and Its Effects

The North American Free Trade Agreement (NAFTA) was created to grow trade among the United States, Canada, and Mexico to help these countries become more competitive in global markets.

Why Do Nations Trade?

No country produces all the goods and services it needs. Because most countries have more than they need of some things but not enough of others, trade is important.

Trade Barriers

Sometimes countries try to protect their industries from competition by setting up trade barriers such as tariffs and quotas. A tariff is a tax on imports. Tariffs raise the prices of imported goods so a country's own industries can produce and sell those goods at competitive prices. A quota restricts the amount of certain goods that can be imported from other countries.

Free Trade

The United States, Mexico, and Canada agreed to free trade, or getting rid of trade barriers, in 1994. The United States also has free trade agreements with 17 other nations, including Australia, Israel, and Peru.

> **"No country produces all the goods and services it needs."**

Disadvantages

Critics of NAFTA say that the agreement has cost U.S. jobs. Workers in Mexico are paid less. As a result, many U.S. industries moved all or part of their production to Mexico. Critics also argue that NAFTA hurt Mexican farmers. Mexico imported more corn and other grains when tariffs on those items were removed. Small farmers in Mexico could not compete with technologically advanced U.S. farms.

Advantages

Supporters of NAFTA say that it creates the largest free trade area in the world. With tariffs removed, the NAFTA countries can trade with one another at lower cost. It allows the 463 million people in the three countries greater choice in the marketplace. The three countries produced an estimated $18 trillion worth of goods and services in 2011.

A worker assembles parts at a U.S. automobile plant. Critics argued that NAFTA resulted in U.S. job losses, especially in the manufacturing industry. ▶

THERE'S MORE ONLINE

SEE the political boundaries of the U.S., Canada, and Mexico • *WATCH* changes in migration patterns

These numbers and statistics can help you learn about the effects of NAFTA.

Growth Triples

In 1993, the year before NAFTA went into effect, U.S. trade with Mexico and Canada totaled $276.1 billion. In 2010 U.S. exports and imports of goods with its NAFTA partners amounted to $918 billion.

U.S. Surplus in Services

A trade deficit occurs when a nation imports (buys) more goods and services than it exports (sells). In 2010 the U.S. experienced a trade deficit of $94.6 billion in *goods* with its NAFTA partners. A trade surplus occurs when a nation exports (sells) more than it buys. In 2009 the United States had a $28.3 billion trade surplus with its NAFTA partners in the value of *services*. The main services exported are financial services and insurance.

$1 million a minute

Almost 400,000 people—truckers, businesspeople, commuters, and tourists—cross the U.S.-Canada border daily. U.S.-Canada two-way trade amounts to $1.4 billion a day. That's almost a million dollars every minute.

One-fifth

Canada is the world's largest supplier of energy to the United States. Canada provides 20 percent of U.S. oil imports and 18 percent of U.S. natural gas imports.

21,444

the number of U.S. Border Patrol agents in 2011. This is double the number of agents in 2003.

700,000 Jobs

Critics of NAFTA say the agreement has cost the jobs of U.S. workers. According to the Economic Policy Institute, the transfer of production to Canada, Mexico, and other countries has resulted in the loss of about 700,000 jobs in the United States since NAFTA began.

FIFTEEN THOUSAND

This is the number of workers at the new Volkswagen plant in Puebla, Mexico, making it one of the country's largest employers.

TOP 10 COUNTRIES IN EXPORTS
Countries with exports in excess of $300 billion

CHINA

U.S.

GERMANY

JAPAN

NETHERLANDS

FRANCE

ITALY

U.K.

SOUTH KOREA

RUSSIA

KEY:
■ $1 trillion or more ■ $500 billion–$999 billion ■ $300 billion–$499 billion

U.S. TRADE WITH OTHER NATIONS
Trade in billions of dollars
(imports and exports combined, 2011)

CANADA **$597***

CHINA **$503** MEXICO **$461***

JAPAN $195

GERMANY $148

SOUTH KOREA $100

NETHERLANDS $66

U.K. $107

BRAZIL $74

SAUDI ARABIA $61

*NAFTA countries

GLOBAL IMPACT

EXPORTS AND IMPORTS Based on 2011 statistics, the exports of three countries—China, the United States, and Germany—exceeded $1 trillion in value. China's major exports to the U.S. include electrical machinery and toys, guns, and sports equipment. Top U.S. exports to China include oil seeds, fruits, vehicles, and aircraft.

The U.S. did more trade, if exports and imports are combined, with Canada in 2011 than with any other nation.

NAFTA Signing 1992

Mexican President Carlos Salinas, U.S. President George H.W. Bush, and Canadian Prime Minister Brian Mulroney look on as the chief trade representatives sign the NAFTA agreement in 1992. NAFTA was ratified by the three countries in 1993.

Thinking Like a
Geographer

1. *Human Geography* What is the purpose of NAFTA?

2. *The Uses of Geography* Find a product in a store or at home that has a label in another language in addition to English. Is that language used in one of the NAFTA countries? Why would a product be labeled in more than one language?

3. *Human Geography* Hold a debate in your class on this statement: NAFTA has been good for U.S. workers and consumers.

Directions: Write your answers on a separate piece of paper.

1 Use your FOLDABLES to explore the Essential Question.

INFORMATIVE/EXPLANATORY WRITING Write a couple of paragraphs explaining how geographical features led the Aztec and then much later the founders of Mexico City to build their cities on the same site.

2 21st Century Skills

INTEGRATING VISUAL INFORMATION Choose a country or one of the islands mentioned in this chapter. Find out more about it by researching it on the Internet. Use the information to create a travel poster or slide show highlighting the country's or island's best features and include any places you would like to see.

3 Thinking Like a Geographer

INTEGRATING VISUAL INFORMATION Draw a graphic organizer like the one shown here and use it to record information about the islands of the Caribbean.

4

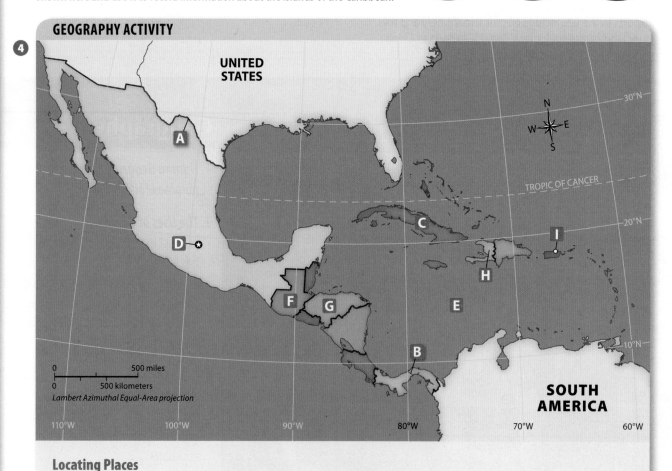

GEOGRAPHY ACTIVITY

Locating Places

Match the letters on the map with the numbered places listed below.

1. Panama Canal   **3.** Mexico City   **5.** Caribbean Sea   **7.** Guatemala   **9.** Cuba

2. Honduras   **4.** Rio Grande   **6.** Haiti   **8.** San Juan, Puerto Rico

REVIEW THE GUIDING QUESTIONS

Directions: Choose the best answer for each question.

1 What are Mexico's two most important natural resources?

A. gold and silver

B. oil and natural gas

C. iron ore and copper

D. bauxite and zinc

2 Which is one of the most important waterways in the world?

F. Lake Nicaragua

G. Río Bravo

H. Panama Canal

I. Caribbean Sea

3 A civilization that flourished in Southern Mexico, Belize, and Guatemala about 3,000 years ago and built pyramids like the Egyptians was the

A. Anasazi.

B. Olmec.

C. Aztec.

D. Maya.

4 Europeans established plantations and brought enslaved people to Cuba, Puerto Rico, and Hispaniola in order to grow

F. tobacco.

G. bananas.

H. sugar.

I. coffee.

5 Which country is Mexico's biggest trading partner?

A. Canada

B. China

C. Venezuela

D. the United States

6 What is the most serious economic challenge facing Central American countries?

F. high rate of population growth

G. fluctuating oil prices

H. food shortages

I. debt

DBQ ANALYZING DOCUMENTS

7 **IDENTIFYING** Read this passage about the Maya.

"*About six million Maya live in Central America. Like their ancestors, many of them survive by growing maize (Indian corn) or other crops on their land, or by producing woven textiles for sale. In some villages, the men have to leave their families to find work in the cities, or on coffee and cotton plantations.*"

—from *How People Live,* DK Publishing

Which Maya activity today is similar to one from ancient times?

A.   working in tourism

B.   working in factories

C.   growing maize

D.   working in cities

8 **DETERMINING CENTRAL IDEAS** Which best explains why some men have to leave their villages?

F.   They leave to seek wives elsewhere.

G.   They're forced to do so by the government.

H.   The villages are overcrowded.

I.   They face a lack of jobs within the villages.

SHORT RESPONSE

"*At the beginning of the 17th century the sweet crystal [sugar] transformed the Caribbean Islands into the Sugar Islands, though the islands did not turn sweet themselves. . . . Entire jungles were leveled; a slave or, later, cheap work force was massively imported from Africa and Asia; [and] a huge wave of European settlers arrived to stay.*"

—from Alfonso Silva Lee, *Natural Cuba/Cuba Natural*

9 **DESCRIBING** In what ways were the Caribbean islands transformed by the spread of sugar farming?

10 **DETERMINING WORD MEANINGS** What does the author of the passage mean by the phrase "the islands did not turn sweet themselves"?

EXTENDED RESPONSE

11 **INFORMATIVE/EXPLANATORY WRITING** Research and then write a brief report comparing and contrasting the cotton and rice plantations of the American South with the sugar plantations of the Caribbean islands.

Need Extra Help?

If You've Missed Question	**1**	**2**	**3**	**4**	**5**	**6**	**7**	**8**	**9**	**10**	**11**
Review Lesson	1	1	2	2	3	3	3	3	2	2	3

SOUTH AMERICA

UNIT **3**

Chapter 8
Brazil

Chapter 9
The Tropical North

Chapter 10
Andes and Midlatitude
Countries

MIKE THEISS/National Geographic Stock

EXPLORE the CONTINENT

SOUTH AMERICA

At nearly 7 million square miles (18 million sq. km) in area, South America is the fourth-largest continent in the world. Two great rivers—the Orinoco and the Amazon—flow through Brazil and the Tropical North. The most distinctive landform in the region is the Andes mountain ranges. The Andes, 4,500 miles (7,242 km) long, is the world's longest continental mountain range.

1 NATURAL RESOURCES Farmers grow a variety of potatoes in plots located in the Peruvian Andes. El Parque de la Papa, also known as "Potato Park," is a bio-reserve that is managed by the local communities surrounding it.

2 BODIES OF WATER Oil tankers travel near the port of Maracaibo, Venezuela. Like a circulatory system, the region's many waterways serve as arteries that transport people and goods throughout the region and to the world.

(bkgd) ©Frank Lukasseck/Corbis; (l) JIM RICHARDSON/National Geographic Stock; (c) JUAN BARRETO/AFP/Getty Images; (r) NOAA National Geophysical Data Center (NGDC)

3 **LANDFORMS** Los Glaciares National Park in Argentina is an area of rugged mountains and many glacial lakes. Its name refers to the glaciers that are part of the Patagonian ice field. The ice field is the largest ice mantle outside Antarctica. Los Glaciares is located along Argentina's border with Chile.

FAST **FACT**

Earth's driest place is in South America.

CENTRAL AMERICA

Caribbean Sea

0   1,000 miles
0   1,000 kilometers
Lambert Azimuthal Equal-Area projection

100°W   80°W   60°W   40°W

Lake Maracaibo

Orinoco R.

Llanos

Guiana Highlands

Gulf of Panama

ANDES

EQUATOR

0°

Galápagos Islands

Rio Negro

Amazon R.

AMAZON BASIN

Purus R.

Madeira R.

Tapajós R.

Xingu R.

Tocantins R.

São Francisco R.

Nevado Huascarán
22,205 ft.
(6,768 m)

PACIFIC OCEAN

Lake Titicaca

Altiplano

BRAZILIAN HIGHLANDS

Atacama Desert

ANDES

Paraguay R.

Paraná R.

20°S

TROPIC OF CAPRICORN

Cerro Aconcagua
22,834 ft.
(6,960 m)

Uruguay R.

PAMPAS

ATLANTIC OCEAN

Colorado R.

Rio de la Plata

40°S

Elevations

10,000 ft. (3,000 m)
5,000 ft. (1,500 m)
2,000 ft. (600 m)
1,000 ft. (300 m)
0 ft. (0 m)
Below sea level

⎯⎯ National boundary
▲ Mountain peak

Valdés Peninsula

PATAGONIA

Falkland Islands

Tierra del Fuego

Strait of Magellan

Cape Horn

South Georgia Island

SOUTH AMERICA

PHYSICAL

MAP SKILLS

1 THE GEOGRAPHER'S WORLD What is the easternmost river in Brazil?

2 PLACES AND REGIONS Describe the elevation differences between western South America and eastern South America.

3 PLACES AND REGIONS Explain why elevation of land in the Amazon Basin makes travel easier.

Galápagos
Islands
(Ecuador)

EQUATOR

CENTRAL
AMERICA

Caribbean Sea

Caracas

VENEZUELA

Orinoco R.

Bogotá

COLOMBIA

Gulf of
Panama

Quito

ECUADOR

PERU

Lima

PACIFIC
OCEAN

SURINAME

Georgetown   Paramaribo

GUYANA   Cayenne

FRENCH GUIANA (Fr.)

Rio Negro

Amazon R.

Purus R.

Madeira R.

Tapajós R.

Xingu R.

Tocantins R.

São Francisco R.

BRAZIL

Brasília

Lake
Titicaca   La Paz

BOLIVIA

Sucre

Paraguay R.

Paraná R.

PARAGUAY

Asunción

CHILE

Uruguay R.

TROPIC OF CAPRICORN

20°S

• National capital
○ Department capital

Santiago

ARGENTINA

Buenos Aires

Colorado R.

URUGUAY

Montevideo

Rio de la Plata

ATLANTIC
OCEAN

40°S

Falkland
Islands (U.K.)

South
Georgia
Island (U.K.)

Strait of
Magellan

0   1,000 miles
0   1,000 kilometers
Lambert Azimuthal Equal-Area projection

N
W   E
S

100°W   80°W   60°W   40°W

0°

POLITICAL

MAP SKILLS

1 PLACES AND REGIONS What is the capital of Uruguay?

2 PHYSICAL GEOGRAPHY Which two countries in South America do not have coastlines?

3 THE GEOGRAPHER'S WORLD Which country in South America shares its border with the most countries?

POPULATION

Per sq. mi.	Per sq. km
1,250 and over	500 and over
250–1,249	100–499
63–249	25–99
25–62	10–24
2.5–24	1–9
Less than 2.5	Less than 1

Cities
(Statistics reflect metropolitan areas.)

- ■ Over 5,000,000
- □ 2,000,000–5,000,000
- ◉ 1,000,000–2,000,000
- • 500,000–1,000,000
- ○ Under 500,000

SOUTH AMERICA

POPULATION DENSITY

MAP SKILLS

1 HUMAN GEOGRAPHY Where do most people in South America live?

2 HUMAN GEOGRAPHY About how many people live in Lima?

3 PLACES AND REGIONS What are the largest cities on South America's eastern coast?

CENTRAL
AMERICA

Caribbean Sea

Gulf of
Panama

EQUATOR

PACIFIC
OCEAN

PACIFIC
OCEAN

TROPIC OF CAPRICORN

ATLANTIC
OCEAN

Resources

- 🚃 Coal
- ⛏ Petroleum
- 💧 Natural gas
- ⚡ Iron ore
- ▼ Tin
- ▣ Zinc
- ✜ Bauxite
- ✳ Uranium
- ● Cobalt
- Ⓝ Nickel
- ◢ Copper
- ✦ Lead
- ✦ Manganese
- ◢ Gold
- ◢ Silver
- ▽ Diamonds

Land Use

- Commercial farming
- Subsistence farming
- Livestock raising
- Primarily forest
- ■ Manufacturing and trade
- 🐟 Commercial fishing
- Little or no activity

ECONOMIC RESOURCES

MAP SKILLS

1 HUMAN GEOGRAPHY Is there more commercial farming or livestock raising in South America?

2 PHYSICAL GEOGRAPHY Where is the greatest concentration of minerals and ores?

3 ENVIRONMENT AND SOCIETY Is South America a manufacturing center?

CENTRAL AMERICA
Caribbean Sea
100°W
80°W
60°W
40°W
0
1,000 miles
0
1,000 kilometers
Lambert Azimuthal Equal-Area projection
N
W
E
S
Caracas
Bogotá
EQUATOR
0°
Manaus
Belém
Lima
PACIFIC OCEAN
La Paz
Brasília
20°S
TROPIC OF CAPRICORN
Asunción
Rio de Janeiro
ATLANTIC OCEAN
Santiago
Buenos Aires
40°S
Tropical rain forest
Tropical savanna
Semi-arid (steppe)
Arid (desert)
Humid subtropical
Marine west coast
Mediterranean
Tundra and high altitude
National capital
City

SOUTH AMERICA

CLIMATE

MAP SKILLS

1. **PHYSICAL GEOGRAPHY** What is the most prevalent climate in South America?

2. **PHYSICAL GEOGRAPHY** Where is South America's desert climate?

3. **PLACES AND REGIONS** In which type of climate is Santiago, Chile, located?

BRAZIL

ESSENTIAL QUESTIONS • *How does geography influence the way people live?*
• *How do governments change?* • *What makes a culture unique?*

Soccer ("football") player
Robinho has many fans
in Brazil and around
the world.

Silvia Izquierdo/AP Images

Lesson 1
*Physical Geography
of Brazil*

Lesson 2
History of Brazil

Lesson 3
Life in Brazil

The Story Matters...

Brazil is located in the eastern half of South America. Brazil's vast land area makes it the giant of South America. Water is also important in defining the country. The great Amazon River flows through Brazil for more than 2,000 miles (3,219 km) and carries as much as one-fourth of the world's freshwater. This river drains the Amazon Basin, which stretches across the northern half of Brazil and contains the world's largest remaining tropical rain forest.

FOLDABLES®
Study Organizer

Go to the Foldables® library in the back of your book to make a Foldable® that will help you take notes while reading this chapter.

Valuable Natural Resources

Urban Population

BRAZIL

Brazil is the largest country in South America with almost 3.3 million square miles (8.5 million sq. km) of land. It accounts for most of the eastern coast of South America. Brazil contains more than 4,665 miles (7,508 km) of coastline along the Atlantic Ocean. The Equator and the Tropic of Capricorn run through the country. As you study the map, look for the geographic features that make this area unique.

Step Into the Place

MAP FOCUS Use the map to answer the following questions.

1 PHYSICAL GEOGRAPHY What is the main river in Brazil?

2 PLACES AND REGIONS How many countries share a border with Brazil?

3 THE GEOGRAPHER'S WORLD Why is it significant that the Equator and the Tropic of Capricorn both run through Brazil?

4 CRITICAL THINKING
ANALYZING Use the scale bar on the map to measure the distance between the cities of Brasília and Rio de Janeiro.

A

RIO, AERIAL VIEW The huge "Christ the Redeemer" statue overlooks Rio de Janeiro. Set between beautiful mountains and the Atlantic coast, Rio de Janeiro was Brazil's capital from 1763 to 1960.

B

BRAZIL'S CAPITAL Brasília is a planned city, built in Brazil's central wilderness area. Brasília has been the country's capital since 1960.

Step Into the Time

ANALYZING Select at least two events on the time line and explain how they illustrate the importance of the Amazon Basin to Brazil's development, as well as the environmental concerns caused by that development.

1500
Cabral is first European to reach Brazil's coast

1800

1822 Brazil gains independence from Portugal

Caribbean Sea

Brazil

National capital
• City

VENEZUELA

GUYANA

FRENCH GUIANA
(France)

ATLANTIC OCEAN

COLOMBIA

SURINAME

ECUADOR

EQUATOR 0°

10°N

PERU

Amazon R.
Madeira R.
Tapajós R.
Purus R.
Xingu R.

BRAZIL

Tocantins R.
São Francisco R.

• Recife

B

10°S

• Salvador

◉ Brasília

BOLIVIA

A

PACIFIC OCEAN

Paraná R.

São Paulo •

• Rio de Janeiro

CHILE

PARAGUAY

20°S

TROPIC OF CAPRICORN

0   500 miles
0   500 kilometers
Lambert Azimuthal Equal-Area projection

ARGENTINA

30°S

URUGUAY

N
W   E
S

1889
Brazil is proclaimed
a republic

2010
Dilma Rousseff
elected president

1960 Capital moves from
Rio de Janeiro to Brasília

1900

2000

1888 Slavery is
abolished in Brazil

2009 Rio de Janeiro chosen
to host 2016 Olympic Games

Reading **HELP**DESK

Academic Vocabulary

- **area**
- **occur**

Content Vocabulary

- **tributary**
- **basin**
- **rain forest**
- **canopy**
- **plateau**
- **escarpment**
- **pampas**
- **Tropics**
- **temperate zone**

TAKING NOTES: *Key Ideas and Details*

Summarize As you read, use a graphic organizer to write a summary sentence about each topic.

Topic	Summary
Waterways	
Climate	
Resources	

Indiana Academic Standards
6.1.18, 6.1.19, 6.3.5, 6.3.7, 6.3.8, 6.3.12, 6.4.7

Lesson 1
Physical Geography of Brazil

ESSENTIAL QUESTION • *How does geography influence the way people live?*

IT MATTERS BECAUSE
Brazil is the world's fifth-largest country in size and population.

Waterways and Landforms

GUIDING QUESTION *What are Brazil's physical features?*

Brazil is the largest country in South America. It occupies about half the continent. Rolling lowland plains and flat highland plateaus cover most of the country.

The Amazon

The Amazon River is one of Brazil's amazing natural features as well as a great natural resource. It begins high in the Andes of Peru and flows east across northern Brazil to the Atlantic Ocean. The river is the Western Hemisphere's longest river and the world's second longest, after the Nile River in Africa.

The Amazon is the largest river in terms of the amount of freshwater it carries. It moves more than 10 times the water volume of the Mississippi River. Of all the water that Earth's rivers empty into the oceans, about 25 percent comes from the Amazon. Its massive flow pushes freshwater more than 100 miles (161 km) out into the Atlantic Ocean. The river's depth allows oceangoing ships to travel more than 2,000 miles (3,219 km) upstream to unload or pick up cargo.

The Amazon Basin

One reason the Amazon carries so much water is that it has more than 1,000 **tributaries**. These smaller rivers feed into the Amazon as it flows from the Andes to the Atlantic Ocean. Several tributaries are more than 1,000 miles (1,609 km) long.

The **area** that a river and its tributaries drain is called a **basin**. The Amazon Basin covers more than 2 million square miles (5.2 million sq. km). Nearly half of Brazil's land lies within this vast region. Its wet lowlands cover most of the country's northern and western areas.

Much of the Amazon Basin is covered by the world's largest **rain forest**. A rain forest is a warm woodland that receives a great deal of rain each year. Tall evergreen trees form a **canopy**, or an umbrella-like covering. The Amazon rain forest is called the Selva. It is the world's richest biological resource. The Selva is home to several million kinds of plants, insects, birds, and other animals.

Only about 6 percent of Brazil's population live in the Amazon Basin. Most of the region contains fewer than two people per square mile. Some are Native Americans who live in small villages and have little contact with the outside world.

Brazilian Highlands

South and east of the Amazon Basin are the Brazilian Highlands. This is mainly a region of rolling hills and areas of high, flat land called **plateaus**. These highlands are divided into western and eastern parts.

Visual Vocabulary

Tributary A tributary is a smaller river or stream that flows into a larger one, or into a lake.

Academic Vocabulary

area a geographic region

South America's Amazon River and North America's Mississippi River cross vast distances and carry enormous amounts of water.
▶ **CRITICAL THINKING**
Comparing How are the Amazon and Mississippi Rivers similar? How are they different?

The western part of the highlands is largely grassland that is partly covered with shrubs and small trees. Farming and ranching are the major economic activities in this part of the highlands. Farther west is the Mato Grosso Plateau, a flat, sparsely populated area of forests and grasslands that extends into Bolivia and Peru.

Low mountain ranges form much of the eastern Brazilian Highlands, although some peaks rise above 7,000 feet (2,134 m). In other places, highland plateaus plunge to the Atlantic coast, forming **escarpments**, or steep slopes. These escarpments, rising from coast to highlands, have hindered development of inland areas.

Brazil's third-largest city, Brasília, is located in the Brazilian Highlands. It was built in the 1950s as Brazil's new capital to encourage settlement in the country's interior. Some 3.5 million people live in and around the city.

About 600 miles (966 km) south of Brasília is São Paulo. This huge city is located on a plateau at the highland's eastern edge, just 30 miles (48 km) from the Atlantic coast. With more than 17 million people, São Paulo is the largest city in the Southern Hemisphere. It is also South America's most important industrial city.

Farther south are grassy, treeless plains called **pampas**. The grass and fertile soil make the pampas one of Brazil's most productive ranching and farming areas.

Atlantic Lowlands

Brazil has one of the longest strips of coastal plains in South America, wedged between the Brazilian Highlands and the Atlantic Ocean. This narrow plains region, called the Atlantic lowlands, is just 125 miles (201 km) wide in the north; it becomes even narrower in the southeast. The rural parts of this region are another important area for farming.

An escarpment slopes down to an Atlantic Ocean beach near the city of São Paulo.

▶ **CRITICAL THINKING**

Describing How have escarpments affected Brazil's development?

Although the coastal lowlands cover only a small part of Brazil's territory, most of the nation's people live here. More than 12 million live in and around Rio de Janeiro, Brazil's second-largest city. Rio's beautiful beaches and vibrant lifestyle make it Brazil's cultural and tourist center.

☑ **READING PROGRESS CHECK**

Analyzing Why do many Brazilians live in the Brazilian Highlands?

A Tropical Climate

GUIDING QUESTION *What are Brazil's climate and weather like?*

Most of Brazil is located in the **Tropics**. This is the zone along Earth's Equator that lies between the Tropic of Cancer and the Tropic of Capricorn. Brazil's climate varies. In fact, the huge country has several different climates.

Wet Rain Forests

The area along the Equator in northern Brazil has a tropical rain forest climate. In this climate, every day is warm and wet. Daytime temperatures average in the 80s Fahrenheit (27°C to 32°C). It feels hotter than this because the wet rain forest makes the air humid.

Ranchers herd cattle on the Mato Grosso Plateau of west-central Brazil.

▶ **CRITICAL THINKING**

Describing What are the main features of the Mato Grosso Plateau?

During periods of drought, the Amazon River carries less water, which exposes sandbars in the river and low-lying areas along the shoreline.

▶ **CRITICAL THINKING**

Identifying What type of climate is found in areas along the Amazon River?

Academic Vocabulary

occur to happen or take place

Areas along the Amazon River have a tropical rain forest climate. They experience winds called monsoons that bring a huge amount of rain—120 inches to 140 inches (305 cm to 356 cm) per year. During the monsoon season, flooding swells the Amazon River in some places to more than 100 miles (161 km) wide. These areas also have a dry season when little rain **occurs**. During the dry season, forest fires are a danger, even in a rain forest.

Tropical Wet/Dry Climate

Tropical wet/dry climates usually exist along the outer edges of tropical rain forest climates. Most of the northern and central Brazilian Highlands has a tropical wet/dry climate. This climate has just two seasons—summer, which is wet, and winter, which is dry. Daily average temperatures change very little. Summers average in the 70°F range (21°C) and winters in the 60°F range (16°C). But even this slight difference is enough to change wind patterns, which affect rainfall. Between 40 inches and 70 inches (102 cm to 178 cm) of rain fall during the summer months. Winters get almost no rain.

Dry and Temperate Climates

The northeastern part of the Brazilian Highlands has a semiarid climate. This region is the hottest and driest part of the country. The daily high temperature during the summer often reaches 100°F (38°C). Frequent and severe droughts have caused many of the region's farms to fail. Even so, the desertlike plant life supports some light ranching.

Southeastern Brazil, including São Paulo and Rio de Janeiro, is located in the **temperate zone**—the region between the Tropic of Capricorn and the Antarctic Circle. It has a temperate climate called humid subtropical. It is the same type of climate that the southeastern United States experiences.

Temperatures vary according to location and elevation in this part of Brazil. Summers are generally warm and humid, and winters are mild. Rainfall occurs year-round. In the southern parts of this climate zone, snow can fall.

☑ **READING PROGRESS CHECK**

Identifying What factors make farming in the northeastern part of Brazil difficult?

Natural Resources

GUIDING QUESTION *What resources are most plentiful and important in Brazil?*

Brazil has some of the world's most plentiful natural resources. Many of the resources have been developed for years, especially in the south and southeast. Recent transportation improvements have made the resources in Brazil's vast interior available to its growing industries and population. Agriculture, mining, and forestry have been important for centuries. The natural riches of Brazil attracted European settlers to the region. They found abundant trees, rich mineral resources, and fertile farmland.

River floodwaters surge through an area of the Amazon rain forest in northwestern Brazil.
▶ **CRITICAL THINKING**
Identifying What yearly natural event causes flooding in the Amazon Basin?

Think Again?

Summer and winter occur at about the same time everywhere.

Not true. While American teens enjoy their summer vacation, young people in Brazil are going to school! That's because south of the Equator, the seasons are reversed. The summer months in the United States are winter months in Brazil.

Kevin Schafer/Photographer's Choice/Getty Images

A worker on a Brazilian coffee plantation picks ripe coffee berries. After picking, the coffee berries are separated for quality and packed in sacks to send to market.

▶ **CRITICAL THINKING**

Identifying Where are Brazil's major coffee-growing areas?

Abundant Forests

Forests cover about 60 percent of Brazil, accounting for about 7 percent of the world's timber resources. Most of the forests in the northeast and south were cleared long ago. Heavy logging continues in the Atlantic lowlands.

Logging in the Amazon Basin is increasing as more roads are built and settlement grows. The rain forest's mahogany and other hardwoods are highly desirable for making furniture. The rain forest is also a source of natural rubber, nuts, and medicinal plants. Logging, mining, and other development have become a major environmental issue. However, the rate of deforestation, or clearing land of forests or trees, has declined in recent years.

Minerals

Brazil has rich mineral resources that are only partly developed. They include iron ore, tin, copper, bauxite, gold, and manganese. At one time, most mining was done in the Brazilian Highlands. Recently, major deposits of minerals have been found in the Amazon basin. The new deposits might make Brazil the world's largest producer of many of the minerals. Brazil also has huge potential reserves of petroleum and natural gas deep under the ocean floor off its coast. Getting to the oil is a challenge, however.

Benjamin Lowy/Getty Images News/Getty Images

Productive Farmland

Brazil is the world's largest producer of coffee, sugarcane, and tropical fruits. The country also produces great amounts of soybeans, corn, and cotton.

Brazilian farmers produce most of their country's food supply. Agriculture is also important in trade, accounting for more than one-third of Brazil's exports. It is a leading exporter of coffee, oranges, soybeans, and cassava. Cassava is used to make tapioca.

Major Crops

Production of coffee throughout the world was estimated to set an all-time high in 2012–2013, up 10 million bags from the previous year. Brazil and Vietnam accounted for most of the increase. The eastern Brazilian Highlands and the Atlantic lowlands are the main coffee-growing areas. Coffee was once Brazil's main export. Today, soybeans provide more income for the country. China is increasing its soybean imports, mostly for animal feed, and much of it comes from Brazil.

Most soybeans are grown in the south, but they are an important crop in the Brazilian Highlands, too. Farming has become easier in the highlands as farmers have begun using tractors and fertilizer to work the savanna soils.

Brazil grows one-third of the world's oranges, making it the world's leading supplier of the citrus fruit. Brazil is also the largest beef exporter in the world. Most of the country's grazing land is in the south and southeast.

In a recent year, Brazil's sugarcane production was more than two and a half times that of India, the second-leading producer. Brazilian sugarcane is used to make ethanol, which is mixed with gasoline and used as fuel for cars and trucks. For many years, the government has required cars to use ethanol. The country's car manufacturers make flexible-fuel vehicles that can use fuel with high levels of ethanol.

☑ **READING PROGRESS CHECK**

Identifying Which two regions are Brazil's most important agricultural areas?

Include this lesson's information in your Foldable®.

LESSON 1 REVIEW

Reviewing Vocabulary
1. How does Brazil's location in the Tropics affect its climate?

Answering the Guiding Questions
2. ***Determining Central Ideas*** Why is the Amazon Basin a unique region?

3. ***Analyzing*** How do a tropical rain forest climate and a tropical wet/dry climate differ?

4. ***Describing*** What resources are important Brazilian exports?

5. ***Informative/Explanatory Writing*** In which of Brazil's physical regions would you most like to live? Write a paragraph to explain why.

14.

Reading **HELP**DESK

Academic Vocabulary

- **comprise**
- **extract**

Content Vocabulary

- **indigenous**
- **slash-and-burn agriculture**
- **emancipate**
- **compulsory**

TAKING NOTES: *Key Idea and Details*

Sequencing As you read about Brazil's history, use the graphic organizer below to note how Brazil became a modern democratic republic.

Portugal's government moves to Brazil.

↓

↓

Indiana Academic Standards
6.1.6, 6.1.10, 6.2.7, 6.4.1

Lesson 2
History of Brazil

ESSENTIAL QUESTION · *How do governments change?*

IT MATTERS BECAUSE
Brazil is one of the world's leading industrial powers.

Early History

GUIDING QUESTION *How did Brazil's early peoples live?*

In 1493 Christopher Columbus returned to Spain with news of his explorations and of new lands. The Spanish worried that neighboring Portugal, a powerful seafaring rival, would try to claim these lands for itself. So they asked the pope to find a solution. The pope decided that all new lands west of a certain line should belong to Spain. Lands east of the line would belong to Portugal. The two countries agreed to this division in 1494 by signing the Treaty of Tordesillas.

Almost nothing was known of the region's geography, so neither side realized how unequal the division was. Almost all of the Americas lay west of the line, which became Spanish territory. The only exception was the eastern part of South America, which became Portuguese territory. Today, this part of South America is Brazil. That is why Brazil is the only South American country that has a Portuguese heritage.

Indigenous Populations

The first Portuguese ships stopped in Brazil in 1500. Their destination was India, so they did not stay in Brazil for long. They had peaceful encounters with some of the **indigenous**, or native, peoples who lived along the coast. The Portuguese commander, Pedro Cabral, claimed the land for Portugal. After just 10 days, the Portuguese left. They had no idea of the vast region and many peoples included in Cabral's claim.

(l to r) Mike Goldwater/Alamy; Diego Frichs Antonello/Getty Images; Eraldo Peres/AP Images

The people the Portuguese met were the Tupi. They lived along the coast and in the rain forests south of the Amazon River, where they grew cassava, corn, sweet potatoes, beans, and peanuts. They hunted fish and other water animals with arrows and harpoons from large log canoes, but they did little hunting on land.

Brazil's native peoples had lived there for more than 10,000 years when the Portuguese arrived. Estimates are that the population was between 2 million and 6 million by 1500. Besides the Tupi, it included the Arawak and Carib people of the northern Amazon and coast, and the Nambicuara in the drier grasslands and highlands. These are not the names of native peoples; they were Brazil's four main language groups. Each group **comprised** many different peoples.

Daily Life

Like the Tupi, Brazil's other lowland and rain forest peoples were mainly farmers. They lived in permanent, self-governing villages and practiced **slash-and-burn agriculture**. This is a method of farming in forests that involves cutting down trees and burning away underbrush to create fields for growing crops. Farther south, most of the Nambicuara of the Brazilian Highlands were nomads, people who move from place to place and have no permanent home. In the dry season, they lived as hunter-gatherers, people who get their food by hunting, fishing, and collecting seeds, roots, and other parts of trees and wild plants. In the wet season, they built temporary villages and practiced slash-and-burn agriculture.

Europeans Arrive

For more than 30 years after Cabral's visit, the Portuguese did not pay much attention to Brazil. Their main focus was on their colonies and trade in Asia. Their trading ships sailed south and east around Africa on their way to Asia. Portuguese sailors established a few trading posts along Brazil's coast and collected brazilwood. The red dye **extracted** from this wood was highly valued in Europe. It was because of this trade that the Portuguese named the region Brazil.

Academic Vocabulary

comprise to be made up of
extract to remove or take out

An Ashaninka family fishes from a boat in Brazil's Amazon rain forest.
▶ **CRITICAL THINKING**
Describing How did indigenous peoples make a living when the first Europeans arrived in Brazil?

The church of São Miguel das Missões was built about 1740 as the center of a Jesuit mission village in southern Brazil.

▶ **CRITICAL THINKING**
Explaining Why did the Jesuits build mission villages in Brazil and other parts of South America?

The valuable brazilwood trade made other Europeans more interested in Brazil. French traders began collecting the wood and shipping it to France. To bring Brazil under tighter Portuguese control, Portugal's King John III established a permanent colony and government there. The first Portuguese settlers arrived in 1533.

✓ **READING PROGRESS CHECK**

Determining Central Ideas Why did the Portuguese colonize Brazil?

Colonial Rule

GUIDING QUESTION *How did the Portuguese colony in Brazil develop?*

Portugal's rule of Brazil lasted more than 300 years. During that time, Portuguese settlements spread all along the coast. Explorers and others traveled up rivers and deep into Brazil's interior. The expansion brought wealth to Portugal, though much of it came at great cost to Brazil's indigenous peoples.

The Portuguese Conquest

King John III gave wealthy supporters huge tracts of land in Brazil. These tracts extended west from the coast about 150 miles (241 km) inland. In return, the people who received a land grant were responsible for developing it. They founded cities and gave land to colonists to farm.

Diego Frichs Antonello/Getty Images

Because the colonists could not do all the work that was required, they soon began enslaving nearby native peoples as laborers. Many of them resisted and were killed. Thousands more died from exposure to European diseases to which they had no natural resistance. Others fled into Brazil's interior. These conditions and other complaints caused King John to end the land-grant system in 1549. He put Brazil under royal control and sent a governor from Portugal to rule the colony.

Spread of Christianity

The new governor brought more colonists with him. They included a number of Jesuit Catholic priests who belonged to a missionary group called the Society of Jesus. The king asked the Jesuits to go to Brazil to help the native peoples and convert them to Christianity. Those who converted were settled in special Jesuit villages and were protected from slavery.

Those Portuguese colonists who held enslaved people complained to the king about the Jesuits' work. In 1574 he ruled that native peoples who did not live in Jesuit villages could be enslaved only if they were captured in war. This ruling sent Jesuits into Brazil's interior to protect and convert peoples there. Slave hunters also moved into the interior to attack and enslave the native peoples. Cattlemen and prospectors followed, slowly spreading development inland.

Sugar and Gold

As Brazil's sugar industry expanded, cattlemen needed new land. The rise of large sugarcane plantations, mainly in the northeast, pushed ranching westward.

Plantation workers carry sugarcane into a Brazilian mill, 1845.
▶ **CRITICAL THINKING**
Identifying Besides sugarcane, what else did large plantations grow?

In the 1600s, sugar became Brazil's main export and Portugal's greatest source of wealth. Coffee and cotton plantations also developed. The discovery of gold in the eastern highlands in the 1690s further boosted the development of the interior. Towns sprang up as thousands of colonists rushed to the area. Large numbers of new colonists arrived from Europe, as well. The discovery of diamonds in the region in the 1720s added to the population boom.

Plantation agriculture and mining required large numbers of workers. This increased the need for enslaved workers. When native populations could not fill the need, the Portuguese began importing large numbers of enslaved Africans. By the 1780s, more than 150,000 enslaved Africans worked in the mining districts. This was twice the size of the Portuguese population. By 1820, some 1.1 million enslaved people accounted for nearly one-third of Brazil's total population.

☑ **READING PROGRESS CHECK**

Determining Central Ideas Why did King John III send Jesuits to Brazil?

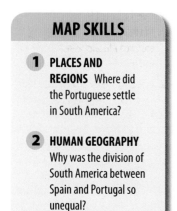

MAP SKILLS

1 **PLACES AND REGIONS** Where did the Portuguese settle in South America?

2 **HUMAN GEOGRAPHY** Why was the division of South America between Spain and Portugal so unequal?

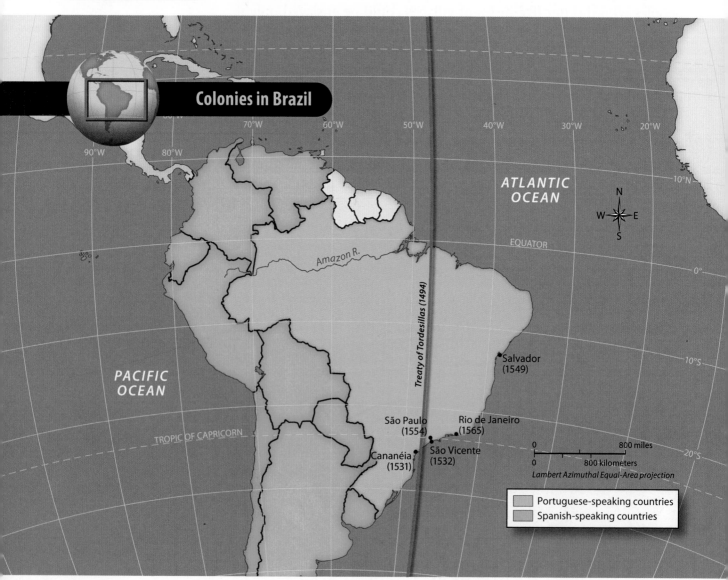

Colonies in Brazil

ATLANTIC OCEAN

EQUATOR

Amazon R.

Treaty of Tordesillas (1494)

Salvador (1549)

PACIFIC OCEAN

São Paulo (1554)

Rio de Janeiro (1565)

TROPIC OF CAPRICORN

Cananéia (1531)

São Vicente (1532)

0   800 miles
0   800 kilometers
Lambert Azimuthal Equal-Area projection

Portuguese-speaking countries
Spanish-speaking countries

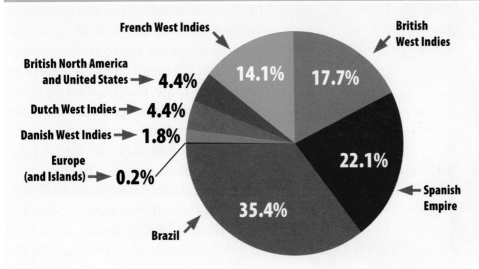

French West Indies

British North America and United States → **4.4%**

Dutch West Indies → **4.4%**

Danish West Indies → **1.8%**

Europe (and Islands) → **0.2%**

Brazil

British West Indies

14.1%

17.7%

22.1%

Spanish Empire

35.4%

THE SLAVE TRADE

More than one of every three enslaved Africans who were transported to the Americas were brought to Brazil.

▶ **CRITICAL THINKING**

1. *Identifying* What percentage of enslaved Africans were transported to the Spanish Empire?

2. *Integrating Visual information* What is the combined percentage of enslaved Africans brought to the West Indies?

Independent Brazil

GUIDING QUESTION *How did Brazil gain independence and become a democracy?*

Brazil gained independence from Portugal in an unusual way. It came gradually, fairly easily, and with little bloodshed. It was also the indirect result of the actions of the French emperor Napoleon Bonaparte.

Independence and Monarchy

In 1805, Britain joined by its allies Russia, Austria and Sweden, went to war with France to crush Napoleon. Instead, Napoleon defeated them and conquered much of Europe. In 1807 Napoleon invaded Portugal. As the French army closed in on Portugal's capital city of Lisbon, ruler Dom João, the royal family, and other government leaders fled to Brazil. Rio de Janeiro became the new capital of the Portuguese Empire. Brazil's status within the empire changed from a colony to a kingdom. This action gave Brazil equal status with Portugal within the empire.

After Napoleon was defeated, the Portuguese people wanted their king back. In 1821 Dom João and the rest of the government returned to Portugal. He left his son Pedro to rule Brazil. In 1822 Portugal's legislature restored Brazil's status as a colony and ordered Pedro to return. Pedro refused to give up the Brazilian throne. He declared independence and crowned himself Emperor Pedro I. Most other independent American nations became republics, but independent Brazil became a constitutional monarchy. In this form of government, a king, a queen, or an emperor acts as head of state.

Most Brazilians had supported independence from Portugal, but they soon tired of Pedro's harsh rule. In 1831 he was forced to turn over the throne to his five-year-old son.

As in the United States, voters in Brazil elect a president every four years.

▶ **CRITICAL THINKING**
Identifying Which group controlled the election of Brazil's president in the early republic?

A series of advisers ruled in the boy's name until he was old enough to rule on his own. In 1840, at age 14, he became Emperor Pedro II.

Pedro II ruled Brazil for nearly 50 years. His reign was marked by great progress. Brazil's population grew from 4 million to 14 million during his rule. He offered land to attract large numbers of Germans, Italians, and other European immigrants to Brazil. Sugar, coffee, and cotton production rose. Brazil's first railroads were built to get these and other products to the coast for export.

In 1850 Brazil stopped importing enslaved people from Africa. In the 1860s, a new movement began to **emancipate**, or free, the enslaved. Pedro II opposed slavery, but he thought it should be ended gradually. An 1871 law granted freedom to all children born to people in slavery. An 1885 law freed enslaved people who were over age 60. Finally, in 1888, all remaining enslaved people were freed.

The Brazilian Republic

Brazil's powerful plantation owners were angered by the loss of their enslaved workers. In 1889 they supported Brazil's army in overthrowing Pedro II. A new government was established, with a constitution based on the Constitution of the United States. Brazil became a republic, a system in which the head of state is an elected ruler instead of a king, a queen, or an emperor. In this republic, the right to vote was limited to wealthy property owners. In 1910, for example, out of a population of 22 million, only 627,000 people could vote.

Most of the power in the early republic was held by the governors of Brazil's southeastern states. Governors were elected by their state's wealthy voters. State governors controlled the election of Brazil's president, who usually came from the highly populated, coffee-rich states of São Paulo and Minas Gerais (General Mines).

These presidents followed economic policies that benefited southeastern Brazil. Coffee became Brazil's main export. By 1902, Brazil was supplying 65 percent of the world's coffee. São Paulo, Minas Gerais, and Rio de Janeiro also became the country's industrial and commercial centers. Over time, some people became unhappy with government policies that continued to favor the coffee growers and other rich Brazilians. In 1930 Getúlio Vargas overthrew the newly elected "coffee president" and seized power. He ruled for the

Eraldo Peres/AP Images

next 15 years. Vargas's reforms made him a hero to most Brazilians. He raised wages, shortened work hours, and let workers form labor unions. Yet for much of his rule, Vargas governed as a dictator. He dissolved the legislature and banned political parties. In 1945 military leaders forced Vargas to resign.

Brazil Under Military Rule

Vargas was elected president again in 1950, but again was forced from office by the military in 1954. For over 30 years, government in Brazil alternated between dictators and elected leaders. Manufacturing thrived throughout this period. Foreign investments brought rapid growth in the steel, auto, and chemical industries.

Industrial growth was accompanied by changes and unrest in Brazilian society. As a result, the military took control of Brazil in 1964, and a series of generals became the heads of government. An elected legislature was allowed, but the army controlled the elections. People who opposed the government were arrested. Many others were frightened into silence. The military gave up power in 1985 and allowed the election of a civilian president.

Modern Brazil

Today Brazil is a democratic republic in which people elect a president and other leaders. In Brazil, voting is **compulsory**. This means that citizens have no choice in deciding whether or not to vote. People from ages 18 to 70 are required by law to vote.

Because Brazil has a high number of well-supported political parties, coalition governments are common. A coalition government is one in which several political parties cooperate to do the work of government. In 2003 a democratically elected president replaced another democratically elected president for the first time in more than 40 years. In 2010 voters elected Dilma Vana Rousseff as the thirty-sixth president of Brazil. She is the first woman president in the country's history.

Include this lesson's information in your Foldable®.

☑ **READING PROGRESS CHECK**

Identifying Central Ideas Why did Brazil's monarchy come to an end?

LESSON 2 REVIEW

Reviewing Vocabulary

1. What kind of agriculture did some *indigenous* farmers practice?

Answering the Guiding Questions

2. *Analyzing* How were the Nambicuara similar to and different from the other main indigenous peoples of early Brazil?

3. *Identifying* Why did African slavery increase in Brazil before it was abolished completely in 1888?

4. *Describing* What were the main steps in Brazil's transition from a colony to a democratic country?

5. *Argument Writing* Take the role of a Brazilian living in 1889. Write a letter to the editor of your local newspaper supporting or opposing the establishment of the republic. Be sure to state the reasons for your opinion.

Reading **HELP**DESK

Academic Vocabulary

• **diverse**
• **unique**

Content Vocabulary

• **hinterland**
• **metropolitan area**
• **central city**
• **favela**

TAKING NOTES: *Key Ideas and Details*

Organize As you read the lesson, use the graphic organizer below to organize information about Brazil by adding one or more facts to each box.

Indiana Academic Standards
6.3.4, 6.3.10

Lesson 3
Life in Brazil

ESSENTIAL QUESTION · *What makes a culture unique?*

IT MATTERS BECAUSE
Brazil's cultures have influenced many people around the world.

People and Places

GUIDING QUESTION *What cultures are represented by Brazilians?*

With some 200 million people, Brazil is the world's fifth-largest country in population. Only China, India, the United States, and Indonesia are home to more people. About half of all South Americans live in Brazil.

Brazil's Diverse Population

Brazil is a mix of several cultures. Many people have a combination of European, African, and native American ancestry. Many are of Portuguese origin or immigrants from Germany and Italy. To a lesser degree, people came from Russia, Poland, and Ukraine. São Paulo, in particular, has a **diverse** population, including a large Japanese community.

Nearly 40 percent of Brazilians have mixed ancestry. This is largely because marriages between people of different ethnic groups have been more acceptable in Brazil than in many other countries. The largest group of multiethnic Brazilians are persons with European and African ancestors. People of European and Native American ancestry are a smaller group.

The smallest multiethnic group is persons of African and Native American descent. About 4 million Africans had been enslaved and brought to Brazil by the 1800s. Many escaped into the **hinterland**, the often remote inland regions, far from the coasts. The Africans lived there with the indigenous Native Americans or formed their own farming communities.

Today, about 80 percent of Brazilians live within 200 miles (322 km) of the Atlantic coast. After slavery ended, many formerly enslaved people left their homes and settled in other agricultural areas or towns. The northeast, however, still has Brazil's highest African and mixed populations. They also form the major population groups in coastal cities and towns north of Rio de Janeiro.

Most Brazilians of European descent live in southern Brazil. Indigenous Native Americans live in all parts of the country. The Amazon rain forest holds the greatest number, but about half of Brazil's Native Americans now live in cities.

Crowded Cities

For most of Brazil's history, the majority of Brazilians lived in rural areas, mainly on plantations, on farms, or in small towns. In the 1950s, millions of people began migrating to cities to take jobs in Brazil's growing industries. By 1970, more Brazilians lived in urban areas than in rural ones. Today, 89 percent of Brazilians live in and around cities.

São Paulo, Brazil's industrial center, is one of the world's largest cities. Some 17 million people live in its **metropolitan area**, or the city and built-up areas around the central city. The **central city** is the largest or most important city in a metropolitan area. São Paulo and Brazil's other large cities look much like cities in the United States. Skyscrapers line busy downtown streets. Cars and trucks jam highways in the mornings and evenings as people travel to and from their jobs. People work in office buildings, shops, and factories. Many own small businesses.

Favelas

Many middle-class urban dwellers live in apartment buildings. Others live in small houses in the suburbs, which are largely residential communities on the outskirts of cities. Wealthy Brazilians live in luxury apartments and mansions.

Most of Brazil's large cities also have shantytowns called **favelas**. Favelas are makeshift communities located on the edges of the cities.

Stuart Dee/Photographer's Choice RF/Getty Images

Academic Vocabulary

diverse differing from one another; varied

Sugarloaf Mountain looms above Rio de Janeiro's Copacabana Beach.
▶ **CRITICAL THINKING**
Explaining What has led to the growth of Brazil's cities since the 1950s?

The Estaiada Bridge, opened in 2008, is one of São Paulo's landmarks. It is known for its curved appearance and X-shaped tower.

▶ **CRITICAL THINKING**

Describing What role does São Paulo play in Brazil's economy?

Favelas arose as millions of poor, rural Brazilians with few skills and little education migrated to cities to seek better lives. These people could not afford houses or apartments. Instead, they settled on land they did not own and built shacks from scraps of wood, sheet metal, cinder blocks, and bricks. Some favelas lack sewers and running water. In many, disease and crime are widespread.

São Paulo and Rio de Janeiro have the most and largest favelas. Rio has about 1,000 of them. About one of every three of the city's residents live in a favela. Rio officials have tried to deal with this problem by offering favela dwellers low-cost housing in the suburbs. Many do not want to move because the long commute from the suburbs to jobs in the city can take hours.

☑ **READING PROGRESS CHECK**

Analyzing Why does Brazil have such a large percentage of people with multiethnic ancestry?

People and Cultures

GUIDING QUESTION *What is it like to live in Brazil?*

Brazilians get along well for a country whose population includes such a variety of racial and ethnic groups. This is largely due to Brazilians' reputation for accepting other people's differences. Personal warmth, good nature, and "getting along" are valued in Brazilian culture. These attitudes and behaviors are an important part of what is known as the "Brazilian Way."

Tensions exist in Brazilian society, but they involve social and economic issues more than ethnic or cultural ones. Ethnicity still plays a factor, though, because Brazilians of European origins have often had better educational opportunities. They hold many of the better jobs as a result.

Ethnic and Language Groups

Until the late 1800s, nearly all European immigrants to Brazil were from Portugal. After slavery ended, large numbers of Italians arrived to work on the coffee plantations.

During the same period, settlers from Germany started farming colonies in southern Brazil. In the early 1900s, the first Japanese arrived to work in agriculture in the Brazilian Highlands. Many of their descendants moved to cities. The first Middle Easterners, mainly Lebanese and Syrians, arrived at about the same time. They became involved in commerce in cities and towns around the country.

The diversity of Brazil's people has given the country a **unique** culture. Portuguese is Brazil's official language. Almost all Brazilians speak it. Brazilian Portuguese is quite different from the language spoken in Portugal. In fact, many Brazilians find it easier to understand films from Spanish-speaking countries in South America than films from Portugal. This is because Brazil's many ethnic groups have introduced new words to the language. Thousands of words and expressions have come from Brazil's indigenous peoples. Dozens of Native American languages are still spoken throughout Brazil.

Religion and the Arts

About two-thirds of Brazilians are Roman Catholics, but only about 20 percent attend services regularly. Women go to church more often than men, and older Brazilians are more active in the Church than the young.

Academic Vocabulary

unique unlike anything else; unusual

Most of the rest of Brazil's population follows the Protestant faith. Those who practice Islam and Eastern religions such as Buddhism are growing in numbers. Many Brazilians blend Christian teachings with beliefs and practices from African religions.

Other African influences on Brazilian culture include foods, popular music, and dance, especially the samba. Brazilians blended samba rhythms with jazz to introduce the world to music called bossa nova. Several Brazilian writers have gained world fame for their books exploring regional and ethnic themes. Brazilian movies and plays also have gained worldwide attention.

Each February, Brazilians celebrate a four-day holiday called Carnival. Millions of working-class and middle-class Brazilians spend much of the year preparing for it by making costumes and building parade floats. Nearly all city neighborhoods are strung with lights. Rio de Janeiro's Carnival is the largest and is world famous. Elaborately costumed Brazilians ride equally elaborate floats in dazzling parades. They are accompanied by thousands of costumed samba dancers moving to the lively music.

Rural Life

Family ties are strong in Brazil. Family members usually live close to one another. They hold frequent reunions or gather at a family farm or ranch on weekends and holidays. Life in rural Brazil has changed little over the years. Most rural families are poor. They work on plantations or ranches or own small farms. They live in one- or two-room houses made of stone or adobe—clay bricks that are dried and hardened in the sun. Their chief foods are beans, cassava, and rice. A stew of black beans, dried beef, and pork is Brazil's national dish.

Urban Life

Many city dwellers are poor, too, and they eat a similar diet. For those who can afford it, U.S. fast-food chains are rapidly expanding in larger Brazilian cities. In general, people in the industrial cities of southern Brazil have a better life than people in the more rural northeast.

Life in Brazil's cities moves at a faster pace. Government services and modern conveniences are available there. Many workers have good jobs and enjoy a decent quality of life. Most middle-class families have cars. Poor families rely on buses to get to work and to the beach or countryside on weekends.

Soccer ("football") is Brazil's most popular sport. It is played nearly everywhere on a daily basis. Matches between professional teams draw huge crowds in major cities. Brazil's national team is recognized as one of the best in the world.

☑ READING PROGRESS CHECK

Describing Describe one element of Brazil's culture. Explain why that element of culture is important to Brazilians.

Buda Mendes/LatinContent WO/Getty Images

Contemporary Brazil

GUIDING QUESTION *What challenges does Brazil face?*

Brazil has the world's seventh-largest economy. It ranks among the leaders in mining, manufacturing, and agriculture. These activities have produced great wealth for some people and a growing middle class. However, only 10 percent of Brazilians receive about half the country's income, while the bottom 40 percent receive only 10 percent of the total income. At the same time, 1 in 10 Brazilians is forced to live on less than $2 a day. About 1 in 5 workers is employed in agriculture, mainly on large farms and ranches owned by corporations or wealthy Brazilians.

Brazil is a member of several organizations designed to promote free trade. MERCOSUR, established in 1991, is South America's leading trading bloc. In 2008 the leaders of 12 South American nations created the Union of South American Nations (UNASUR).

Education and Earning a Living

Education is an important key to success in Brazil. College graduates earn twice as much as high school graduates do, and high school graduates earn four times as much as those with little or no schooling.

Soccer ("football") players scramble for the ball during a match at a Rio de Janeiro stadium.
▶ **CRITICAL THINKING**
Describing How important is football to Brazilians?

Boys read in front of the class at a public school in Brazil's Amazon area.

▶ **CRITICAL THINKING**

Describing How well educated are most Brazilians?

School is free up to age 17. Yet 60 percent of Brazilians have only four years of schooling or less. These people have a hard life. They work long hours for low pay. In 2011 the government launched "Brazil Without Poverty," a program aimed at raising the standard of living and improving access to education and health care.

Seeking to create a skilled workforce, Brazil's government is trying to improve education at all levels. It has increased funds to build better primary and secondary schools. At the university level, Brazil has introduced the "Science Without Borders" program, which aims to send thousands of students to universities abroad, including to colleges in the United States.

Connections and Challenges

Improving citizens' quality of life is just one of the challenges facing Brazil. The government is sponsoring a program to colonize the country's sparsely populated interior. Several highways have been built across the country. The most important is the Transamazonica Highway, from the coastal city of Recife to the border with Peru. To relieve poverty and overcrowding, poor rural Brazilians have been offered free land in the Amazon if they will develop it. Thousands have followed new roads into the Amazon Basin to take advantage of this offer.

Brazilians also have worked to develop the energy resources the country needs for continued economic development. Large power plants along several major rivers use water power to produce most of Brazil's electricity. In the 1970s, the high cost of oil caused the government to develop a program that substitutes ethanol, a fuel made from sugarcane, for gasoline. Recent discoveries of oil and natural gas off Brazil's coast provide the country with the energy it needs.

Environmental Concerns

Programs to develop Brazil's interior have resulted in great concern for the future of the Amazon rain forest. Logging has long been a problem, as trees are cut down to sell as wood. The Transamazonica Highway and other new roads have increased this destruction by making it easier to get into the rain forest and to get the logs out.

The farmers, ranchers, miners, and other settlers the roads have brought into the region have become cause for even greater concern. About 15 percent of the rain forest is already gone, and the rate of its destruction has attracted worldwide attention.

It is easy to think that good soils must lie underneath tropical rain forests. However, this is often not true. The heat and moisture of the area keep the nutrients in the biosphere, that is, in the living organisms, particularly the plants. As a result, the soil is poor. When the forest is cleared for farming, the soil cannot support crops.

✓ READING PROGRESS CHECK

Identifying What are reasons for allowing development in the rain forest?

A highway cuts through Brazil's Amazon rain forest.

▶ CRITICAL THINKING
Explaining How do new roads benefit and harm Brazil's development, especially in rain forest areas?

FOLDABLES®
Study Organizer

Include this lesson's information in your Foldable®.

LESSON 3 REVIEW

Reviewing Vocabulary
1. What is Brazil doing to develop some of its *hinterlands*?

Answering the Guiding Questions
2. *Identifying* In what parts of Brazil do most of its population live?

3. *Determining Central Ideas* How has Brazil's African heritage affected its culture today?

4. *Analyzing* How do education issues contribute to economic inequalities in Brazil?

5. *Argument Writing* Choose one challenge Brazil faces today and write a short essay suggesting how to solve it.

©Paulo Fridman/Sygma/Corbis

Rain Forest Resources

Many medicines that we use today come from plants found in rain forests. From these plants, we derive medicines to treat or cure diabetes, heart conditions, glaucoma, and many other illnesses and physical problems.

Largest Rain Forests The world's largest rain forests are located in the Amazon Basin in South America, the Congo Basin in Africa, and the Indonesian Archipelago in Southeast Asia. The Amazon rain forest makes up more than half of Earth's remaining rain forest.

The Planet's Lungs Rain forests are often called the "lungs of the planet" for their contribution in producing oxygen, which all animals need for survival. Rain forests also provide a home for many people, animals, and plants. Rain forests are an important source of medicine and foods.

> **Every year, less and less of the rain forest remains. Human activity is the main cause of this deforestation.**

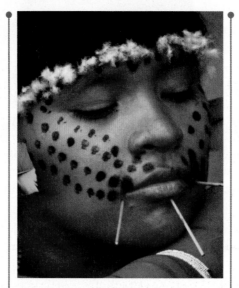

The Yanomami People
An ancient indigenous people, the Yanomami live in the Amazon rain forest regions of Brazil and Venezuela. For many years, the Yanomami lived in isolation. They rely on their environment for their food, shelter, and medicine.

Deforestation Every year, less and less of the rain forest remains. Human activity is the main cause of this deforestation. Humans cut rain forests for grazing land, agriculture, wood, and the land's minerals. Deforestation harms the native peoples who rely on the rain forest. The loss of rain forests also has an extreme impact on the environment because the rich biological diversity of the rain forest is lost as the trees are cut down.

Preserving Rain Forests More and more people realize that keeping the rain forests intact is critical. Groups plant trees on deforested land in the hope that forests will eventually recover. More companies are operating in ways that minimize damage to rain forests.

More Research Thirty years ago, very little research on the medicines of the rain forest was being done. Today, many drug companies and several branches of the U.S. government, including the National Cancer Institute, are taking part in research projects to find medicines and cures for viruses, infections, cancer, and AIDS.

Ashaninka children are at play in the rain ▶ forest. The Ashaninka comprise one of the largest indigenous groups in South America.

(c) Claudia Andujar/Photo Researchers; (bl, tr) Brand X Pictures/PunchStock;

THERE'S MORE ONLINE

HEAR why the rain forest is important • SEE the loss of the rain forest • WATCH plants become medicine

These numbers and statistics can help you learn about the resources of the rain forest.

1.4 Billion Acres

The Amazon rain forest covers 1.4 billion acres (2,187,500 sq km). If the rain forest were a nation, it would be the 13th-largest country in the world.

80%

About 80 percent of the diets of developed nations of the world originated in tropical rain forests. Included are such fruits as oranges and bananas; corn, potatoes, and other vegetables; and nuts and spices.

EIGHTY PERCENT

For centuries, people who live in rain forests have used the plants and trees to meet their health needs. The World Health Organization (WHO) estimates that about 80 percent of the indigenous peoples still rely on traditional medicine.

OVER SEVEN PERCENT

Tropical rain forests make up about 7 percent of the world's total landmass. But found within the rain forest are half of all known varieties of plants.

120

Today, 120 prescription drugs sold worldwide are derived from rain forest plants. About 65 percent of all cancer-fighting medicines also come from rain forest plants. An anticancer drug derived from a special kind of periwinkle plant has greatly increased the survival rate for children with leukemia.

ONE PERCENT

Although ingredients for many medicines come from rain forest plants, less than 1 percent of plants growing in rain forests have been tested by scientists for medicinal purposes.

40 Years

In 1950 rain forests covered about 14 percent of Earth's land. Rain forests cover about 7 percent today. Scientists estimate that, at the present rate, all rain forests could disappear from Earth within 40 years.

50,000 Square Miles

When rain forests are cleared for land, animal and plant life disappears. Almost half of Earth's original tropical forests have been lost. Every year, about 32 million acres—50,000 square miles (129,499 sq. km)—of tropical forest are destroyed. That's roughly the area of Nicaragua or the state of Alabama.

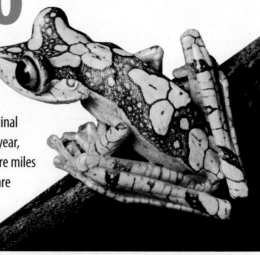

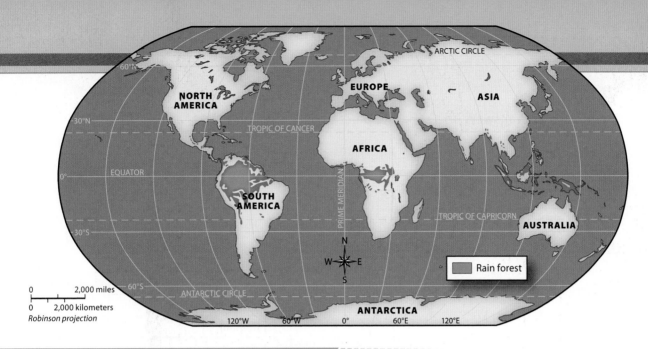

Rain forest

0 2,000 miles
0 2,000 kilometers
Robinson projection

THE WORLD'S RAIN FORESTS Rain forests are located in a belt around Earth near the Equator. Abundant rain, relatively constant temperatures, and strong sunlight year-round are ideal conditions for the plants and animals of the rain forest.

Rain forests cover only a small part of Earth's surface. The Amazon Basin in South America is the world's largest rain forest area.

Rain Forest Research

Laboratories provide a research base for scientists to conduct environmental research. This laboratory in Mumbai attracts rain forest scientists from around the world.

Thinking Like a
Geographer

1. **Environment and Society** Why do you think scientists only know about a small fraction of potential medicines from the rain forest?

2. **Environment and Society** How do you think native doctors in the Amazon rain forest discovered medical uses for plants?

3. **Human Geography** List two reasons to explain why some people support saving rain forests. List two reasons to explain why some people support cutting down rain forests. Write a paragraph to state which position you support. Include facts to support your position.

Directions: Write your answers on a separate piece of paper.

1 Use your **FOLDABLES** to explore the Essential Question.

INFORMATIVE/EXPLANATORY Write an essay explaining how the Brazilians' conversion of rain forest land to farmland may affect the environment of the rest of the world.

2 **21st Century Skills**

IDENTIFYING POINT OF VIEW Given what you have learned about the benefits of rain forests, do you think Brazil has an obligation to maintain what remains of them? Write two or three paragraphs to explain your viewpoint.

3 **Thinking Like a Geographer**

DESCRIBING On a graphic organizer, note important differences between Brazilians who live in the major cities and those who do not. Add other categories you think are important in describing the differences.

	City dwellers	Country dwellers
Wealth		
Housing		
Work		

4 **GEOGRAPHY ACTIVITY**

Locating Places

Match the letters on the map with the numbered places below.

1. Brasília
3. Amazon River
5. Recife
7. São Francisco River

2. Atlantic Ocean
4. São Paulo
6. Pacific Ocean
8. Amazon Basin

Chapter 8 ASSESSMENT

REVIEW THE GUIDING QUESTIONS

Directions: Choose the best answer for each question.

1 Brazil is the world's largest exporter of
A. beef.
B. clocks.
C. peanut butter.
D. tropical plants.

2 Most of Brazil's population lives in
F. the rain forest.
G. the coastal lowlands.
H. the Amazon Basin.
I. northeastern Brazil.

3 The first Portuguese explorer to lay claim to Brazil was
A. Ferdinand Magellan.
B. a Jesuit priest.
C. Getúlio Vargas.
D. Pedro Cabral.

4 When Napoleon invaded Portugal in 1807 and the Portuguese royal family and government leaders fled to Brazil, which city became the new capital of the Portuguese Empire?
F. São Paulo
G. Campinas
H. Rio de Janeiro
I. Brasília

5 Brazil's largest metropolitan area, or city and surrounding suburbs, is
A. Brasília.
B. Buenos Aires.
C. São Paulo.
D. Rio de Janeiro.

6 The official language of Brazil is
F. Spanish.
G. Brazilian.
H. Portuguese.
I. English.

DBQ ANALYZING DOCUMENTS

7 IDENTIFYING POINT OF VIEW Read the following news report:

"*Brazilian farmers meanwhile have been demanding the country's Congress ease environmental laws in the Amazon region. They support a bill that would let them clear half the land on their properties in environmentally sensitive areas. Current law allows farmers to clear just 20 percent of their land in the Amazon zone.*"

—from Marco Sibaja, "Amazon Deforestation in Brazil Increases"

Why do Brazilian farmers want to be able to clear more land?

A. They oppose any environmental protection laws.

B. They want more land for crops to increase their profits.

C. They plan to sell the cleared land for new housing developments.

D. They hope to drive Native Americans from the land.

8 ANALYZING What is likely to happen if the bill the farmers support becomes law?

F. Brazil's economy will suffer from too much emphasis on agriculture.

G. Agriculture in Brazil will decline because the land is unproductive.

H. Deforestation in the Amazon will increase at a faster rate.

I. Environmentalists will stop fighting over deforestation.

SHORT RESPONSE

"*Exploiting vast natural resources and a large labor pool, [Brazil] is today South America's leading economic power. . . . Highly unequal income distribution and crime remain pressing problems.*"

—from *CIA World Factbook*

9 ANALYZING How has Brazil's economy benefited from the nation's large size?

10 IDENTIFYING Why is unequal income distribution in Brazil a problem?

EXTENDED RESPONSE

11 INFORMATIVE/EXPLANATORY WRITING Compare and contrast the cultures of the United States and Brazil, including political, economic, and social factors.

Need Extra Help?

If You've Missed Question	❶	❷	❸	❹	❺	❻	❼	❽	❾	❿	⓫
Review Lesson	1	1	2	2	1	3	3	3	1	3	3

THE TROPICAL NORTH

ESSENTIAL QUESTIONS • *How does geography influence the way people live?*
• *Why does conflict develop?* • *What makes a culture unique?*

©Pablo Corral V/Corbis

Farmer Juan Lucas harvests palms in Ecuador's tropical coastal lowlands.

Lesson 1
Physical Geography of the Region

Lesson 2
History of the Countries

Lesson 3
Life in the Tropical North

The Story Matters...

The countries of the Tropical North are home to some of the most ethnically diverse populations in the world. Native Americans, Europeans, Africans, and Chinese are among those who live in this subregion of South America. It is also home to some of the most diverse environments in the world. Landscapes include jungles, towering mountain ranges, broad river plains, plunging waterfalls, and an archipelago renowned for its unique animal life.

FOLDABLES®
Study Organizer

Go to the Foldables® library in the back of your book to make a Foldable® that will help you take notes while reading this chapter.

○	Trade
○	Foreign Influences and Resources
○	Geography
○	The Tropical North

THE TROPICAL NORTH

Ecuador, Colombia, Venezuela, Guyana, Suriname, and French Guiana are the lands that make up South America's Tropical North.

Step Into the Place

MAP FOCUS Use the map to answer the following questions.

1 **THE GEOGRAPHER'S WORLD** Which country in the Tropical North is connected to Central America?

2 **THE GEOGRAPHER'S WORLD** In which direction would you go if you were traveling from French Guiana to Suriname?

3 **PLACES AND REGIONS** Why do you think the Galápagos Islands belong to Ecuador?

4 **CRITICAL THINKING DESCRIBING** Use the map to help you describe how Guyana and Suriname are similar geographically.

A

RIVER TRAVEL Indigenous peoples, such as the Makushi, live in small villages on the banks of rivers that wind their way through Guyana's rain forests.

B

URBAN CENTER With about 4 million people, Caracas is the capital and largest city of Venezuela.

Step Into the Time

DESCRIBING Choose one event from the time line and write a paragraph describing the social, political, or environmental effect that event had on the region and the world.

1821 Simón Bolívar frees Venezuela from Spanish rule

1667 The Netherlands acquires Suriname from Britain

1800

1835 English naturalist Charles Darwin arrives on Galápagos Islands

The Tropical North

networks
There's More Online!

TROPIC OF CANCER

20°N

ATLANTIC OCEAN

ANTIGUA AND BARBUDA

ST. KITTS AND NEVIS

DOMINICA

Caribbean Sea

ST. LUCIA
BARBADOS
GRENADA
ST. VINCENT AND THE GRENADINES

B

TRINIDAD AND TOBAGO

10°N

GUATEMALA
HONDURAS

EL SALVADOR
NICARAGUA

COSTA RICA

PANAMA

Gulf of Panama

Barranquilla

Maracaibo
Valencia
Caracas
Maracay

Lake Maracaibo

Orinoco R.

Georgetown
Paramaribo
Cayenne

VENEZUELA
GUYANA
SURINAME
FRENCH GUIANA (Fr.)

Atrato R.
Cauca R.
Magdalena R.

Medellín

Bogotá

Cali

COLOMBIA

A

Galápagos Islands (Ecuador)

EQUATOR

Quito

ECUADOR

Guayaquil
Guayas R.

BRAZIL

0°

PACIFIC OCEAN

PERU

✪ National capital
○ Department capital
● City

N
W   E
S

0   400 miles
0   400 kilometers
Lambert Azimuthal Equal-Area projection

10°S

90°W   80°W   70°W

BOLIVIA

1935
Ecuador declares part of Galápagos Islands a wildlife sanctuary

1998
Hugo Chávez is elected president of Venezuela

1966 Guyana gains independence from Britain

1900

1978 UNESCO adds the Galápagos Islands to the World Heritage List

1990s Ecuadoran Indians protest for rights

2000

networks

There's More Online!

☑ **GRAPHIC ORGANIZER**

☑ **MAP** Tropical North

☑ **SLIDE SHOW** Emeralds

☑ **VIDEO**

Reading **HELP**DESK

Academic Vocabulary

- **exceed**
- **despite**

Content Vocabulary

- **elevation**
- **trade winds**
- **cash crop**

TAKING NOTES: *Key Ideas and Details*

Identify As you read the lesson, use a graphic organizer like this one to record the important resources of each of these countries.

Country	Resources
Ecuador	
Colombia	
Venezuela	
Guyana	
Suriname	

Indiana Academic Standards
6.1.18, 6.1.19, 6.3.12, 6.4.7

Lesson 1
Physical Geography of the Region

ESSENTIAL QUESTION • *How does geography influence the way people live?*

IT MATTERS BECAUSE
The land and waters of the Tropical North provide oil, bauxite, and emeralds, along with shrimp and other food products that people and industries in the United States and around the world need or want.

Landforms and Waterways

GUIDING QUESTION *What are the major physical features of the Tropical North?*

South America's Tropical North consists of five countries and a colony. From west to east, they are Ecuador, Colombia, Venezuela, Guyana, Suriname, and French Guiana.

Colombia is the Tropical North's largest country, and Venezuela is the second largest. Each is more than twice the size of California. Ecuador and Guyana, the third and fourth largest, are about the size of Colorado and Kansas, respectively. Suriname is about the size of Washington State; French Guiana, the smallest, is the size of Maine. Together, the countries of the Tropical North total only about one-third the size of nearby Brazil.

Landforms of the Tropical North

Ecuador, Colombia, and Venezuela have the region's most diverse physical geography. The Andes mountain ranges, which extend the length of western South America, run through each country. Some of the peaks have **elevations**, or height above the level of the sea, that **exceed** 18,000 feet (5,486 m)—almost 3.5 miles (5.6 km) high. Many peaks are covered with snow year-round. About 40 peaks are volcanoes.

(l to r) Fabio Filzi/Vetta/Getty Images; Tips Images/Tips Italia Srl a socio unico/Alamy; ©Last Refugee/Robert Harding World Imagery/Corbis

Cotopaxi in Ecuador, at 19,347 feet (5,897 m), is the world's highest active volcano. In Colombia, the Sierra Nevada de Santa Marta mountains along the Caribbean coast are the world's highest coastal range.

Colombia is the only country in South America with coastlines on both the Pacific Ocean and the Caribbean Sea. The mountains make travel between the coasts difficult. So does the Darién, a wilderness region of deep ravines, swamps, and dense rain forest along Colombia's border with Panama.

West of the Andes, Colombia and Ecuador have narrow lowlands that border their Pacific coasts. East of the mountains, more lowlands extend into Peru, Brazil, and Venezuela. The southern half of the lowlands is part of the Amazon Basin. The northern half is a grassy plain called the Llanos. This plain also covers most of northern Venezuela.

Southern Venezuela contains a heavily forested region of rolling hills, low mountains, and plateaus called the Guiana Highlands. Along the border with Brazil, groups of forest-covered mesas called *tepuis* rise to heights of 9,000 feet (2,743 m) in places. The Guiana Highlands extend east into Guyana, Suriname, and French Guiana. Rain forest covers most of this region except for a narrow band of low and sometimes swampy plains along the Atlantic coast.

Abundant Waterways

Rivers flow across much of northern South America. The 1,300-mile-long (2,092 km) Orinoco River is the continent's third-longest river. Its more than 400 tributaries form the north's largest river system. The Orinoco crosses Venezuela in a giant arc, dropping from the Guiana Highlands through the Llanos to the Atlantic Ocean. One of its tributaries flows over Angel Falls, the world's highest waterfall. Angel Falls is more than 20 times higher than Niagara Falls. From the top of a *tepui*, the water plunges more than a half-mile to the fall's base.

Academic Vocabulary

exceed to be greater than; to go beyond a limit

Visual Vocabulary

Mesa A mesa is a small, elevated area of land that has a flat top and sides that are usually steep cliffs.

A shepherd tends alpaca below the western slope of the Cotopaxi volcano in Ecuador.
▶ **CRITICAL THINKING**
Analyzing Why is there snow on Cotopaxi even though the volcano lies close to the Equator?

Academic Vocabulary

despite in spite of

The waters of Venezuela's Angel Falls drop from such a height that they are vaporized by the wind and turn into mist before reaching the ground.

▶ **CRITICAL THINKING**

Identifying Angel Falls is part of what major river system?

The Tropical North region has coastlines on three bodies of ocean water. Ecuador and western Colombia lie along the Pacific Ocean. Northern Colombia and Venezuela lie along the Caribbean Sea. The Atlantic Ocean washes the shores of Guyana, Suriname, and French Guiana.

Colombia's two main rivers, the Magdalena and the Cauca, flow north across Andes plateaus and valleys to the Caribbean Sea. These rivers form important routes into the country's agricultural and industrial interior. Both can be navigated by commercial ships for much of their length.

Other rivers that begin in the Andes flow west to the Pacific. Of these, Ecuador's Guayas River is the most important because it has made Guayaquil the country's largest city and a major port.

Rivers in Guyana, Suriname, and French Guiana flow north and empty into the Atlantic. Most are shallow, slow moving, and responsible for the region's swampy coastline. They are not useful for long-distance transportation into the interior.

Galápagos Islands

The Galápagos Islands lie in the Pacific, about 600 miles (966 km) west of Ecuador. They consist of 13 major islands, six smaller ones, and many tiny islands called islets. These rocky islands, which were formed by underwater volcanoes, are owned by Ecuador. Most have no human population.

The islands' isolation makes them home to many unusual animals, such as lizards that swim and birds with wings although they do not fly. In the 1800s, British scientist Charles Darwin studied the islands' animals to develop his theory of evolution. Today, the islands are tourist attractions. Many are protected as national parks.

☑ **READING PROGRESS CHECK**

Analyzing How do Colombia's rivers help the nation's economy?

Climates

GUIDING QUESTION *How and why do climates vary in the Tropical North?*

South America's Tropical North lies along the Equator. **Despite** its location, the region has a variety of climates. Many of the variations result from differences in elevation and location, and from the influence of ocean currents and winds.

Tropical Climates

The region's coasts, interior lowlands, plains, and highlands all have some type of tropical climate. This means warm temperatures throughout the year.

Much of the coastal and eastern lowlands of Ecuador and Colombia have a tropical monsoon climate, with a short, dry season and a long, wet season of heavy rainfall. In Colombia's coastal Chocó region, which includes the rugged Darién, it rains more than 300 days per year. This produces more than 400 inches (1,016 cm)—about 33 feet (10 m)—of rainfall each year, making it one of the wettest places on Earth.

The Llanos of Colombia and Venezuela have a tropical wet-dry climate, with an annual rainfall of 40 inches to 70 inches (102 cm to 178 cm). Most rain falls between May and October. Average daily temperatures are above 75°F (24°C) throughout the year. The Guiana Highlands have a tropical monsoon climate in some places. In other areas, a tropical rain forest climate (which has no dry season) is normal.

Guyana, Suriname, and French Guiana have the same climate as Venezuela's highlands. Yearly rainfall ranges from 70 inches to 150 inches (178 cm to 381 cm). Their coasts are not as hot as might be expected because of the **trade winds**, steady winds that blow from higher latitudes toward the Equator. The Caribbean coast of Venezuela and Colombia is also cooler. It has a semiarid climate, receiving less than 20 inches (51 cm) of rain per year.

Cooler Highlands

Mountain climates depend on elevation. From 3,000 to 6,500 feet (914 m to 1,981 m) is the *tierra templada,* or "temperate land." This zone has moderate rainfall and temperatures with daily averages between 65°F (18°C) and 75°F (24°C). Next is the *tierra fria,* or "cold land," reaching to about 10,000 feet (3,048 m). A colder zone called the *páramo* begins at about 10,000 feet (3,048 m); daily average temperatures in this zone are below 50°F (10°C). Wind, fog, and light drizzle are common in this zone. Vegetation is mainly grasses and hardy shrubs. Above 15,000 feet (4,572 m), the ground is permanently covered with snow and ice.

✔ **READING PROGRESS CHECK**

Identifying How do the climates of the Pacific coast, the Atlantic coast, and the Caribbean coast differ?

Natural Resources

GUIDING QUESTION *Which natural resources are most important to the economies of the Tropical North's countries?*

Tropical rain forests cover much of the North, but lack of roads and the region's physical geography have made it difficult for any of its countries to exploit this natural resource. The North's largest countries, Venezuela and Colombia, are its richest and most diverse in other resources, as well.

Think Again?

Angel Falls was named after a pilot.

True. The falls are called Salto Ángel in Spanish. They are named for Jimmie Angel, a Missouri-born pilot who was the first person to fly over the falls in a plane in 1933. In 2009, however, Venezuelan president Hugo Chávez declared that the falls should be known as Kerepakpai Merú, which means "waterfall of the deepest place" in the language of the local Pemón people. He believed that Venezuela's most famous natural wonder should have an indigenous name. At the time of Chávez's death in early 2013, the name of the falls remained in dispute.

EMERALD MINING
Emeralds are one of the most valued gems in the world.

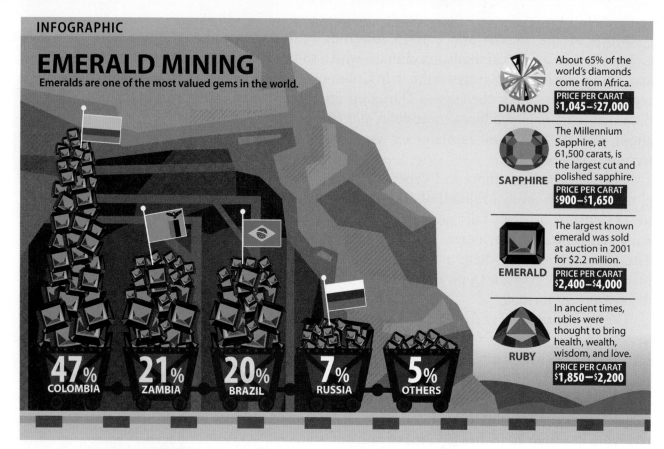

DIAMOND About 65% of the world's diamonds come from Africa.
PRICE PER CARAT $1,045–$27,000

SAPPHIRE The Millennium Sapphire, at 61,500 carats, is the largest cut and polished sapphire.
PRICE PER CARAT $900–$1,650

EMERALD The largest known emerald was sold at auction in 2001 for $2.2 million.
PRICE PER CARAT $2,400–$4,000

RUBY In ancient times, rubies were thought to bring health, wealth, wisdom, and love.
PRICE PER CARAT $1,850–$2,200

47% COLOMBIA
21% ZAMBIA
20% BRAZIL
7% RUSSIA
5% OTHERS

Emeralds were first mined in South America by indigenous peoples centuries ago. Later, the Spanish mined emeralds and shipped them to Europe as part of the valuable treasure from their American empire.

▶ **CRITICAL THINKING**

Analyzing Why do you think emeralds, gold, and diamonds are considered to be valuable?

Fossil Fuels

Oil is found across much of the Tropical North. Venezuela is South America's top producer of oil and ranks eleventh in the world. Some of the world's largest known reserves are in the Llanos, at the mouth of the Orinoco River, and offshore in the Caribbean. Large amounts also exist around Lake Maracaibo, South America's largest lake, along the country's northwestern coast. Venezuela has some of the world's largest natural gas deposits and is South America's second-largest coal producer. Most of the coal lies along the country's southwestern border with Colombia.

Colombia is South America's largest coal producer, with major deposits in its lowlands. It is also South America's third-largest oil producer (Brazil is second), with deposits in the Amazon lowlands, the Llanos, and the Magdalena River valley.

Ecuador produces less oil than Colombia, but it accounts for 40 percent of Ecuador's exports. It is piped over the Andes from oil fields in the east. Suriname and Guyana also have oil resources, but they do not produce enough to even meet their needs.

Minerals and Gems

Gold is found throughout the Tropical North. The largest deposits are in Colombia's mountains, eastern Ecuador, and Venezuela's

Guiana Highlands. In Ecuador, thousands of miners live in remote jungle regions and do dangerous work in tunnels that sometimes collapse in heavy rains.

Diamonds are mined from Colombia to Suriname, but Colombia is better known for high-quality emeralds and is the world's leading emerald producer. Guyana is one of the world's largest producers of bauxite, a mineral used to make aluminum. Venezuela and Suriname also have major bauxite deposits. In addition, the four countries have important deposits of copper, iron ore, and other minerals. Except for gold, Ecuador's mineral resources are limited, and French Guiana has no important mining industries.

Agriculture and Fishing

The differing elevations and climates in Ecuador and Colombia allow farmers to grow a variety of crops. Both countries export bananas from their tropical lowlands and coffee from the *tierra templada*. Ecuador's agriculture, however, is not well developed. The amount of farmland is limited, and most rural Ecuadorans grow only enough to feed their families. Corn, potatoes, beans, and cassava are common crops in both countries. Colombia produces rice, wheat, sugarcane, and cattle for sale, as well as cotton for the country's large textile industry.

Coffee is Venezuela's main **cash crop**, a product raised mainly for sale. Venezuela's main food crops are corn and rice. Most farming takes place in the northwest, and most ranching happens on the Llanos. Only about 10 percent of Venezuelans are farmers or ranchers, and much the same is true of Venezuela's neighbors to the east. Guyana, Suriname, and French Guiana have little farming because much of the land is covered by rain forest. Any farming takes place mainly along the coast.

Fishing is not a major economic activity in the Tropical North, which is unusual for countries that border the sea. The region's people do not eat much fish. The major catch of its small fishing industry is shrimp, most of which is exported.

☑ **READING PROGRESS CHECK**

Identifying Which fossil fuel, mineral, and gem are most widespread in the Tropical North?

Include this lesson's information in your Foldable®.

LESSON 1 REVIEW

Reviewing Vocabulary
1. What *cash crops* are important to the economy of the Tropical North?

Answering the Guiding Questions
2. *Analyzing* Why are Venezuela's Orinoco and Colombia's Magdalena rivers so important?

3. *Identifying* How are climate and elevation related in the Tropical North?

4. *Analyzing* Why is agriculture more important in Colombia than elsewhere in the Tropical North?

5. *Informative/Explanatory Writing* Which of the Tropical North's countries would you most like to visit? Write a paragraph to explain why.

Reading **HELP**DESK

Academic Vocabulary

- **conflict**
- **stable**

Content Vocabulary

- **immunity**
- *encomienda*
- **hacienda**

TAKING NOTES: *Key Ideas and Details*

Analyze As you read the lesson, write summary sentences about five important events in the history of the Tropical North on a graphic organizer like the one below.

Important Events
• Native Americans settle in villages along the region's coast.
•
•
•
•

Indiana Academic Standards
6.1.6

Lesson 2
History of the Countries

ESSENTIAL QUESTION • *Why does conflict develop?*

IT MATTERS BECAUSE
The countries of the Tropical North export products that are sought after and highly valued by the rest of the world.

Early History and Colonization

GUIDING QUESTION *How did Europeans colonize the Tropical North?*

The Tropical North's indigenous peoples lived there for thousands of years before encountering Spanish explorers. These explorers invaded the region in the early 1500s. Less than 50 years later, the Spanish had conquered and colonized most of the region.

Early Peoples of the Tropical North
The Native Americans of the Tropical North included Carib, Arawak, and other hunter-gatherer peoples. They settled in villages along the Caribbean and Atlantic coasts.

To the west, the Cara and other peoples built fishing villages along the Pacific coast. Over time, groups like the Chibcha and Quitu moved inland to mountain valleys in the Andes. There they created advanced societies that farmed, made cloth from cotton and ornaments of gold, and traded with the Inca, an advanced civilization that developed to the south. In the late 1400s, some of the groups were conquered by the Inca and became part of the Inca Empire.

Arrival of the Europeans
In the early 1500s, Spanish adventurers landed on the Caribbean and Atlantic coasts, seeking gold and enslaving native peoples. When they met resistance and found no gold, they lost interest. The first Spanish settlements did not appear

on the Caribbean coast—in Venezuela and Colombia—until 1523 and 1525. The Spanish made no effort to colonize east of Venezuela.

On the Pacific coast, the Spanish conquered the Inca in 1530 and seized their silver and gold. Driven by hunger for more wealth, they invaded Ecuador in 1534. By the mid-1500s, the conquest of the area that is now Ecuador, Colombia, and Venezuela was complete.

Spanish Colonies

To control their new colonies, the Spanish set up governments. Bogotá, which the Spanish founded in 1538, became the capital of Colombia in 1549. The Spanish placed Ecuador's government at the native town of Quito in 1563. Caracas, which the Spanish founded in 1567, eventually became Venezuela's capital. The Spanish located these cities where Native Americans already had settlements. Most were located inland, in the higher elevations where climates are milder than on the tropical coasts. For many years, Venezuela was ruled from Peru. In the 1700s, Spain placed Venezuela, Ecuador, and Colombia under a single government located at Bogotá.

Native American peoples suffered greatly under the Spanish. As in Brazil, thousands died from European diseases to which they had no natural **immunity**, or protection against illness. Others found themselves forced to work for the Spanish under a system called **encomienda**. This system allowed Spanish colonists to demand labor from the Native Americans who lived in a certain area.

The *encomienda* provided workers for Spanish mines and for the large estates, called **haciendas,** that developed in some rural areas. Native Americans in remote regions, such as Venezuela's Llanos and the rain forests of eastern Ecuador and Colombia, came under the control of Roman Catholic missionaries who were trying to convert them to Christianity.

Under Spanish rule, the Native American village of Teusaquillo became known as Bogotá. Life in Spanish Bogotá focused on the main plaza, or square, surrounded by a cathedral and government buildings.

▶ **CRITICAL THINKING**
Describing What role did location play in the selection of Bogotá and other cities as centers of government in northern South America?

Most haciendas became plantations that grew coffee, tobacco, sugarcane, or other cash crops. Others, mostly on the Llanos, were cattle ranches. As the hacienda system grew, the Spanish brought in thousands of enslaved Africans to provide more labor. African slavery was most common in Venezuela.

European Colonization

The French, British, and Dutch fought over and colonized Guyana, Suriname, and French Guiana. The British and the Dutch established sugar plantations and brought the first enslaved Africans to the area. Control of these colonies changed hands several times in the 1600s and 1700s. Eventually, what is now Guyana became British Guiana. Suriname was called Dutch Guiana, and French Guiana became a colony of France.

 **READING PROGRESS CHECK**

Identifying Which European nations founded colonies in the Tropical North, and which countries did each nation colonize?

Simón Bolívar, also known as "the Liberator," led the movement that won freedom for several countries in the Americas.

282 *Chapter 9*

Independence

GUIDING QUESTION *How did Spain's colonies become independent countries?*

By the late 1700s, many Spanish colonists who were born in the Americas wanted independence from their Spanish rulers. Their chance came in 1808, when the French ruler Napoleon invaded and conquered Spain. Spain found it difficult to fight the French in Europe and to rule its colonies. Some of the colonists in the Americas took this opportunity to fight for independence from Spain.

Overthrow of Colonial Rule

Ecuadorans rose up against Spanish rule in 1809. Colombians and Venezuelans soon followed. A long war began, at first mainly between groups who remained loyal to Spain and those who favored independence. After the Spanish expelled the French from Spain in 1814, Spain's king sent troops to South America to try to restore Spanish control. In the south, resistance to the Spanish was led by Argentine general José de San Martín. In the north, Venezuela's Simón Bolívar led the revolt.

Spanish forces were not finally defeated until 1823. In 1819, however, Bolívar united Venezuela, Colombia, Panama, and Ecuador to form an independent republic called Gran Colombia. He became its first president.

Independent Countries

Gran Colombia broke apart after Bolívar's death in 1830. Ecuador and Venezuela formed independent countries. Colombia and Panama remained united as one country. In the early 1900s, Panama separated from Colombia and became independent.

Independence and self-government did not bring democracy and peace. Wealthy landholders competed with wealthy city businesspeople for control of the government. **Conflict** over the Catholic Church's role in society added to the unrest. The tensions resulted in civil wars in Colombia and Venezuela. Throughout the history of Ecuador, Colombia, and Venezuela, military or civilian leaders often ruled as dictators.

Labor and Immigration

While Ecuador, Colombia, and Venezuela struggled with self-government, British, Dutch, and French Guiana remained colonies. The British abolished slavery in their colony in 1838. The French and the Dutch followed in 1863.

To replace the once-enslaved workers, British and Dutch plantation owners recruited laborers from India and China. The Dutch also imported workers from their colony in Indonesia. The immigrants had to work on their colony's sugar, rice, coffee, or cacao plantations for a required length of time. At the end of their contract, they were free. Many stayed in the colony and, like the formerly enslaved people they replaced, founded towns along the coast.

In 1852 France began sending convicted criminals to its colony. More than 70,000 convict laborers arrived between 1852 and 1939. The worst convicts were imprisoned off the coast on notorious Devil's Island.

✓ **READING PROGRESS CHECK**

Determining Central Ideas How did British, Dutch, and French colonists find workers after slavery ended in their colony?

British and Dutch colonial rulers recruited foreign workers from India, China, and other parts of Asia to harvest various tropical crops in their South American colonies. The prison on Devil's Island (above) housed convict laborers.

▶ CRITICAL THINKING

Describing How did the arrival of foreign workers in British Guiana and Dutch Guiana change these territories in a way that made them different from other parts of South America?

Academic Vocabulary

conflict a serious disagreement

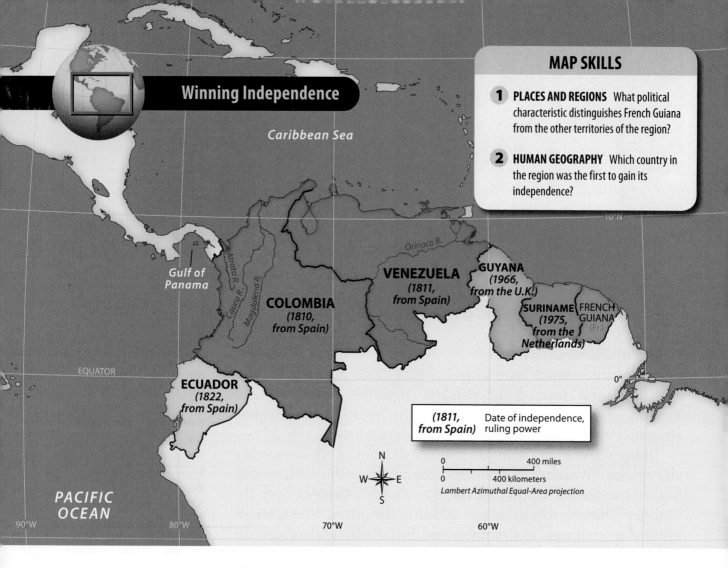

Caribbean Sea

MAP SKILLS

1 **PLACES AND REGIONS** What political characteristic distinguishes French Guiana from the other territories of the region?

2 **HUMAN GEOGRAPHY** Which country in the region was the first to gain its independence?

Orinoco R.

Gulf of Panama

Atrato R.

Cauca R.

Magdalena R.

VENEZUELA
(1811, from Spain)

GUYANA
(1966, from the U.K.)

COLOMBIA
(1810, from Spain)

SURINAME
(1975, from the Netherlands)

FRENCH GUIANA
(Fr.)

EQUATOR

ECUADOR
(1822, from Spain)

0°

| *(1811, from Spain)* | Date of independence, ruling power |

N S E W

0 400 miles
0 400 kilometers
Lambert Azimuthal Equal-Area projection

PACIFIC OCEAN

90°W   80°W   70°W   60°W

10°N

Challenges and Change

GUIDING QUESTION *What challenges do the countries of the Tropical North face?*

The political and social problems that plagued Ecuador, Colombia, and Venezuela after independence continued through most of the twentieth century. Venezuela, for example, did not achieve a peaceful transfer of power between opposing groups until 1969. Meanwhile, the region's other countries, which gained independence in the twentieth century, experienced similar issues and challenges.

Gaining Independence

Independence came slowly for Guyana and Suriname. The British granted their colony limited self-government in 1891. In 1953 all colonists were given the right to vote and allowed to elect a legislature. Guyana finally gained independence in 1966.

Colonists in Dutch Guiana obtained the right to vote in 1948 and self-government in 1953. The colony became the independent country of Suriname in 1975.

The people of French Guiana became French citizens and gained the right to vote in 1848. In 1946 French Guiana's status changed from a colony to an overseas department, or district, of the country of France. French Guiana remains part of France and has representatives in France's national legislature.

Revolutions and Borders

Academic Vocabulary

stable staying in the same condition; not likely to change or fail

The Tropical North's lack of strong, **stable** governments has resulted in major unrest in its countries, as well as conflicts between them. In Colombia, assassinations and other violence between feuding political groups took as many as 200,000 lives between 1946 and 1964. In the 1960s and 1970s, small rebel groups began making attacks throughout the country in hopes of overthrowing the government.

Ecuador's government has not maintained control over its remote region, which lies in the Amazon Basin, to the east of the Andes. In the 1940s, Peru seized some of this land. The two countries often clashed, until a settlement was finally reached in 1968. In 2008 tensions between Ecuador and Colombia were strained after Colombian forces attacked a Colombian rebel camp in Ecuador's territory. In 2010 Colombia accused Venezuela of allowing Colombian rebels to live in its territory. War was narrowly avoided.

Guyana's independence renewed an old border dispute with Venezuela that arose when Guyana was a British colony. The dispute was not settled until 2007. Another dispute arose on Guyana's eastern border after Suriname gained independence in 1975. Several clashes took place before that boundary was settled in 2007. Guyana also experienced years of social and political unrest as its African and South Asian populations competed for power.

Like Guyana, Suriname has faced internal unrest since independence. The military removed civilian leaders in 1980 and again in 1990. Meanwhile, rebel groups of Maroons, the descendants of escaped slaves, disrupted the country's bauxite mining in an effort to overthrow the government. The army responded by killing thousands of Maroon civilians. Thousands more fled to safety in French Guiana.

Include this lesson's information in your Foldable®.

☑ **READING PROGRESS CHECK**

Identifying Which of the region's nations have experienced serious internal unrest since gaining independence?

LESSON 2 REVIEW

Reviewing Vocabulary

1. How were the *encomienda* and the *hacienda* related?

Answering the Guiding Questions

2. *Analyzing* Why were the Spanish more interested in colonizing Ecuador, Colombia, and Venezuela than Guyana, Suriname, and French Guiana?

3. *Identifying* How did conflicts in society lead to independence for Spain's colonies and cause unrest afterward?

4. *Determining Central Ideas* Why do the Tropical North's nations have a history of tense relations and internal unrest?

5. *Argument Writing* Write a short speech calling for or opposing independence for French Guiana. Support your view.

Lesson 3
Life in the Tropical North

(l to r) John Coletti/AWL Images/Getty Images; Kristin Piljay/Lonely Planet Images/age fotostock; Kymri Wilt/DanitaDelimont.com "Danita Delimont Photography"/Newscom

Reading **HELP**DESK

Academic Vocabulary

- **ratio**
- **migrate**

Content Vocabulary

- **Creole**
- **tariff**

TAKING NOTES: *Key Ideas and Details*

Organize As you read the lesson, use a graphic organizer to list information about life in the Tropical North.

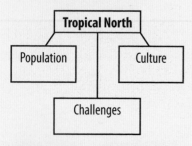

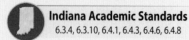

Indiana Academic Standards
6.3.4, 6.3.10, 6.4.1, 6.4.3, 6.4.6, 6.4.8

ESSENTIAL QUESTION • *What makes a culture unique?*

IT MATTERS BECAUSE
Many nations of the world, including the United States, have important trade relations with countries of the Tropical North.

People and Places

GUIDING QUESTION *What ethnic groups populate the Tropical North, and where do they live?*

People of European, African, Native American, and mixed descent are the major population groups of the countries that border the Pacific and Caribbean coasts. African, South Asian, and ethnically mixed peoples form the majority in the Atlantic coast countries.

Population Groups

Ecuador has the Tropical North's greatest indigenous population. About one in four Ecuadorans is Native American. If mestizos, or people of white and Native American descent, are added, the **ratio** becomes 9 of every 10 Ecuadorans.

Venezuela and Colombia have the opposite distribution. Some 20 percent of their populations are white, and 1 to 2 percent are Native American. Colombia's native population is the lowest of any Andean country. Mestizos are the largest group, accounting for more than two-thirds of Colombians and Venezuelans. The African populations of Venezuela, Ecuador, and Colombia are small, although some 15 percent of Colombians have mixed African and European ancestry.

The descendants of contract laborers from India are Suriname's largest group, making up nearly 40 percent of the population. An equal number are people of African and mixed-African descent. A large Indonesian population is also present. Whites and Native Americans total less than 5 percent of Surinam's population.

Neighboring Guyana is home to more Native Americans; this group makes up almost 10 percent of the country's population. Ethnic Africans make up one-third of the population, and East Indians account for more than 40 percent. The country has no significant white population. About one in six Guyanese is of mixed ancestry.

People of mixed descent make up most of French Guiana's population. Small groups of French, Native Americans, Chinese, East Indians, Laotians, Vietnamese, Lebanese, Haitians, and Africans also live in French Guiana.

Where People Live

Guyana's population remains largely rural. Elsewhere in the Tropical North, most people live in cities. Bogotá, Colombia's capital on a high Andes plateau, is home to almost 5 million people. It is the north's largest city and the fifth largest in South America.

Colombia's Caribbean lowlands are home to about 20 percent of its people, mainly in Cartagena and other port cities along the coast.

Academic Vocabulary

ratio the relationship in amount or size between two or more things

The Iglesia de San Francisco, built by Catholic missionaries in about 1560, is the oldest restored church in Bogotá, Colombia.
▶ **CRITICAL THINKING**
Describing How important is the city of Bogotá in the Tropical North region today?

The country's Pacific coast is sparsely settled. Most of the people there are descendants of enslaved Africans who worked on plantations near the Caribbean Sea. As they were freed or they escaped, they migrated into remote areas in western Colombia. The Llanos, where cattle ranching is the main activity, is another area with few people.

Quito, Ecuador, is another mountain capital city, with nearly 2 million people. Most of Ecuador's Native Americans live in or around Quito, or they farm rural mountain valleys nearby. Most other Ecuadorans live along the coast. Guayaquil, the country's largest city and major port, is located there.

Most Venezuelans live along the coast. As in Colombia and Ecuador, Venezuelans began **migrating** to cities in the mid-1900s for the jobs and opportunities they offered. Today, more than 90 percent of the country's people live in Caracas, the capital city of 3 million, and other cities on or near the coast.

The countries of Guyana, Suriname, and the territory of French Guiana are sparsely populated. The population of the three combined totals only about half the population of Caracas. The interior of French Guiana has few roads and is largely uninhabited. In Suriname, small groups of Native Americans live in the Guiana Highlands. Nearly everyone else lives along the coast. Suriname's capital, Paramaribo, a city of 260,000, is home to more than half the country's population.

Most Guyanese also live on the coast, mainly in small farm towns. Each town's farmlands extend inland for several miles. The country's interior is home to a few groups of Native Americans and scattered mining and ranching settlements.

☑ **READING PROGRESS CHECK**

Determining Central Ideas Where do the greatest number of people in the Tropical North live?

Academic Vocabulary

migrate to move from one place to another

People and Cultures

GUIDING QUESTION *What is the Tropical North's culture like?*

Despite its largely Spanish heritage, no one culture unifies the Tropical North. Instead, its culture can be defined by the wide variety of ethnic groups that populate the region.

Language Groups

Spanish is the official language of Ecuador, Colombia, and Venezuela. There are differences in Ecuadoran Spanish because of the influence of Native American languages in each region of the country. More than 10 native languages are spoken in Ecuador. More than 25 native languages are spoken in Venezuela and some 180 in Colombia. Colombians, however, have taken great care to preserve the purity of the Spanish language.

Languages in Guyana, Suriname, and French Guiana reflect their colonial heritage as well as their ethnic populations. **Creole**, a group of languages that enslaved people from various parts of Africa developed to communicate on colonial plantations, is widely spoken. Most people in Guyana speak English. In Suriname, the official language, Dutch, is spoken only as a second language. Native American languages, Hindi, and other South Asian languages are heard in both countries.

Whether they come from the region's rural or urban areas, many people enjoy a celebration called Carnival. This festival is celebrated just before the beginning of Lent, the Christian holy season that comes before Easter.

Religion, Daily Life, and the Arts

Ecuador, Colombia, and Venezuela are overwhelmingly Roman Catholic. No more than 10 percent of the people in these countries practice other religions. The religions practiced in Guyana, French Guiana, and Suriname reflect the variety of ethnic groups that live there. Suriname's population is made up of about equal numbers of Roman Catholics, Protestants, Hindus, and Muslims. Guyana's population is largely Protestant and Hindu, with sizable Catholic and Muslim minorities. In all countries, some Native Americans practice indigenous religions.

Each country's foods, music, and other cultural elements reflect its ethnic and religious makeup. Venezuela, Colombia, and Ecuador celebrate Carnival, though the festivities are not as colorful or as lively as those in Brazil. Regional religious festivals are celebrated in many Andes communities.

A girl in the Andes of Ecuador plays a pan flute, one of the most popular instruments of Andean music.

▶ **CRITICAL THINKING**

Describing In addition to music, what other cultural traditions are practiced by people in the region?

Culture often differs by geographic area. Native Americans in mountain regions weave baskets and cloth using designs that are hundreds of years old. They play Andean music using drums, flutes, and other traditional instruments. Along the coast of Colombia and Venezuela, a dance called the *cumbia* blends the region's Spanish and African heritages. Other Venezuelan coastal music and dance, such as salsa and merengue, show Caribbean island influences. Maracas and guitars are used to perform the music of the Llano.

☑ **READING PROGRESS CHECK**

Analyzing What language and religion are most common in the region?

Ongoing Issues

GUIDING QUESTION *What challenges do the countries of the Tropical North face?*

Many people who live in the Tropical North are poor, although natural resources in the region are plentiful. For generations, those resources have mostly benefited only a wealthy few. This situation has created tensions within and between countries.

Trade Relations

Many South American leaders believe that one way to strengthen their countries' economies is to expand trade. In 2008 the countries in the Tropical North joined with the rest of South America to form the Union of South American Nations (UNASUR). One of the organization's goals is ending **tariffs**—taxes on imported goods—on trade between member nations. Another goal is adopting a uniform currency, similar to the euro.

A Northern Neighbor

Another challenge is improving the region's relationship with the United States. The relationship has sometimes been rocky in the past, as when the United States helped Panama gain independence from Colombia in the early 1900s. Relations between the United States and Colombia have improved greatly. The United States and the Colombian government are working together to stop the flow of illegal drugs.

Challenges in Venezuela

In 1998 Venezuelans elected Hugo Chávez, a former military leader, as president. Chávez frequently criticized the United States and became friendly with anti-U.S. governments in Cuba and Iran.

After his election, Chávez promised to use Venezuela's oil income to improve conditions for the country's poor. Among other actions that angered U.S. leaders, in 2009 he seized control of U.S. companies that were developing oil resources in Venezuela. His strong rule split Venezuela into opposing groups. Working-class people supported Chávez, but middle-class and wealthy Venezuelans opposed his policies.

Struggles in Colombia and Ecuador

Colombia has undergone a long and bitter struggle between the country's government and a Colombian organization called the Revolutionary Armed Forces of Colombia (FARC). One of FARC's goals is to curtail the role of foreign governments and businesses in Colombia's affairs. Another goal is to provide help and support for the nation's poor farmers. FARC is funded through various means, including the production and sale of illegal drugs.

In Ecuador, indigenous peoples protested for rights and blamed President Rafael Correa for not keeping his promises. Correa had promised to rewrite Ecuador's constitution. Among other things, he pledged to extend the rights of the people. Disappointed when Correa did not act, indigenous peoples organized to win rights for access to land, basic services, and political representation.

☑ **READING PROGRESS CHECK**

Determining Central Ideas How did Hugo Chávez increase tensions between Venezuela and the United States?

Include this lesson's information in your Foldable®.

LESSON 3 REVIEW

Reviewing Vocabulary
1. Why did enslaved Africans create *Creole*?

Answering the Guiding Questions
2. *Identifying* Why is the Tropical North home to so many ethnic groups?

3. *Analyzing* Why are there Hindu and Muslim populations in northern South America?

4. *Analyzing* How and why is UNASUR likely to affect the economies and people of the Tropical North's countries?

5. *Informative/Explanatory Writing* Choose one of the challenges the Tropical North faces, and write a short essay suggesting how to solve it.

Directions: Write your answers on a separate piece of paper.

1 Use your **FOLDABLES** to explore the Essential Questions.

INFORMATIVE/EXPLANATORY WRITING Choose one of the region's countries or colonies. Compare the physical and population maps found at the beginning of the unit. Then write at least two paragraphs explaining how the physical geography affects where people live and work.

2 **21st Century Skills**

ANALYZING Working in small groups, choose one of the countries or the colony found in the region and research the most common occupations practiced by its people. Are any of those jobs unique to that country and its culture? Present your findings to the class in a slideshow or a poster.

3 **Thinking Like a Geographer**

IDENTIFYING Create a two-column chart. List the name of the Tropical North country or colony in the first column. List the primary languages spoken in the second column.

4 **GEOGRAPHY ACTIVITY**

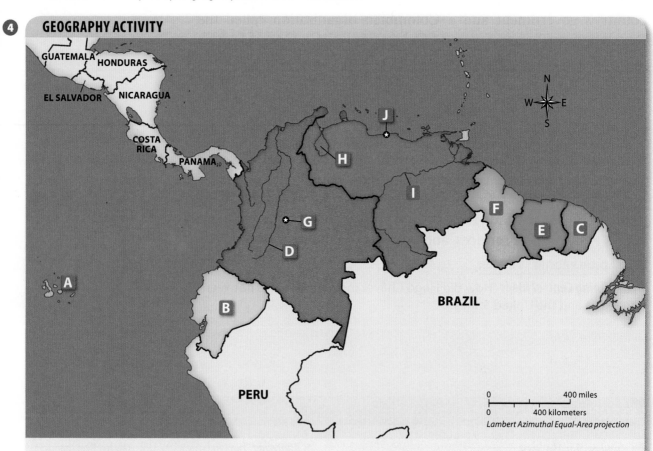

Locating Places

Match the letters on the map with the numbered places listed below.

1. Lake Maracaibo
2. Ecuador
3. French Guiana
4. Caracas
5. Orinoco River
6. Guyana
7. Bogotá
8. Galápagos Islands
9. Suriname
10. Magdalena River

REVIEW THE GUIDING QUESTIONS

Directions: Choose the best answer for each question.

1 The Galápagos Islands were the site of
   A.   Christopher Columbus's second landing in the Americas.
   B.   an outpost of the Incas.
   C.   Charles Darwin's study that resulted in the theory of evolution.
   D.   Ecuador's largest volcanic eruption.

2 The world's leading producer of emeralds is
   F.   Venezuela.
   G.   Colombia.
   H.   Ecuador.
   I.   French Guiana.

3 South America's native populations were forced into laboring for the Spanish under a system called
   A.   *encomienda.*
   B.   immunity.
   C.   hacienda.
   D.   Quito.

4 The transition from colonial governments to independence in the Tropical North of South America
   F.   happened suddenly in 1550.
   G.   took place slowly over a period of more than 100 years.
   H.   began in Brazil.
   I.   was a result of the War of 1812.

5 The most populous spot in the Tropical North is
   A.   Ecuador.
   B.   Bogotá.
   C.   the Llanos.
   D.   Cartagena.

6 Of the languages used in the Tropical North countries, Spanish is used in its purest form in
   F.   Ecuador.
   G.   French Guiana.
   H.   Venezuela.
   I.   Colombia.

DBQ ANALYZING DOCUMENTS

7 **CITING TEXT EVIDENCE**   Read the following passage:

"*With one of the highest deforestation rates in Latin America, Ecuador is losing 200,000 hectares (494,211 acres) of forest every year . . . Although most of the forests of the country are public lands, an important percentage of what is left is in the hands of indigenous people and farmers, among the country's poorest citizens.*"

—from Steve Goldstein, "A Grand Plan: Ecuador and 'Forest Partners'" (2008)

What group in Ecuador controls the forests on public lands?

A.   indigenous people

B.   farmers

C.   the government

D.   businesses

8 **ANALYZING**   Which most likely explains why the indigenous people and the farmers might be willing to sell their land?

F.   to enjoy a better way of life

G.   so they can buy more productive land

H.   so they can move to the city

I.   to rid themselves of the burden of the land

SHORT RESPONSE

"*Latin American pop superstar Shakira will be at this weekend's gathering of the Western Hemisphere's leaders advocating for [promoting] her favorite issues: early childhood development and universal education.*"

—from Gregory M. Lamb, "Shakira Advocates for Children at the Summit of the Americas" (2012)

9 **IDENTIFYING POINT OF VIEW**   How would improving child development and education benefit the countries of the region?

10 **ANALYZING**   What step do you think nations in this region should take to improve child development? Why?

EXTENDED RESPONSE

11 **INFORMATIVE/EXPLANATORY WRITING**   What is school like for children who live in the Tropical North of South America? Research and then describe the educational system in this part of the world in a detailed report. Find out how many months of the year school is in session, how long the typical day is, and what kinds of subjects are taught. You might also want to find out how education is funded. Then compare the results with your school experience.

Need Extra Help?

If You've Missed Question	**1**	**2**	**3**	**4**	**5**	**6**	**7**	**8**	**9**	**10**	**11**
Review Lesson	1	1	2	2	3	3	1	3	3	3	3

ANDES AND MIDLATITUDE COUNTRIES

ESSENTIAL QUESTIONS · *How does geography influence the way people live?* · *Why do civilizations rise and fall?* · *What makes a culture unique?*

©iStockphoto.com/hadynyah

Woman from the town of Tarabuco in south central Bolivia

Lesson 1
Physical Geography of the Region

Lesson 2
History of the Region

Lesson 3
Life in the Region

The Story Matters...

Running the length of the Pacific coast of South America, the Andes define the countries of Peru, Bolivia, and Chile. For the people who have made the Andes their home, the grandeur of the high mountain peaks often contrasts sharply with the challenges of life in such a rugged location. And yet, it was in this very location of mountains, high plateaus, plains, and deserts that the ancient Inca built their powerful and highly developed civilization.

FOLDABLES®
Study Organizer

Go to the Foldables® library in the back of your book to make a Foldable® that will help you take notes while reading this chapter.

Geography	Culture	Economy

ANDES AND MIDLATITUDE COUNTRIES

The world's longest mountain system runs parallel to the Pacific coast of South America. The Andes stretch about 4,500 miles (7,242 km) and include many high mountain peaks. As you study the map, look for other geographic features that make this area unique.

Step Into the Place

MAP FOCUS Use the map to answer the following questions.

1 **PLACES AND REGIONS**
Which country has two capitals? Name them.

2 **THE GEOGRAPHER'S WORLD** Which two bodies of water does the Strait of Magellan connect?

3 **THE GEOGRAPHER'S WORLD** What is Uruguay's capital city?

4 **CRITICAL THINKING**
ANALYZING Why do you think Chile has such a unique shape?

A

RUGGED TERRAIN Gigantic cacti dot the hilly landscape on Bolivia's Incahuasi Island. The island is in the middle of the world's largest salt flats.

B

HANDICRAFTS The people of the small island of Taquile on Lake Titicaca are known for making some of Peru's highest-quality handwoven clothing.

Step Into the Time

TIME LINE Which event on the time line discusses natural resources? Write a paragraph explaining positive and negative effects on a country's colony that holds valuable resources.

1533 Spanish conquer Inca Empire

1811 Paraguay gains independence from Spain

1800

1545 Silver is discovered in Potosí, Bolivia

1808 Rebellion against Spanish rule grows

Andes and Midlatitude Countries

EQUATOR

0°

ECUADOR

PERU

BRAZIL

Lima

Cuzco

Ucayali R.

Beni R.

Mamoré R.

Lake Titicaca

La Paz

BOLIVIA

Sucre

A

B

PARAGUAY

Paraná R.

N
W — E
S

Asunción

Paraguay R.

TROPIC OF CAPRICORN

20°S

CHILE

ARGENTINA

Elqui R.

PACIFIC OCEAN

ATLANTIC OCEAN

Santiago

Uruguay R.

URUGUAY

Buenos Aires

Montevideo

Río de la Plata

Bío-Bío R.

Colorado R.

40°S

0   1,000 miles
0   1,000 kilometers
Albers Equal-Area Conic projection

✪	National capital
○	Territorial capital
●	City

Falkland Islands (U.K.)
Stanley

South Georgia Island (U.K.)

Strait of Magellan

2006
Michelle Bachelet elected first woman president of Chile

1946 Juan Perón becomes president of Argentina

2007 Earthquake in southwest Peru leaves 200,000 people homeless

1900

1900s Foreign companies run mining operations in the region

2000

2009 Bolivia's new constitution empowers indigenous peoples

There's More Online!

☑ **SLIDE SHOW** Landforms: Andean Region

☑ **IMAGES** Uses of Wool

☑ **MAP** Comparing Mountain Ranges

☑ **VIDEO**

Reading **HELP**DESK

Academic Vocabulary

- **isolate**

Content Vocabulary

- **cordillera**
- **altiplano**
- **pampas**
- **estuary**
- **altitude**

TAKING NOTES: *Key Ideas and Details*

Integrate Visual Information
As you read, use a graphic organizer like this one to identify significant physical features of the region.

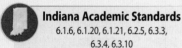

Indiana Academic Standards
6.1.6, 6.1.20, 6.1.21, 6.2.5, 6.3.3, 6.3.4, 6.3.10

Lesson 1
Physical Geography of the Region

ESSENTIAL QUESTION · *How does geography influence the way people live?*

IT MATTERS BECAUSE
Much of the terrain of the southern and western part of South America is extremely rugged. The geography of the area presents unique challenges to the people who live there.

Andes Countries

GUIDING QUESTION *What are the physical features of the Andean region?*

Three countries make up the bulk of the Andean region in South America. They are Peru, Bolivia, and Chile. From north to south, these countries span from the Equator to the southern tip of the continent of South America. The physical landscape includes towering mountains, sweeping plains, and significant waterways.

(l to r) Duane Miller/Flickr/Flickr/Getty Images; Steve Allen Travel Photography/ Alamy; ©VICTOR ROJAS/dea/Corbis; Imágenes del Perú/Flickr/Getty Images;

The Andes

On a map of South America, one of the first features you might notice is the system of mountain ranges running parallel to the continent's Pacific coast. These are the Andes, the longest continuous group of mountain ranges in the world and the tallest in the Western Hemisphere. The Andes include high plateaus and high plains, with even higher mountain peaks rising above them. The entire series of the Andes range stretches for 4,500 miles (7,242 km).

The peaks that make up the Andes are not arranged in one neat line. Instead, they form a series of parallel mountain ranges. The parallel ranges are called **cordilleras**. The rugged terrain of the cordilleras makes travel difficult. These ranges **isolated** human settlements from one another for centuries.

In Peru and Bolivia, the two main branches of the Andes border a high plain called the **altiplano**. In fact, *altiplano* means "high plain" in Spanish. About the size of Kentucky, the altiplano has an elevation of 11,200 feet to 12,800 feet (3,414 m to 3,901 m) above sea level.

The Andes mountain ranges are the result of collisions between tectonic plates. This kind of geologic activity comes as no surprise. After all, the Andes are part of the Ring of Fire. All around the rim of the Pacific Ocean, plates are colliding, separating, or sliding past each other. Those forces make earthquakes and volcanic eruptions a part of life throughout much of the Andes.

Plains and Deserts

The Andes run parallel to the Pacific coast but lie 100 miles to 150 miles (161 km to 241 km) inland from the coast. The land between the Andes and the coast averages more than 3,500 feet (1,067 m) above sea level. In most places, the land rises steeply from the ocean. The area has tall cliffs and almost no areas of coastal plain. In Peru and northern Chile, the area between the Pacific and the Andes is a coastal desert. On the Atlantic side of South America, broad plateaus and valleys spread across Uruguay and eastern Argentina.

Academic Vocabulary

isolate to separate

Spectacular mountains are part of Torres del Paine National Park in southern Chile.

▶ **CRITICAL THINKING**
Describing How were the Andes mountain ranges formed?

The Illimani mountains tower over a small village in the plains of western Bolivia.

▶ CRITICAL THINKING

Describing What are the main economic activities in the area?

This plain is called the **pampas**. Its thick, fertile soils come from sediments that have eroded from the Andes. The pampas, like North America's Great Plains, provide land for growing wheat and corn and for grazing cattle.

Coastal Peru and Chile and most of southern Argentina have deserts. Wind patterns, the cold Peru Current, and high elevations are the causes of the low precipitation. The Atacama Desert in Peru and northern Chile is so arid that in some places no rainfall has ever been recorded. The Patagonia Desert in Argentina lies in the rain shadow of the Andes.

Waterways

The Paraná, the Paraguay, and the Uruguay rivers combine to create the second-largest river system in South America, after the Amazon. This river system drains much of the eastern half of South America. The system is especially important to Paraguay, because Paraguay is a landlocked country. The river system provides transportation routes and makes possible the production of hydroelectric power. Along the Paraguay River is the Pantanal, one of the world's largest wetlands. This area produces a diverse ecosystem of plants and animals.

Steve Allen Travel Photography/Alamy

The Paraná-Paraguay-Uruguay river system flows into the Río de la Plata (Spanish for "river of silver"). This river then empties into the Atlantic Ocean on the border of Argentina and Uruguay. The Río de la Plata meets the ocean in a broad estuary. An **estuary** is an area where the ocean tide meets a river current.

South America has few large lakes. The largest lake in the Andean region is Lake Titicaca. It lies on the border between Bolivia and Peru. Lake Titicaca is on the altiplano at 12,500 feet (3,810 m) above sea level. It is the world's highest lake that is large enough and deep enough to be used by small ships.

☑ **READING PROGRESS CHECK**

Analyzing How do you think the geography of the Andean region affects the lives of the people who live there?

Climate Diversity

GUIDING QUESTION *How does climate affect life in the Andean region?*

Climate is part of a region's physical geography. The varying mountains, plains, and other landforms in the Andean region and midlatitude countries of South America mean that the region's climate is extremely diverse.

The Effect of Altitude

The main factor that determines climate in the Andes is **altitude**, or height above sea level. The higher the altitude, the cooler the temperatures are. This is true even in the warm tropics. Conditions in the region can range from hot and humid at lower elevations to freezing in the mountain peaks.

Farming is a challenge in the rugged Andean region. Farmers have successfully terraced the hillsides to grow crops such as potatoes, barley, and wheat.

Visitors to the Andes may find the altitude at the higher elevations hard to handle. Oxygen is thin. This results in heavy breathing and tiring easily. The region's inhabitants are adapted to the thinner air, as are various native species of plants and animals.

Think **Again**

The Atacama Desert is unpopulated.

Not true. More than a million people live in this region. They live in mining towns, fishing villages, and coastal cities. Farmers use irrigation to grow olives and tomatoes. The Atacama is also a favorite place for teams of astronomers. They take advantage of the area's crystal-clear night skies to probe the secrets of the universe.

Midlatitude Variety

Climates of the midlatitude countries of South America are quite different from the Andean region. These countries enjoy a generally temperate, or moderate, climate. In Uruguay, for example, the average daily temperature in the middle of winter are a mild 50°F to 54°F (10°C to 12°C). In mid-summer, the average daily temperatures reach a comfortable 72°F to 79°F (22°C to 26°C). There is no wet or dry season—rainfall occurs throughout the year. Inland areas, however, are drier than the coast.

Argentina is much larger than Uruguay and includes a greater variety of landforms. As you might expect, Argentina's climate is extremely diverse. It varies from subtropical in the north to tundra in the far south. Northern Argentina has hot, humid summers. Southern Argentina has warm summers and cold winters with heavy snowfall, especially in the mountains.

Paraguay presents yet a different climate. Paraguay is a landlocked country. The climate is generally temperate or subtropical. Strong winds often sweep the pampas in Paraguay because the country lacks mountain ranges to serve as wind barriers.

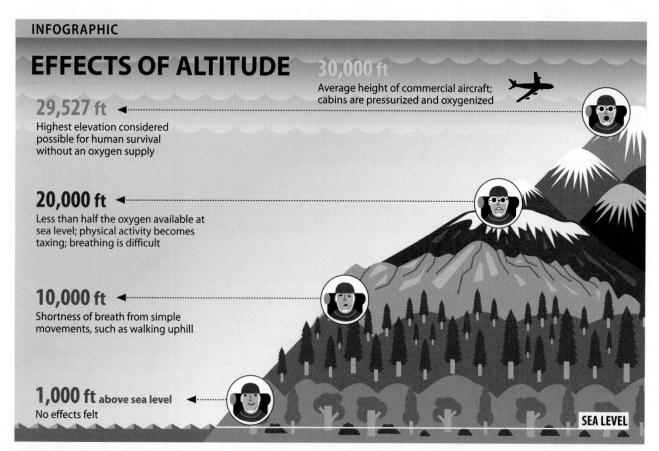

INFOGRAPHIC

EFFECTS OF ALTITUDE

30,000 ft
Average height of commercial aircraft; cabins are pressurized and oxygenized

29,527 ft
Highest elevation considered possible for human survival without an oxygen supply

20,000 ft
Less than half the oxygen available at sea level; physical activity becomes taxing; breathing is difficult

10,000 ft
Shortness of breath from simple movements, such as walking uphill

1,000 ft above sea level
No effects felt

SEA LEVEL

At all latitudes, altitude, or the height above sea level, influences climate. Earth's atmosphere thins as altitude increases. Thinner air retains less heat.

► CRITICAL THINKING
Describing What happens to temperatures as altitude increases? How do higher elevations affect humans?

Climate Extremes

Extremes of climate can be experienced in the Andean countries without changing latitude. Altitude is all that has to change. The climate changes tremendously in the Andes from the lower to the higher elevations. The *tierra caliente,* or "hot land," is the land near sea level. The hot and humid conditions do not change much from month to month. In this zone, farmers grow bananas, sugarcane, cacao, rice, and other tropical crops.

From 3,000 feet to 6,000 feet (914 m to 1,829 m), the air is pleasantly cool. Abundant rainfall helps forests and a great variety of crops grow. This zone is the *tierra templada,* or "temperate land." It is the most densely populated area. Here, farmers grow a variety of crops, such as corn, coffee, cotton, wheat, and citrus fruits.

Higher up, the climate changes. This is the *tierra fría,* or "cold land." It extends from 6,000 feet to 10,000 feet (1,829 m to 3,048 m). The landscape is a combination of forests and grassy areas. Farmers here grow crops that thrive in cooler temperatures, including potatoes, barley, and wheat.

The land at the highest altitude is the *tierra helada,* or "frozen land." Here, above 10,000 feet (3,048 m), conditions can be harsh. The winds blow cold and icy, and temperatures fall well below freezing. Vegetation is sparse, and few people live in this zone.

Mining (left), especially copper mining, is an important part of the Chilean economy. Mining can be dangerous. Miners celebrate their rescue on October 12, 2010 (right), after being trapped underground for 69 days.

Identifying Besides minerals and metals, what natural resources are abundant in the Andean and midlatitude countries?

(l to r) ©VICTOR ROJAS/dpa/Corbis; ©CEZARO DE LUCA/epa/Corbis

Llamas are camelids, relatives of camels. They are bred and raised for food, wool, and as pack animals for pulling carts.

Identifying What other camelids are common in the region?

El Niño and La Niña

Every few years, changes in wind patterns and ocean currents in the Pacific Ocean cause unusual and extreme weather in some places in South America. One of these events is called El Niño. During an El Niño, the climate along the Pacific coast of South America becomes much warmer and wetter than normal. Floods occur in some places, especially along the coast of Peru.

El Niños form when cold winds from the east are weak. Without these cold winds, the central Pacific Ocean grows warmer than usual. More water evaporates, and more clouds form. The thick band of clouds changes wind and rain patterns. Some areas receive heavier-than-normal rains. Other areas, however, have less-than-normal rainfall.

Scientists have found that El Niños occur about every three years. They also found that in some years, the opposite kind of unusual weather takes place. This event is called La Niña. Winds from the east become strong, cooling more of the Pacific. When this happens, heavy clouds form in the western Pacific.

✔ **READING PROGRESS CHECK**

Identifying Why is the *tierra templada* the most populated climate zone by altitude in the Andean region?

Imágenes del Perú/Flickr/Getty Images

Natural Resources

GUIDING QUESTION *Which natural resources are important to the region?*

The Andean and midlatitude countries are rich in natural resources. Energy sources are especially important. Bolivia holds the second-largest reserves of natural gas in South America, trailing behind only Venezuela. Bolivia also has extensive deposits of petroleum. Paraguay's hydroelectric power plants produce nearly all the country's electricity. The governments want to use these resources to develop and strengthen their economies.

Minerals and Metals

Besides energy resources, the region has a number of mineral resources. Most of the area's mines are in the Andes. Chile leads the world in exports of copper. Tin production is important to the Bolivian economy. Bolivia and Peru have deposits of silver, lead, and zinc, and Peru also has gold.

Wildlife

The region's varied geography and climate support a variety of wildlife, including many species of birds and butterflies. The ability of plants and animals to thrive in the region varies with altitude.

A group of mammals called camelids is especially important in this region. Camelids are relatives of camels, but they do not have the typical humps of camels. Two kinds of camelids are the llama and the alpaca. The llama is the larger of the two. Llamas serve as pack animals and are a source of food, wool, and hides. Native Americans throughout the Andes tend herds of llamas. These animals are used to carry goods or pull carts. They are also raised for food, hides, and wool.

Alpacas are found only in certain parts of Peru and Bolivia. The animal's thick, shaggy coat is an important source of wool. Alpaca wool is strong yet soft and repels water. It is used for all kinds of clothing and as insulation in sleeping bags.

Include this lesson's information in your Foldable®.

✔ **READING PROGRESS CHECK**

Analyzing What metal is important to Chile's economy?

LESSON 1 REVIEW

Reviewing Vocabulary
1. Why is *altitude* an important feature in the Andean region?

Answering the Guiding Questions
2. ***Determining Central Ideas*** Why do earthquakes and volcanoes occur in the Andes?

3. ***Analyzing*** Why is the climate wet on the western slopes of the Andes in southern Argentina but dry on the eastern slopes?

4. ***Describing*** How does the climate of the Andes countries compare with that of the midlatitude countries?

5. ***Identifying*** Give a specific example of how a family living in the Andes might use llamas.

6. ***Narrative Writing*** You are living with a relative for a month somewhere in the Andes. Choose the country and the area where you are staying. Then write a letter to a friend describing where you are and what you did yesterday.

Reading **HELP**DESK

Academic Vocabulary

- **hierarchy**

Content Vocabulary

- **smallpox**
- **guerrilla**
- **multinational**
- **coup**

TAKING NOTES: *Key Ideas and Details*

Describe As you read this lesson, use a graphic organizer like this one to write an important fact about each topic.

Topic	Fact
Inca Empire	
Spanish rule	
Independence movements	

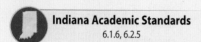

Indiana Academic Standards
6.1.6, 6.2.5

Lesson 2
History of the Region

ESSENTIAL QUESTION · *Why do civilizations rise and fall?*

IT MATTERS BECAUSE
In the Andean and midlatitude countries of South America, history and government have developed in very different ways.

Early History and Conquest

GUIDING QUESTION *How has history influenced the region?*

Native Americans and European colonizers made major contributions to the history of this region of South America. The actions and achievements of both groups continue to influence life today. Almost all the countries in the region have been independent for nearly two centuries. Still, their history continues to influence their culture.

Rise and Fall of the Inca Empire

Before the rise of the Inca in the 1100s, the Andean region was the home of small Native American societies such as the Moche, the Mapuche, and the Aymara. These societies were based primarily on agriculture. The Moche settled on the arid coastline of northern Peru. Archaeological finds show that they were talented at engineering and irrigation. They used a complex irrigation system to grow corn (maize), beans, and other crops.

The Inca developed a highly sophisticated civilization. They first settled in the Cuzco Valley in what is now Peru. It was not until their fifth emperor, Capac Yupanqui, that they began to expand outside the valley. In the 1400s, under the rule of Pachacuti Inca Yupanqui, the Inca made extensive conquests. By the early 1500s, the Inca ruled a region stretching from northern Ecuador through Peru and then southward into Chile. Historians estimate that the area was home to 12 million people. This population included dozens of separate cultural groups, who spoke many different languages.

The Inca state was called Tawantinsuyu. The name means "the land of the four quarters." The imperial capital was located where the four quarters, or provinces, of the Inca Empire met, at Cuzco. Inca society was highly structured. At the top of the **hierarchy** were the emperor, the high priest, and the commander of the army. The nobility served the emperor as administrators. At the bottom of the social pyramid were farmers and laborers.

Inca technology and engineering were highly advanced. The Inca built extensive irrigation systems, roads, tunnels, and bridges that linked regions of the empire to Cuzco. Today you can still see the remains of Inca cities and fortresses. Some of the most impressive ruins are at Machu Picchu, located about 50 miles (80 km) northwest of Cuzco.

The Inca had no written language. Instead, they created a counting system called quipu for record keeping. A quipu was a series of knotted cords of various colors and lengths.

Messengers carrying quipu could travel as far as 150 miles (241 km) per day on the roads. The Inca became extremely wealthy because of their vast natural resources of gold and silver.

Spain Conquers Peru

Unfortunately for the Inca, however, their advanced culture could not turn back the invasion that led to the empire's downfall. The Spanish conquests in Mexico encouraged them to move into South America. In 1532 a Spanish adventurer named Francisco Pizarro landed in Peru with a small band of soldiers.

(tr) ©Werner Forman/Corbis; (b) HUGHES Hervé©/hemis.fr/Getty Images

Visual Vocabulary

quipu The quipu was an Inca counting device.

The ruins of the city of Machu Picchu are high in the Andes.

▶ **CRITICAL THINKING**
Describing How did the Inca build a large empire?

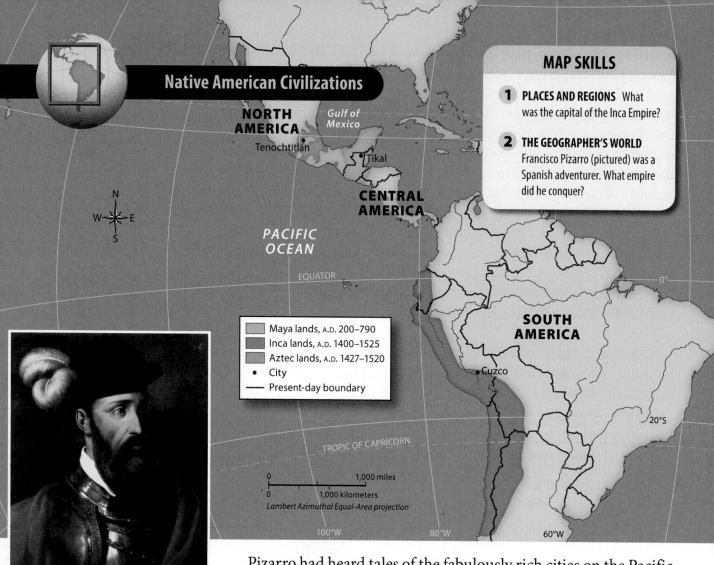

Native American Civilizations

NORTH AMERICA

Gulf of Mexico

Tenochtitlán

Tikal

CENTRAL AMERICA

PACIFIC OCEAN

EQUATOR

SOUTH AMERICA

Cuzco

20°S

0°

TROPIC OF CAPRICORN

Maya lands, A.D. 200–790
Inca lands, A.D. 1400–1525
Aztec lands, A.D. 1427–1520
• City
— Present-day boundary

0   1,000 miles
0   1,000 kilometers
Lambert Azimuthal Equal-Area projection

100°W   80°W   60°W

MAP SKILLS

1 PLACES AND REGIONS What was the capital of the Inca Empire?

2 THE GEOGRAPHER'S WORLD Francisco Pizarro (pictured) was a Spanish adventurer. What empire did he conquer?

Spanish conquistador Francisco Pizarro conquered the empire of the Inca.

Pizarro had heard tales of the fabulously rich cities on the Pacific coast of South America. He also learned that the Inca empire had been badly weakened by a civil war from which the new emperor, Atahualpa, had emerged victorious. When the Spanish confronted Atahualpa, the Inca army, unprepared to face Spanish artillery, was nearly destroyed. Within a few years, the Spanish controlled the entire empire. The Spanish also seized control of the region's precious metals. The road system that the Inca developed became an important transportation route for Spanish goods. The Inca are the ancestors of the Native Americans who live in the Andes region today.

The Spanish also branched out from Peru to create colonies in Argentina, Chile, and other parts of South America. The Spanish military victors were called conquistadors, from the Spanish word for "conqueror." The Spanish colonies became sources of wealth for Spain. Some Spanish settlers prospered from gold and silver mining. Spanish rule of the Andean and midlatitude areas in the region continued for nearly 300 years.

After the Inca lost their empire to the Spanish, they and other Native Americans in the region endured great hardships. Their numbers declined drastically as a result of **smallpox**, a highly infectious disease introduced by the Europeans. The introduction of epidemic diseases also affected numerous Native American groups in North America.

✓ **READING PROGRESS CHECK**

Analyzing How were the Spanish, under conquistador Francisco Pizarro, able to conquer such a mighty empire as the Inca?

Independent Countries

GUIDING QUESTION *How did the countries of the Andean region gain their independence?*

By the early 1800s, most of South America had been under Spanish control for nearly 300 years. History was about to change once again. The reasons for this shift were local as well as international.

Overthrow of Spanish Rule

In the early 1800s, revolution and liberation movements were occurring around the world. The United States threw off British rule, and the French replaced the monarchy with a republic. Struggles for independence also occurred in Mexico and the Caribbean. People in South America were encouraged by these events. It was exactly the right time for two South American revolutionary leaders to lead the fight against Spanish rule. These two leaders were Simón Bolívar and José de San Martín.

The two leaders were able to rally support for independence. San Martín pioneered many elements of **guerrilla** warfare—the use of troops who know the local landscape so well that they are difficult for traditional armies to find. By the mid-1800s, many South American countries had gained independence.

Power and Governance

After the Spanish left South America, several different countries formed. Their borders mostly followed the divisions set in place by the Spanish colonizers. But despite gaining independence, political and economic hardships continued on the continent.

In contrast to the United States, there was no strong momentum for unity in South America. In fact, the rulers of many of the newly independent states were wealthy aristocrats, powerful landowners, or military dictators. Their mindset was more European than South American. In addition, communication between countries was difficult because of the mountainous terrain.

The new countries lacked a tradition of self-government. The British colonies in North America had elected representatives in their colonial legislatures. The new states of South America, however, did not have a structure in place for a government to function. The newly independent countries drafted constitutions.

Thinking Like a Geographer

A Land of Two Capitals

A map of Bolivia reveals an unusual feature. Bolivia has two capitals: Sucre and La Paz. When Bolivia gained its independence, the question of which city would serve as the nation's capital was never resolved. Today, La Paz is the home of the country's executive and legislative branches, and Sucre serves as the center of the country's judicial system.

Eva Perón was married to Juan Perón, president of Argentina. She won admiration for her efforts to support the woman suffrage movement and to improve the lives of the poor.

▶ **CRITICAL THINKING**

Identifying How was Juan Perón removed from office? Who ruled Argentina after Perón?

The enormously uneven distribution of wealth between rich and poor, however, resulted in social and economic instability. Several countries engaged in bloody conflicts over boundary disputes and mineral rights. These conflicts led to much loss of life and weakened economies.

☑ **READING PROGRESS CHECK**

Describing Why are Simón Bolívar and José de San Martín important in the history of the region?

History of the Region in the Modern Era

GUIDING QUESTION *What challenges did the countries of the region face in the late 1800s and 1900s?*

The Andean and midlatitude countries continued to face challenges during the late 1800s and 1900s. With military backing, dictators seized power, and they ignored democratic constitutions. Economies were still dependent on outside powers.

Economic Challenges

The countries of the region faced economic challenges. Among the challenges were developing and controlling resources, building roads and railroads, and establishing trade links. Before independence, the countries of the region depended economically on Spain and Brazil. After independence, the economies of the region remained tied to countries outside South America.

Rapidly industrializing countries in Europe exploited the region for its raw materials. Wealthy landowners, cattle grazers, and mining operators refused to surrender their ties to European investors. Beginning in the early 1900s, large U.S. and European **multinational** firms—companies that do business in several countries—started mining and smelting operations in the region. As the economies expanded, profits grew for wealthy landowners and multinational companies. But many workers and farmers and their families remained mired in poverty.

Political Instability

Economic woes led to calls for reform. Political leaders promised changes for the better. In 1946 Argentinians elected General Juan Perón as the nation's president. Perón and his wife, Eva, were popular with the people. The new government enacted economic reforms to benefit the working people. However, the Perón government limited

Keystone-France/Gamma-Keystone/Getty Images

free speech, censored the press, and added to the country's debt. After Perón was overthrown in 1955, the military government ruled Argentina.

The new government moved to put an end to unrest. The rulers imprisoned thousands of people without trial. Some were tortured or killed. Others simply "disappeared." Argentina was also troubled by conflict over the Falkland Islands. Argentina and Great Britain both claimed the Falklands. After a brief war in 1982, Argentina was defeated, and the Falklands remain a British territory.

Significant changes were also taking place in the country of Chile. In the presidential election of 1970, Chileans elected a socialist candidate named Salvador Allende. Allende took action to redistribute wealth and land. The government took over Chile's copper industry and banking system. Allende's economic reforms were popular with workers but angered the upper classes. In 1973 Chilean military officers staged a **coup**, an illegal seizure of power, and killed Allende. A military dictatorship, headed by General Augusto Pinochet, ruled Chile for the next 16 years.

Movements for Change

In recent years, democracies have replaced dictatorships. Yet the countries in the region are still struggling to end corruption in government, shrink the gap between rich and poor, provide jobs, and protect human rights.

Voters also have elected new leaders. In 2005 Bolivians elected Evo Morales, the country's first indigenous president. Morales introduced a new constitution and land reforms, brought industries under government ownership, and moved to limit U.S. corporate involvement in the country's politics. In 2006 Chileans elected the country's first female president, Michelle Bachelet. A year later, Cristina Fernández de Kirchner became Argentina's first elected female president. Both female leaders started efforts to improve human rights and equal opportunity.

Include this lesson's information in your Foldable®.

☑ **READING PROGRESS CHECK**

Determining Central Ideas After independence, why did the countries in this region continue to experience economic hardship?

LESSON 2 REVIEW

Reviewing Vocabulary

1. What is one advantage of *guerrilla* warfare?

Answering the Guiding Questions

2. *Identifying* What were some of the strengths and achievements of the Inca culture?

3. *Describing* What were two events in North America and Europe that set the stage for the independence movement in the Andes and midlatitude countries?

4. *Determining Central Ideas* What can you infer about democratic government based on what you learned about Juan Perón in Argentina and Salvador Allende in Chile?

5. *Argument Writing* You are either Simón Bolívar or José de San Martín. The year is 1822, when the two men met face to face in Guayaquil (now located in Ecuador). Write a paragraph or two in which you urge your fellow leader to pursue the struggle for independence from Spain.

— Inca road system
— Pan-American highwa

Lesson 3
Life in the Region

ESSENTIAL QUESTION · *What makes a culture unique?*

Reading **HELP**DESK

Academic Vocabulary

- **impact**
- **contemporary**

Content Vocabulary

- **pueblo jóven**

TAKING NOTES: *Key Ideas and Details*

Describe As you read this lesson, use a graphic organizer like this one to write an important fact about each topic.

Topic	Fact
People and Places	
People and Cultures	
Ongoing Issues	

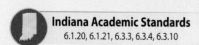

Indiana Academic Standards
6.1.20, 6.1.21, 6.3.3, 6.3.4, 6.3.10

IT MATTERS BECAUSE
The population of the Andean and midlatitude countries is ethnically diverse. People, places, and cultures have been shaped by physical geography, urban growth, migration, and immigration.

People and Places

GUIDING QUESTION *What are the major population patterns in the Andean region?*

The population of the Andean region is not evenly distributed. Population patterns in this region reflect the changing **impact** of politics, economics, and the availability of natural resources and jobs.

Population Density and Distribution

The people of the region came from many different places. Three centuries of colonization by Spain have left their mark on the population. Enslaved Africans were brought as laborers, especially in Peru; this was less common in other countries of the region. After independence, immigrants from many European countries traveled to settle in South America. Immigrants also came from Asia.

Today, Bolivia and Peru have large Native American populations. Argentina and Chile have many people of European ancestry, including Spanish, Italian, British, and German backgrounds. Peru has descendants of people from Europe, but also people with origins in Japan and Southwest Asia.

In the Andean and midlatitude countries, as with the whole continent, the population is densest in the coastal areas. This area is sometimes called the "population rim." The rugged, mountainous areas and the tropical rain forest have discouraged settlement in many inland areas. Transportation is difficult, and communication can be slow or nonexistent. Coastal regions offer fertile land, favorable climates, and easy transportation.

Large Cities and Communities

The largest city in the region is Buenos Aires, the capital of Argentina. This is a bustling port and cultural center. It resembles a European city with its parks, buildings, outdoor cafes, and wide streets. About 2.8 million people live in the central city of Buenos Aires, but the metropolitan area includes 11.5 million people. This is more than one-fourth of Argentina's entire population. Although Argentina as a whole is not densely populated, the area in and around Buenos Aires is.

Buenos Aires and many other large cities in the region have shantytowns. These makeshift communities often spring up on the outskirts of a city. Poor people migrate here from remote inland areas to seek a better life. They cannot afford houses or apartments in the city or suburbs, so they settle in the shantytowns. They build shacks from scraps of sheet metal, wood, and other materials. Shantytowns often lack sewers, running water, and other services. They tend to be dangerous places with widespread crime.

In Lima, the capital and largest city in Peru, the shantytowns are called **pueblos jóvenes**. The name means "young towns." One pueblo jóven was home to María Elena Moyano. She worked to improve education, nutrition, and job opportunities in the pueblo jóven. She refused to give in to the demands of the government or to the communist rebels, who assassinated her in 1992. She has become recognized as a national hero for her courage.

✔ READING PROGRESS CHECK

Analyzing What limits the population in many inland areas?

WINFIELD PARKS/National Geographic Stock

Shacks made of metal, wood, and other materials stand in front of high-rise apartments in Buenos Aires.

▶ CRITICAL THINKING

Describing What challenges do people living in the shantytowns face?

People and Cultures

GUIDING QUESTION *How do ethnic and religious traditions influence people's lives?*

The Andean and midlatitude regions are home to a wide range of ethnic groups. Although many people trace their ancestry back to Europe, Asia, and Africa, Native American groups still thrive in parts of the region.

Ethnic and Language Groups

The Guarani is a Native American group that lives mainly in Paraguay, but people of Guarani descent also can be found in Argentina, Bolivia, and Brazil. The Guarani lived in tropical forests and practiced slash-and-burn agriculture—cutting and burning small areas of forest to clear land for farming. After a few years, the soil's nutrients were used up, and the people moved to a new area.

Today, Guarani customs and folk art are an important part of the culture in Paraguay. Guarani is one of the country's official languages. A related language is Sirionó. The Sirionó live in eastern Bolivia. In Peru, Quechua, a surviving language of the central Andes, is still widely spoken, along with Spanish.

Traditional Medicine

In Bolivia, the custom of Kallawaya medicine is widespread. The word *Kallawaya* might come from the Aymara word for "doctor." Kallawaya healers use traditional herbs and rituals in their cures. Many Bolivians seek the help of Kallawayas when they get sick, either because they prefer these healers or because they cannot afford other doctors. In fact, 40 percent of the Bolivian population relies on the natural healers, who travel from place to place.

Religion and the Arts

During the centuries of Spanish colonization, the Roman Catholic Church was one of the region's most important institutions. The influence of the Catholic Church continues. Millions of people practice mixed religions. Many of the native peoples of the Andes combine their indigenous rituals and beliefs with Roman Catholicism. Others have adopted Protestant religions.

Traditional arts, crafts, music, and dance thrive in the Andean and midlatitude countries. In literature, two Chilean poets, Gabriela Mistral and Pablo Neruda, have won the Nobel Prize for Literature. The works of writers from Argentina,

Kallawaya healers use traditional methods in efforts to cure the sick. The customs and languages of the indigenous peoples are an important part of the culture of the region.

▶ **CRITICAL THINKING**
Identifying What are some of the languages that are spoken in the region?

©David Mercado/Reuters/Corbis

including Jorge Luis Borges and Manuel Puig, are popular with readers around the world. Isabel Allende from Chile, a cousin of the country's former president Salvador Allende, is a **contemporary** writer of great distinction. Many writers have been praised for their use of magic realism. This style combines everyday events with magical or mythical elements. It is especially popular in Latin America.

Academic Vocabulary

contemporary living or happening now

Daily Life

In large cities and towns and in wealthier areas, family life revolves around parents and children. In the countryside, extended families are more common. In the region, the *compadre* relationship is still valued. This relationship is a strong bond between a child's parents and other adults who serve as the child's godparents.

In a megacity like Buenos Aires, people can wander through large, modern shopping malls. They may dine in outdoor cafes or fancy restaurants. They may work in modern office buildings. Traditional Andean foods of the countryside include *pachamanca* in Peru. This is a mixture of lamb, pork, and chicken baked in an earthen oven. Pachamanca dates back to the time of the Inca Empire.

Soccer, or football, is the most popular sport in the region. Football is Argentina's national game, and its teams have won several World Cup titles. Equestrian sports, or sports featuring riders on horseback, are popular in the region. Argentina's polo teams have long dominated international competition. The Argentina team won the first ever Olympic polo gold medal in 1924.

The Larcomar Shopping Mall in Lima, Peru, is a popular attraction for international tourists.

Football is also the national sport for many of the other countries in the region. The top professional league, the American Football Confederation, is made up of teams from 10 South American nations. League teams are eligible for the World Cup and the America Cup. Other popular sports include basketball, golf, boxing, and rugby. Social life focuses on family visits, patriotic events, religious feast days, and festivals.

✓ **READING PROGRESS CHECK**

Analyzing How is Kallawaya medicine different from the modern medicine that is practiced in most Western countries?

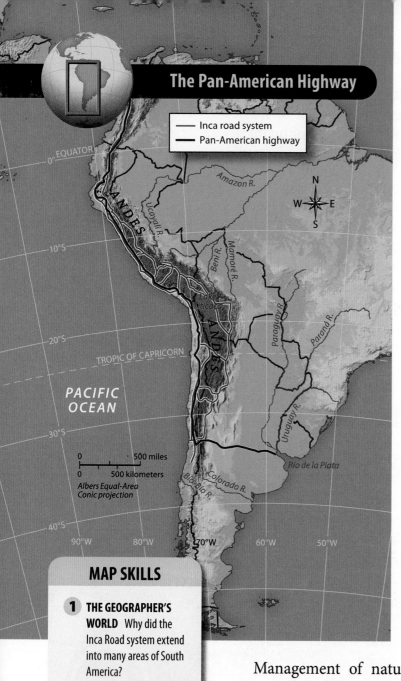

The Pan-American Highway

Legend:
— Inca road system
— Pan-American highway

EQUATOR

Amazon R.

Ucayali R.

ANDES

Beni R.

Mamoré R.

Lake Titicaca

Paraguay R.

Paraná R.

TROPIC OF CAPRICORN

PACIFIC OCEAN

Uruguay R.

Río de la Plata

0   500 miles
0   500 kilometers
Albers Equal-Area
Conic projection

Bío Bío R.

Colorado R.

90°W   80°W   70°W   60°W   50°W

N W E S

MAP SKILLS

1 THE GEOGRAPHER'S WORLD Why did the Inca Road system extend into many areas of South America?

2 THE GEOGRAPHER'S WORLD Why do you think the Pan-American Highway links many capital cities?

Ongoing Issues

GUIDING QUESTION *How are economic and environmental issues affecting the region?*

The population growth rate in the Andean and midlatitude countries is generally not high. However, population is growing enough in some places to add to today's challenges of earning a living.

Earning a Living

It is difficult to build strong economies in the Andean region largely because of the rugged terrain. Many countries rely heavily on agriculture, which is limited and difficult in the mountains. About one-third of Peruvians, for example, are farmers. They grow potatoes, coffee, and corn on terraces built into the mountain slopes. Farms in the valleys along the coast produce sugarcane, asparagus, mangoes, and many other crops.

Mining and fishing are other important economic activities in Peru. Mines in the mountains produce silver, zinc, copper, and other minerals. Peru also produces oil and natural gas. The country's coastal waters are rich fishing grounds. Much of the fish catch is ground into fishmeal for animal feed and fertilizer.

Management of natural resources presents many important issues and challenges. An example is conflict between countries over gas reserves. Bolivia has the second-largest reserve of natural gas in Latin America. Bolivia is landlocked. So, to export the gas, it must move through Peru or Chile.

In 2003 the Bolivian government proposed moving the natural gas through Chile, because it would be cheaper than an alternate plan to go through Peru. The Bolivian people turned out in huge numbers to protest. In Bolivia, suspicion and anger against Chile are widespread. These feelings date back to the Pacific War of the early 1880s, when Chile took over Bolivia's former coastal lands.

As the economies in the region develop, the primary economic activities of agriculture, mining, and fishing remain important. Other activities have become important as well. About 20 percent of the workers in the region are employed in the secondary, or manufacturing, sector. They work in factories making products.

About 65 percent find jobs in the tertiary, or services, sector. They work in a wide variety of occupations, ranging from transportation and retail sales to banking and education.

Transportation and Trade

This region has many geographic and regional barriers. The Andes limit construction of roads and railroads. Yet highways do link large cities. The Pan-American Highway, for example, runs from Argentina to Panama, then continues northward after a break in the highway. A trans-Andean highway connects cities in Chile and Argentina. Peru and Brazil are building the Transoceanic Highway. Parts of it opened in 2012. Eventually, this road will link Amazon River ports in Brazil with Peruvian ports on the Pacific Ocean. Unlike other countries, Argentina has an effective railway system.

Trade also connects countries. In 1991 a trade agreement, known as MERCOSUR, was signed by Argentina, Paraguay, Uruguay, and Brazil. In 2011 MERCOSUR merged into a new organization—the Union of South American Nations (UNASUR). The Union set up an economic and political zone modeled after the European Union (EU). Its goals are to foster free trade and closer political unity.

Addressing Challenges

Looking toward the future, the Andean and midlatitude countries must address many challenges. Environmental issues are among the most important. Air and water pollution is a major problem, especially in the shantytowns of urban areas, where the lack of sewage systems and garbage collection increases disease.

Disputed borders have presented challenges for years. For example, Bolivia and Paraguay long disputed rights to a region thought to be rich in oil. In 1998 Peru and Ecuador finally settled a territorial dispute after years of tensions marked by episodes of armed conflict. Border wars use up resources of people, money, time, and brainpower that could be used to address economic development and environmental concerns.

✔ **READING PROGRESS CHECK**

Determining Central Ideas How does the physical landscape hamper transportation? What actions are being taken to improve transportation?

FOLDABLES®
Study Organizer

Include this lesson's information in your Foldable®.

LESSON 3 REVIEW

Reviewing Vocabulary
1. Why might someone live in a *pueblo jóven*?

Answering the Guiding Questions
2. ***Describing*** How are the populations of Peru and Bolivia different from the populations of Argentina and Chile?

3. ***Analyzing*** Why might the Kallawaya of Bolivia have an important influence on people's lives?

4. ***Identifying*** What are two important industries in Peru today?

5. ***Informative/Explanatory Writing*** Write a paragraph or two to explain what you think is the most pressing problem facing the people who live in shantytowns.

What Do You Think?

Is Globalization Destroying Indigenous Cultures?

Globalization makes it easier for people, goods, and information to travel across borders. Customers have more choices when they shop. Costs of goods are sometimes lower. However, not everyone welcomes these changes. Resistance is particularly strong among indigenous peoples. They see the expansion of trade and outside influences as a threat to their way of life. Is globalization deadly for indigenous cultures?

No!

PRIMARY SOURCE

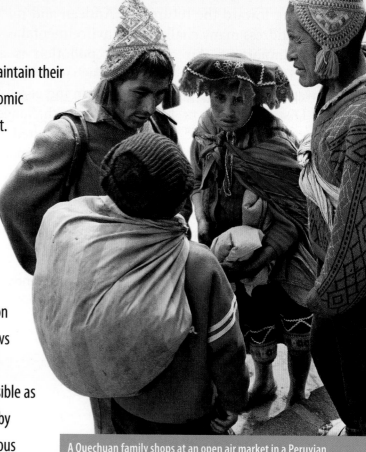

❝ Indigenous people have struggled for centuries to maintain their identity and way of life against the tide of foreign economic investment and the new settlers that often come with it. ... But indigenous groups are increasingly assertive. Globalization has made it easier for indigenous people to organize, raise funds and network with other groups around the world, with greater political reach and impact than before. The United Nations declared 1995–2004 the International Decade for the World's Indigenous People, and in 2000 the Permanent Forum on Indigenous Issues was created. ... Many states have laws that explicitly recognize indigenous people's rights over their resources. ... Respecting cultural identity [is] possible as long as decisions are made democratically—by states, by companies, by international institutions and by indigenous people. ❞

—Report by the United Nations Development Programme (UNDP)

A Quechuan family shops at an open air market in a Peruvian village. The Quechua is the term for several native groups who speak a common language. They live in many countries throughout the region.

PHOTO: Martin Mejia/AP Images

A woman carries water home in a mining town in central Peru. Protesters contend that aggressive mining practices contaminate the environment and destroy the way of life of indigenous people.

Yes !

PRIMARY SOURCE

" Globalization . . . is a multi-pronged attack on the very foundation of [indigenous people's] existence and livelihoods. . . . Indigenous people throughout the world . . . occupy the last pristine [pure and undeveloped] places on earth, where resources are still abundant [plentiful]: forests, minerals, water, and genetic diversity. All are ferociously sought by global corporations, trying to push traditional societies off their lands. . . . Traditional sovereignty [control] over hunting and gathering rights has been thrown into question as national governments bind themselves to new global economic treaties. . . . Big dams, mines, pipelines, roads, energy developments, military intrusions all threaten native lands. . . . National governments making decisions on export development strategies or international trade and investment rules do not consult native communities. . . . The reality remains that without rapid action, these native communities may be wiped out, taking with them vast indigenous knowledge, rich culture and traditions, and any hope of preserving the natural world, and a simpler . . . way of life for future generations. "

—**International Forum on Globalization (IFG), a research and educational organization**

What Do You Think? DBQ

1 *Citing Text Evidence* According to the IFG, why are indigenous people at risk?

2 *Describing* According to the United Nations report, how has globalization given indigenous people more power?

Critical Thinking

3 *Identifying* One effect of globalization is that more tourists are visiting remote places such as rain forests in South America and wildlife areas in Africa. How do you think indigenous peoples feel about the growth of tourism in their communities?

Directions: Write your answers on a separate piece of paper.

1 Use your **FOLDABLES** to explore the Essential Questions.

INFORMATIVE/EXPLANATORY WRITING Write a short essay to answer the question: Why are people who live in the Andes more likely to follow a traditional way of life than those who live in cities?

2 **21st Century Skills**

DESCRIBING Write a radio script for a two-minute segment about what people do for recreation in one of this region's countries. Research the topic, and outline the information you want to include in your script. Share your notes with an adult, and ask: (a) Is my outline clear and well-organized? (b) Does the outline have too much or too little information? Revise your outline as needed, and then use it to write the script. Exchange scripts and compare your script with a classmate's. Discuss strong and weak points in each script.

3 **Thinking Like a Geographer**

ANALYZING Think about why industrialization requires good transportation and communication systems. Then, describe obstacles that slow the development of these systems.

4 **GEOGRAPHY ACTIVITY**

Locating Places

Match the letters on the map with the numbered places below.

1. Rio de la Plata	**3.** Lima	**5.** Santiago	**7.** Paraguay	**9.** Uruguay River
2. Bolivia	**4.** Falkland Islands (Malvinas)	**6.** Chile	**8.** Uruguay	**10.** Buenos Aires

REVIEW THE GUIDING QUESTIONS

Directions: Choose the best answer for each question.

1 The ocean tide meets a river current at a(n)

A. cordillera.

B. altiplano.

C. estuary.

D. *tierrra templada.*

2 Chile leads the world in exports of

F. emeralds.

G. copper.

H. lead.

I. gold.

3 The Inca became extremely wealthy because of

A. the system called quipu.

B. rich farmland.

C. huge deposits of gold and silver.

D. a lucrative fur trade.

4 The broad plain that spreads across Uruguay and eastern Argentina is called

F. the pampas.

G. the Atacama.

H. the Rio Blanco.

I. the Río de la Plata.

5 The group of South American animals called camelids includes

A. horses and donkeys.

B. sheep and goats.

C. llamas and alpacas.

D. mules and llamas.

6 The Spanish explorer who took the Inca emperor Atahualpa hostage was

F. Pizarro.

G. Columbus.

H. Magellan.

I. Fernandez.

DBQ ANALYZING DOCUMENTS

7 **CITING TEXT EVIDENCE**   Read the following passage:

"*[Argentina's] President [Cristina] Kirchner . . . announce[d] the nationalization of the Argentine oil company. . . . The move . . . raised concerns that this may be the first of many expropriations [government takeovers] of privately run companies.*"

—from Jonathan Gilbert, "The Next Venezuela?" (2012)

What does the writer say that Argentina's government might do in the future?

A.   break the nation's dependence on imported oil

B.   take over other companies

C.   replace nuclear power with oil as the main source of energy

D.   end high unemployment in Argentina

8 **IDENTIFYING**   What impact will Argentina's action probably have on foreign companies doing business there?

F.   They probably will try to sell their businesses to the government.

G.   They will expect to have lower costs when oil prices fall.

H.   They are likely to fear that their companies will be taken over, too.

I.   They will seek to buy the oil company from the government.

SHORT RESPONSE

"*Visitors to modern Cuzco frequently marvel at the exquisite [very fine] workmanship of its many Inca walls. The . . . impression is that a great deal of Inca Cuzco has survived. . . . [Actually,] new streets have been created, ancient ones lost, and the bulk of the city's former palaces, halls, temples, and shrines [holy places] have been demolished.*"

—from Brian S. Bauer, *Ancient Cuzco: Heartland of the Inca* (2004)

9 **DETERMINING CENTRAL IDEAS**   Is the impression visitors have that much of Inca Cuzco has survived correct? Why or why not?

10 **IDENTIFYING**   Who do you think was responsible for these changes to ancient Cuzco? Why?

EXTENDED RESPONSE

11 **INFORMATIVE/EXPLANATORY WRITING**   Think about what you have read about the physical geography of the Andes and midlatitude countries. What geographic factors influence where people have settled in the region?

Need Extra Help?

If You've Missed Question	**1**	**2**	**3**	**4**	**5**	**6**	**7**	**8**	**9**	**10**	**11**
Review Lesson	1	1	2	1	1	2	3	3	2	2	1

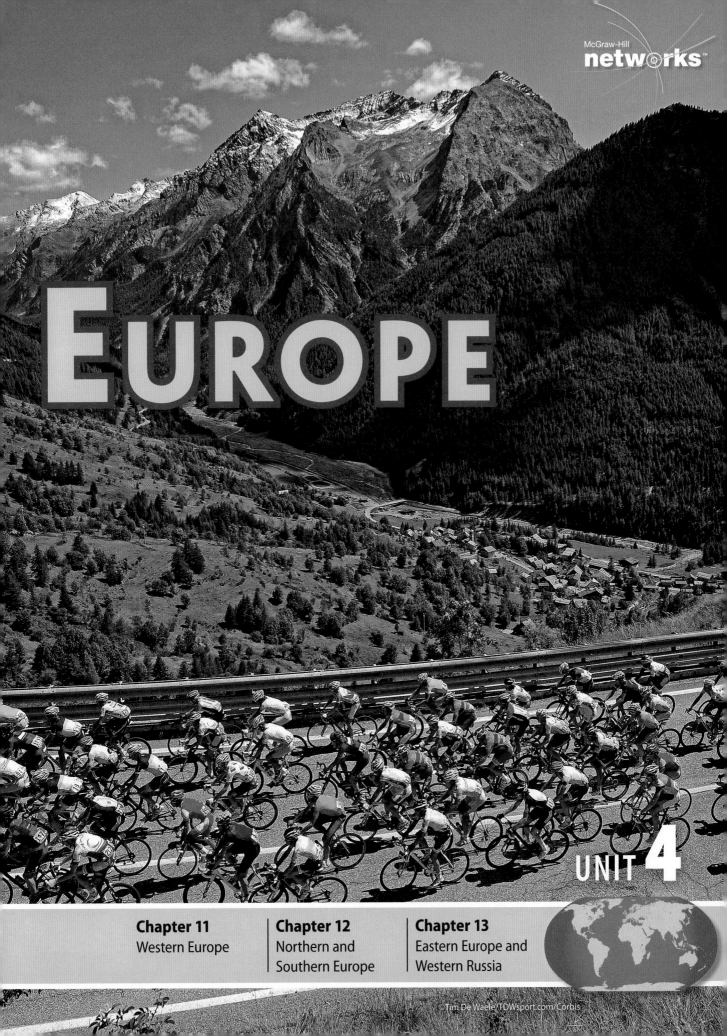

EUROPE

UNIT 4

Chapter 11
Western Europe

Chapter 12
Northern and
Southern Europe

Chapter 13
Eastern Europe and
Western Russia

EXPLORE the CONTINENT

EUROPE

Europe is a relatively small continent with a long, jagged coastline. Fertile plains extend across much of northern Europe. Farther south, the plains become rugged hills, and then mountains. Great rivers wind their way through Europe's landscape, linking inland areas with the seas.

(1) BODIES OF WATER A fisher uses a net to catch salmon on Loughros Bay in northwest Ireland. Europe's deep bays and well-protected inlets shelter fine harbors. For centuries, the rivers of Europe have provided links between coastal ports and inland population centers.

(2) LANDFORMS A herder tends to cattle on the Great Hungarian Plain in the eastern part of Hungary. Farmers raise livestock and cultivate grains, fruits, and vegetables here. In the northern part of the region, the North European Plain stretches from France to Russia. The plain's fertile soil and wealth of rivers originally drew farmers to the area.

3 **NATURAL RESOURCES** Located in west central Italy, the region of Tuscany is known for its natural resources as well as its beautiful landscape. In the mountains, iron and magnesium are produced, and marble is quarried. Carrara marble is in high demand for building and statues worldwide. Along the northern coast, pine forests are abundant.

FAST **FACT**

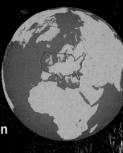

Europe is only about 10 percent larger than the United States.

ARCTIC OCEAN

Barents
Sea

Iceland

ARCTIC CIRCLE

KOLA
PENINSULA

RUSSIA

Faeroe
Islands

Norwegian
Sea

SCANDINAVIA

N. Dvina R.

Europe/Asia
boundary

URAL MOUNTAINS

Shetland
Islands

Hebrides

Orkney
Islands

ATLANTIC
OCEAN

Ireland

North
Sea

Baltic Sea

NORTHERN EUROPEAN PLAIN

Ural R.

Irish Sea

Great
Britain

Celtic
Sea

English Channel

Thames R.

Elbe R.

Rhine R.

Oder R.

Vistula R.

Dnieper R.

Volga R.

Don R.

Seine R.

Loire R.

Danube R.

Dniester R.

Sea of
Azov

Caspian
Sea

Bay of
Biscay

Matterhorn
14,691 ft.
(4,478 m)

ALPS

Mont Blanc
15,781 ft.
(4,810 m)

Po R.

HUNGARIAN
PLAIN

Carpathian
Mts.

Danube R.

Black Sea

Caucasus Mts.

Mt. Elbrus
18,510 ft.
(5,642 m)

MESETA

Pyrenees Mts.

Ebro R.

Corsica

Apennines

Adriatic Sea

BALKAN
PENINSULA

Bosporus

Tagus R.

IBERIAN
PENINSULA

Balearic
Islands

Sardinia

Tyrrhenian
Sea

Ionian
Islands

Aegean Sea

ASIA

Strait of
Gibraltar

Sicily

Mt. Etna
10,902 ft.
(3,323 m)

Crete

Cyprus

Mediterranean Sea

Elevations

10,000 ft. (3,000 m)
5,000 ft. (1,500 m)
2,000 ft. (600 m)
1,000 ft. (300 m)
0 ft. (0 m)
Below sea level

—— National boundary
▲ Mountain peak

AFRICA

0   400 miles
0   400 kilometers
Lambert Azimuthal
Equal-Area projection

EUROPE

PHYSICAL

MAP SKILLS

1 PHYSICAL GEOGRAPHY How would you describe the landforms of central Europe?

2 PLACES AND REGIONS How high is the Meseta in the Iberian Peninsula?

3 PLACES AND REGIONS What body of water separates Great Britain from Europe's mainland?

ARCTIC OCEAN

Barents Sea

Murmansk

Europe/Asia boundary

Reykjavík
ICELAND

Norwegian Sea

FINLAND

SWEDEN

NORWAY

Helsinki

Oslo Stockholm

Tallinn St. Petersburg

RUSSIA

ATLANTIC OCEAN

EST.

LATVIA

Riga

Moscow

Ural R.

IRELAND

Dublin

UNITED KINGDOM

North Sea

DENMARK

Copenhagen

LITH.

Vilnius

Kaliningrad Minsk

BELARUS

KAZAKHSTAN

Volga R.

Volgograd

NETH.

Amsterdam

London Brussels

English Channel

Thames R.

Berlin

GERMANY

Warsaw

POLAND

Kiev (Kyiv)

UKRAINE

Don R.

Elbe R. *Rhine R.* *Oder R.* *Vistula R.*

Seine R. Paris

BELG.

Luxembourg

LUX.

Prague

CZECH REP.

Dnieper R.

Loire R.

Danube R.

SLOVAKIA

Bratislava

MOLDOVA

Chişinău

Dniester R.

Sea of Azov

FRANCE

Bern

SWITZ.

LIECH.

Vaduz

Vienna

AUST.

SLOV.

Ljubljana

Budapest

HUNG.

Zagreb

ROMANIA

Bucharest

Caspian Sea

Bay of Biscay

Andorra la Vella

CROATIA

B.&H.

Sarajevo

Belgrade

SERB.

KOS.

Priština

Black Sea

PORTUGAL

SAN MARINO

Madrid

ANDORRA

MONACO

ITALY

Rome

Corsica (France)

Podgorica

MONT.

Sofia

BULGARIA

TURKEY

Lisbon

Tagus R.

Ebro R.

Rhône R. *Po R.*

Adriatic Sea

Skopje

MAC.

SPAIN

Strait of Gibraltar

Sardinia (Italy)

VATICAN CITY (within Rome)

Tiranë

ALBANIA

GREECE

Aegean Sea

Balearic Islands (Spain)

Sicily (Italy)

Athens

Nicosia

CYPRUS

○ National capital
● City

MALTA Valletta

Mediterranean Sea

Crete (Greece)

30°N

0   400 miles

0   400 kilometers

Lambert Azimuthal Equal-Area projection

AFRICA

0°   10°E   20°E   30°E

Abbreviations	
AUST.	Austria
B.&H.	Bosnia & Herzegovina
BELG.	Belgium
CZECH REP.	Czech Republic
EST.	Estonia
HUNG.	Hungary
KOS.	Kosovo
LIECH.	Liechenstein
LITH.	Lithuania
LUX.	Luxembourg
MAC.	Macedonia
MONT.	Montenegro
NETH.	Netherlands
SERB.	Serbia
SLOV.	Slovenia
SWITZ.	Switzerland

POLITICAL

MAP SKILLS

1 **THE GEOGRAPHER'S WORLD** Where is Monaco located?

2 **PLACES AND REGIONS** What country lies between Belarus and Romania?

3 **PLACES AND REGIONS** Which countries border the Baltic Sea?

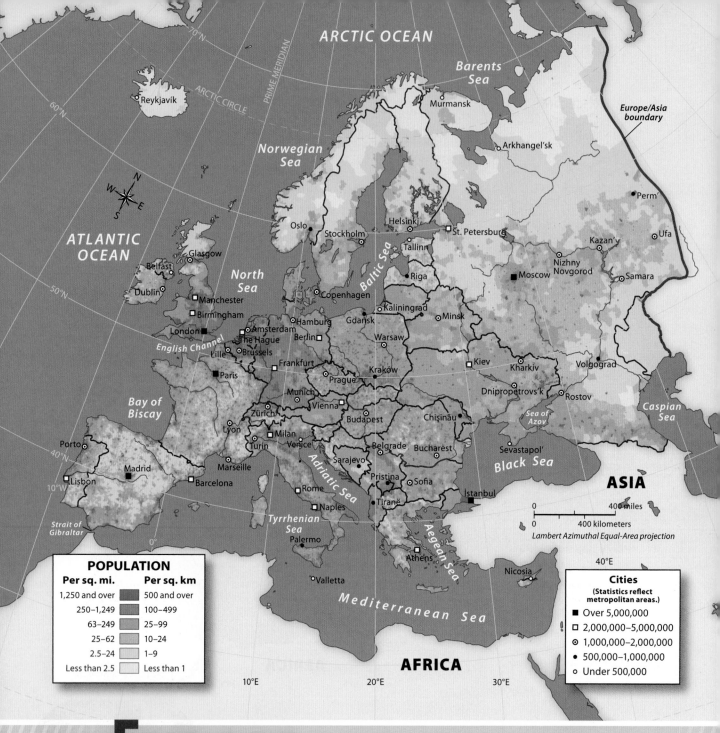

POPULATION

Per sq. mi.	Per sq. km
1,250 and over	500 and over
250–1,249	100–499
63–249	25–99
25–62	10–24
2.5–24	1–9
Less than 2.5	Less than 1

Cities
(Statistics reflect metropolitan areas.)

■ Over 5,000,000
□ 2,000,000–5,000,000
◉ 1,000,000–2,000,000
• 500,000–1,000,000
○ Under 500,000

Lambert Azimuthal Equal-Area projection

EUROPE

POPULATION DENSITY

MAP SKILLS

1 **PLACES AND REGIONS** How would you describe the population in the area north of St. Petersburg?

2 **THE GEOGRAPHER'S WORLD** What cities are located in Europe's most densely populated area?

3 **PLACES AND REGIONS** What part of Europe has a low population density?

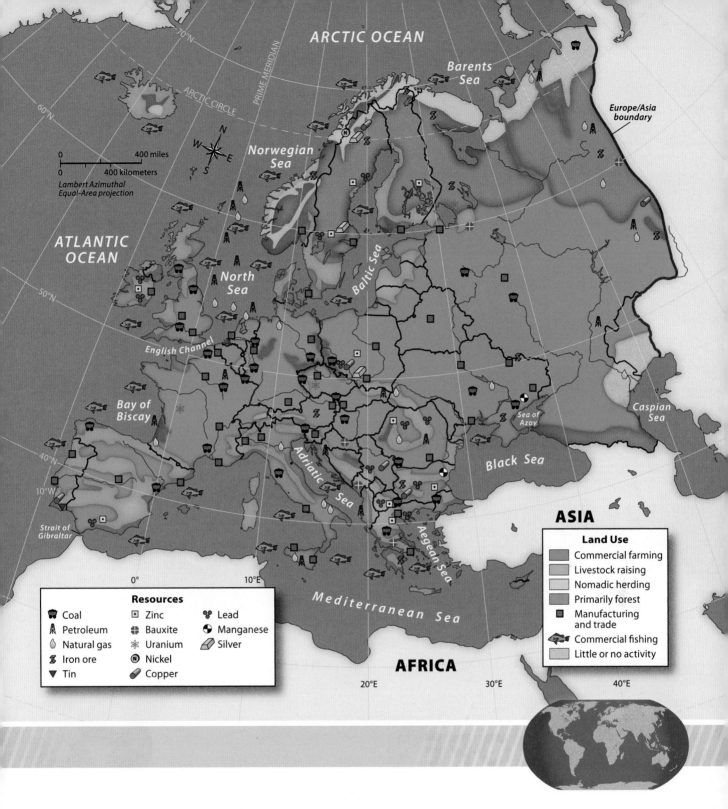

ARCTIC OCEAN

Barents Sea

Europe/Asia boundary

Norwegian Sea

ATLANTIC OCEAN

0 — 400 miles
0 — 400 kilometers
Lambert Azimuthal Equal-Area projection

North Sea

Baltic Sea

English Channel

Bay of Biscay

Adriatic Sea

Sea of Azov

Caspian Sea

Black Sea

ASIA

Strait of Gibraltar

Aegean Sea

Mediterranean Sea

AFRICA

20°E   30°E   40°E

Resources

Coal	Zinc	Lead
Petroleum	Bauxite	Manganese
Natural gas	Uranium	Silver
Iron ore	Nickel	
Tin	Copper	

Land Use

- Commercial farming
- Livestock raising
- Nomadic herding
- Primarily forest
- Manufacturing and trade
- Commercial fishing
- Little or no activity

ECONOMIC RESOURCES

MAP SKILLS

1 **PLACES AND REGIONS** In what part of Europe is bauxite found?

2 **ENVIRONMENT AND SOCIETY** How is land used in the far northern areas of Europe?

3 **PLACES AND REGIONS** What mineral resources are located under the North Sea?

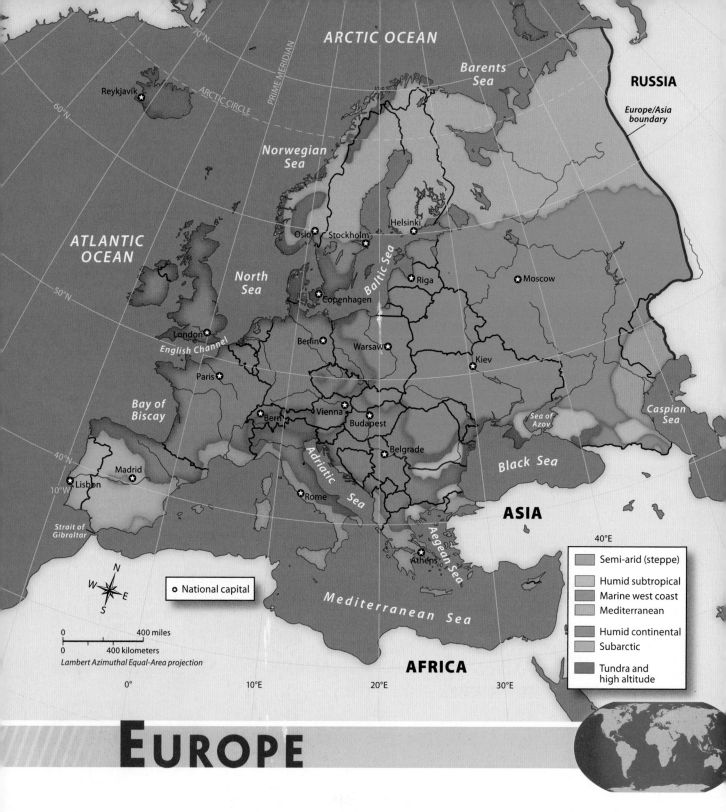

ARCTIC OCEAN

Barents Sea

RUSSIA

Europe/Asia boundary

Reykjavík

Norwegian Sea

ATLANTIC OCEAN

Helsinki

Oslo   Stockholm

Riga   Moscow

North Sea   Copenhagen

London   Berlin   Warsaw
English Channel   Kiev

Paris

Bay of Biscay   Bern   Vienna
Budapest

Sea of Azov   Caspian Sea

Madrid   Belgrade

Lisbon   Rome   Black Sea

Strait of Gibraltar

ASIA

Adriatic Sea   Aegean Sea   Athens

National capital

Mediterranean Sea

AFRICA

0   400 miles
0   400 kilometers
Lambert Azimuthal Equal-Area projection

0°   10°E   20°E   30°E   40°E

	Semi-arid (steppe)
	Humid subtropical
	Marine west coast
	Mediterranean
	Humid continental
	Subarctic
	Tundra and high altitude

EUROPE

CLIMATE

MAP SKILLS

1 THE GEOGRAPHER'S WORLD Where are Europe's driest areas located?

2 PLACES AND REGIONS What type of climate do people in Paris experience?

3 PLACES AND REGIONS What area of Europe has a largely humid continental climate?

WESTERN EUROPE

ESSENTIAL QUESTIONS · *How does geography influence the way people live?*
· *Why do civilizations rise and fall?* · *How do governments change?*

©Alex Masi/Corbis

Coal mine worker in South Wales

Lesson 1
Physical Geography of Western Europe

Lesson 2
History of Western Europe

Lesson 3
Life in Western Europe

The Story Matters...

Western Europe has always been a crossroad of cultures. Today, this is reflected in the diversity of cultures represented in the population, particularly in large cities, such as Paris, France. Long before modern times, however, the geography and resources of Western Europe influenced early civilizations and their rise to economic and political power, shaping the history and governments of this region.

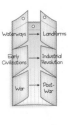

FOLDABLES
Study Organizer

Go to the Foldables® library in the back of your book to make a Foldable® that will help you take notes while reading this chapter.

Islands and peninsulas branch out from Western Europe into the various oceans, seas, bays, and channels that border the region. In the past, these landforms and waterways isolated groups of people. As a result, many different cultures developed. As you study the map, look for other geographic features that make this area unique.

Step Into the Place

MAP FOCUS Use the map to answer the following questions.

1 THE GEOGRAPHER'S WORLD Which two countries does the English Channel separate?

2 PLACES AND REGIONS Which island country in Western Europe lies farthest west?

3 THE GEOGRAPHER'S WORLD What is the capital city of Switzerland?

4 CRITICAL THINKING
ANALYZING In what ways might the large number of rivers and waterways be important to the people of Western Europe?

A

SWISS ALPS A train begins its climb to the summit. The railway, completed in the early 1900s, makes this scenic trip popular with visitors.

B

BRANDENBURG GATE Completed in the late 1700s, the Brandenburg Gate is one of the well-known landmarks in Berlin, Germany.

Step Into the Time

TIME LINE Choose an event from the time line and predict its long-term political, cultural, or geographical consequences.

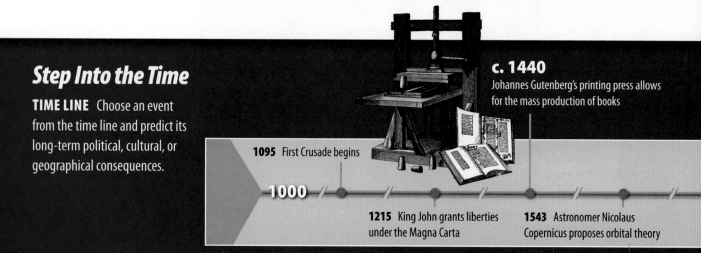

c. 1440 Johannes Gutenberg's printing press allows for the mass production of books

1095 First Crusade begins

1000

1215 King John grants liberties under the Magna Carta

1543 Astronomer Nicolaus Copernicus proposes orbital theory

ages;

Western Europe

Legend:
- ✪ National capital
- • City

60°N

50°N

40°N

ATLANTIC OCEAN

PRIME MERIDIAN

North Sea

Baltic Sea

NORWAY **SWEDEN**

ESTONIA

LATVIA

LITHUANIA

DENMARK

RUSSIA

B Berlin ✪

POLAND

IRELAND
Dublin ✪

UNITED KINGDOM

Thames R.

London •

English Channel

NETHERLANDS
Amsterdam ✪

GERMANY

Elbe R.

Rhine R.

CZECH REPUBLIC

SLOVAKIA

Brussels •
BELGIUM

Luxembourg ✪
LUXEMBOURG

Paris ✪
Seine R.

Danube R.

Vienna ✪

AUSTRIA

HUNGARY

LIECHTENSTEIN

Bern ✪ Vaduz ✪

FRANCE

Loire R.

Rhone R.

A

SWITZERLAND

SLOVENIA

CROATIA

ITALY

BOSNIA & HERZEGOVINA

MONTENEGRO

Monaco ✪
MONACO

Bay of Biscay

PORTUGAL

SPAIN

Ajaccio •
Corsica (France)

Mediterranean Sea

10°W *0°* *10°E*

0 — 400 miles
0 — 400 kilometers
Lambert Azimuthal Equal-Area projection

1804
Napoleon crowned emperor of France

1993
European Union created

1 EURO

1900

2000

1918 World War I ends

1945 Japan surrenders, ending World War II

1989 Berlin Wall torn down as East and West Germany reunite

There's More Online!

☑ **IMAGE** How Windmills Work

☑ **SLIDE SHOW** Agriculture in
Western Europe

☑ **VIDEO**

Reading **HELP**DESK

Academic Vocabulary

- **adapt**

Content Vocabulary

- **dike**
- **polder**
- **estuary**
- **Westerlies**
- **deciduous**
- **coniferous**

TAKING NOTES: *Key Ideas and Details*

Identify Use a graphic organizer like this one to list important resources in Western Europe.

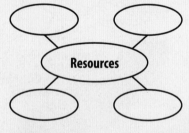

Resources

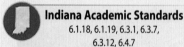

Indiana Academic Standards
6.1.18, 6.1.19, 6.3.1, 6.3.7,
6.3.12, 6.4.7

Lesson 1

Physical Geography of Western Europe

ESSENTIAL QUESTION • *How does geography influence the way people live?*

IT MATTERS BECAUSE

The geography of Western Europe has provided the people who live there with several advantages: closeness to the sea, abundant resources, and temperate climates.

Landforms and Waterways

GUIDING QUESTION *How do the physical features of Western Europe make the region unique?*

You probably have seen images of Europe on television or in photographs. Maybe you have seen a movie showing a Dutch windmill on a windswept plain. You might have watched news clips showing cyclists racing through the rolling French countryside during a yearly sports event called the Tour de France. These images show just a part of Western Europe's varied landscape.

Western Europe comprises the following nations: Ireland, the United Kingdom, France, Luxembourg, Germany, the Netherlands, Belgium, Austria, and Switzerland. Belgium, the Netherlands, and Luxembourg are often referred to as the Benelux Countries. Western Europe also includes the tiny countries of Monaco and Liechtenstein, which have a combined population of fewer than 70,000 people.

Low-Lying Plains

Western Europe's landscape consists of plains with mountains cutting through some places. Shaped by wind, water, and ice, these landforms have influenced the lives of the region's peoples. Much of Western Europe lies within an area called

(l to r) TADAO YAMAMOTO/a.collectionRF/Getty Images; Willfried Gredler/age fotostock; Arterra Picture Library/Alamy; Philippe Hugueo/AFP/Getty Images

the Northern European Plain. France, Belgium, the Netherlands, Luxembourg, and most of Germany are located on this plain.

Massive sheets of ice shaped the Northern European Plain during the last ice age, which ended about 11,000 years ago. In some places, the melting glaciers left behind fertile soil, but also thick layers of sand and gravel. Ocean waves, currents, and winds have eroded these deposits into sand dunes along some of the North Sea coastline. The glaciers also left behind areas of poorly drained wetlands along the North Sea and Atlantic coasts in Ireland and the United Kingdom.

Mountains and Highlands

Two mountain ranges—the Pyrenees (PIR•eh•NEES) and the Alps—separate Western Europe from neighboring nations in Southern Europe. Both mountain ranges divide the cooler climates of the north from the warm, dry climate of the Mediterranean region to the south. To the southwest, the Pyrenees form a natural barrier between France and Spain. This mountain range stretches 270 miles (435 km) from east to west. Pico de Aneto—the tallest mountain in the Pyrenees—reaches a height of 11,169 feet (3,404 m). The average height of the Pyrenees, however, is only about 5,300 feet (1,615 m).

To the east of the Pyrenees lie the Alps. Like the Pyrenees, the Alps were created by the folding of rocks as a result of plate tectonics. Then these mountains were further shaped by glaciers. The Alps extend about 750 miles (1,207 km) along the southeastern border of France through Switzerland, Austria, and Germany. The Alps are much larger and higher than the Pyrenees. The tallest mountain in the Alps is Mont Blanc, at 15,771 feet (4,807 m).

The Alps and Pyrenees are geologically younger than other mountainous areas in Western Europe that have been worn down by glaciers. The most extensive is a plateau called the Middle Rhine Highlands in Germany, Belgium, Luxembourg, and France. Its highest elevations are about 3,000 feet (914 m).

Seas, Islands, and Waterways

Western Europe has long, irregular coastlines that touch many bodies of water, including the Atlantic Ocean and the North, Baltic, and Mediterranean seas. This closeness to the sea has long shaped European life.

The North Sea is part of the Atlantic Ocean but shallower. Separating the island of Britain from the rest of Europe, the North Sea is a rough, dangerous body of water to navigate. It is also a rich fishing ground for the Netherlands and the United Kingdom. The North Sea has long been important for trade between the United Kingdom and the rest of Europe. It is also the location of large oil and natural gas reserves.

Windmills are used to pump water into a reservoir for this village in the plains of the Netherlands.

▶ **CRITICAL THINKING**
Identifying What is the major plain in the region?

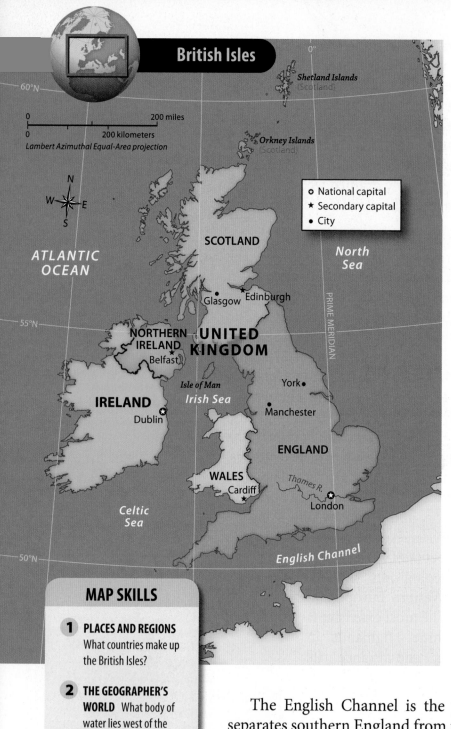

The North Sea has helped but also hindered the Dutch, the people of the Netherlands. The Netherlands sits at a low elevation—25 percent of the country is below sea level. To **adapt**, the Dutch have built **dikes**, which are walls or barriers to hold back the water. The Dutch call the land they reclaim from the sea **polders**. This land is used for farming and settlement. Stormy seas, however, have broken dikes and caused flooding in recent times.

Just off the northern coast of France lie the British Isles, or the Atlantic Archipelago—the islands of Britain and Ireland. You might think each main island is its own country, but that is not the case. There are four parts of the United Kingdom: England, Wales, Scotland, and Northern Ireland. Three parts—England, Wales, and Scotland—occupy the island of Britain. The fourth part, Northern Ireland, is the northeastern section of Ireland. The country of Ireland, or Eire, occupies all but the northeastern section of the island of Ireland.

The English Channel is the part of the Atlantic Ocean that separates southern England from northern France. The channel is a busy sea route connecting the North Sea with the Atlantic Ocean. In addition to the ships that travel the route, trains pass beneath this waterway. A tunnel runs through the rock under the water. Called the Chunnel, it houses a high-speed train that connects Britain to mainland Europe.

Western Europe has a wealth of rivers and smaller waterways. The region's rivers have played an important role in how Europe developed over the centuries. Rivers determined the locations of cities, such as London, Paris, and Hamburg. Rivers provide transportation routes for goods and people. They also form some political borders. Rivers linked by canals provide transportation

MAP SKILLS

1 **PLACES AND REGIONS** What countries make up the British Isles?

2 **THE GEOGRAPHER'S WORLD** What body of water lies west of the United Kingdom?

Academic Vocabulary

adapt to change a trait to survive

networks from inland areas to the sea. They also provide water for farming and electric power.

The Thames River in southern England is one of the most well-known rivers in the world. The Thames flows for 205 miles (330 km) from the Cotswold Hills to the city of London. When it reaches London, the Thames becomes an **estuary** and extends for another 65 miles (105 km) before it enters the North Sea. An estuary is where part of the sea connects to the lower end of a river.

The Rhine River is the busiest waterway in Europe. It begins high in the Swiss Alps and empties into the North Sea. Along its course, it plunges over waterfalls, cuts deeply into the Middle Rhine Highlands, and connects industrial areas to the port in Rotterdam. The Rhine meanders lazily across the plains of the Netherlands to the North Sea. The river serves as part of the political boundary between France and Germany. It runs through the most populated region of Europe.

Another important waterway in Germany is the Elbe River, which also empties into the North Sea. Historically, the Elbe formed part of the border between West Germany and East Germany when that country was split into two from 1945 to 1990 (after World War II). Another important river in Germany, the Danube, is different from the other rivers because it flows toward the east. It passes through southern Germany and Austria on its way into Eastern Europe. The Danube is Eastern Europe's most important waterway. The Main (MINE) River, a tributary of the Rhine, is linked to the Danube by the Main-Danube Canal. The canal links the North Sea with the Black Sea at the southeastern end of Europe.

This seaport in Dover, England, is one of the busiest in the world.

▶ **CRITICAL THINKING**

Analyzing What are some of the important waterways in Western Europe?

Think Again ?

London has black fogs that can kill people.

Not true. At one time, London had "pea soupers," thick, black and greenish fog. The fogs were caused in part by the burning of soft coal in homes and industry, which produced soot and poisonous sulfur dioxide. This mixed with the mist and fogs of the Thames Valley. In 1952 the Great Smog killed 4,000 London residents. Coal burning has been outlawed in London, but fumes from automobiles still produce smog in the city.

Roadways run along the Danube River in Vienna, Austria.

▶ **CRITICAL THINKING**

Describing How is the Danube different from other rivers in the region?

In France, the longest river is the Loire River, at a length of 634 miles (1,020 km). It passes through the Loire Valley, the most important agricultural region in France. It is the Seine River, however, which runs through the capital city of Paris, that carries far more of the country's inland water traffic.

☑ **READING PROGRESS CHECK**

Analyzing How did the rivers in Western Europe affect its economic development?

Climate

GUIDING QUESTION *Why is the climate mild in Western Europe?*

Western Europe is located at northern latitudes, but it has a milder climate than other places at the same latitudes. For example, southern France is at roughly the same latitude as Halifax, Nova Scotia, in Canada, but southern France experiences a milder climate. Why? Western Europe is located near the Atlantic Ocean. Warm winds off the ocean are the primary factor that shapes the region's climate.

Temperate Lowlands

Most of Western Europe lies in the path of the **Westerlies**, strong winds that travel from west to east. The Westerlies blow a constant stream of relatively warm air from the sea to the land. Why is the air so warm this far north? The answer is found on the other side of the Atlantic Ocean, in the tropical waters of the Caribbean Sea. Here, an

Willfried Gredler/age fotostock

ocean current called the Gulf Stream moves warm tropical water north along the eastern coast of North America. The Gulf Stream then flows across the Atlantic Ocean, where its eastern extension, the North Atlantic Current, approaches the European coast. The current's warm water heats the air above it. This warm, moist air moves inland on the Westerlies and brings mild temperatures and rain to most of Western Europe throughout the year. Summers are cool, and winters are mild. This climate is known as a marine west coast climate.

Most places that have a marine west coast climate have mountain ranges running north-south along the coast. The mountains block the warm, moist air from moving farther inland. Western Europe, however, does not have coastal mountain ranges, so the Westerlies blow farther across the European continent.

Mediterranean Climate Area

Other areas of Western Europe, such as southern France, have a drier climate. A high-pressure system called the Azores High travels north over the Atlantic Ocean during the summer months. This pressure system pushes moist air northward. As a result, summers in southern France are hot and dry. In winter, the nearby Mediterranean Sea moderates the climate so that winters are mild or cool. Most of the rainfall occurs in spring and autumn. This is called a Mediterranean climate.

✔ **READING PROGRESS CHECK**

Describing How do the Westerlies affect the climate of Western Europe?

Walkers take an autumn stroll through a forest in southern Belgium. In this area, autumns start mild but soon become cool. Winters are cold and snowy.

Natural Resources

GUIDING QUESTION *How do the people of Western Europe use the region's natural resources?*

Western Europe has many important natural resources. Layers of coal lie beneath rich, fertile soils. Beneath the North Sea are large pockets of oil and natural gas. Western Europeans use these natural resources to support their economies and populations.

Energy Sources

Deposits of coal are plentiful throughout Britain, Belgium, the Netherlands, France, and Germany. Coal was used to fuel machines invented during the Industrial Revolution of the 1800s. Today, coal production is declining in Northern Europe. Much less coal is available than in the past, so the region is importing more coal. Coal has also become less important as a source of energy as people rely more on other energy sources to meet their needs. In places where other fuels are scarce, Europeans burn peat. A peat bog is a wetland in which large masses of vegetable matter decay in the poorly drained soil. Peat—the name for the decaying vegetable matter—can be used as fuel for heating. The peat is dug up, cut into blocks, and dried so that it can be burned.

In 1959 oil and natural gas were discovered under the North Sea. Since then, the North Sea has become the region's most important source for these resources. The United Kingdom and the Northern European nation of Norway are the leading producers of oil and natural gas from North Sea oil fields. The Netherlands and Germany also produce oil and natural gas from the North Sea.

Other countries in Western Europe use their rivers to supply energy. Switzerland, for example, uses its fast-flowing rivers to produce electricity. Hydroelectricity supplies more than half of Switzerland's electricity needs.

Rich Soils

The Northern European Plain has some of the richest soils in Europe so people can farm there. Soils of the Northern European Plain contain humus. Humus is decomposed plant and animal material that makes soils rich and fertile, and is good for growing crops and raising livestock.

France is Western Europe's leading agricultural producer. France devotes more surface area to agriculture than any other country in the region. Large wheat fields stretch across northern France. Orchards and vineyards are common in the central and southern parts of the country. On the Northern European Plain, farmers grow a variety of crops and raise cattle and hogs. In the Netherlands, dairy farming is important for the country's economy.

Plants and Wildlife

The moderate climate and abundance of rainfall in most of Western Europe support a wide variety of plant and animal life. The British Isles have dense forests, grasslands, scrublands, and wetlands. The natural vegetation in most of the British Isles is **deciduous** forest, which includes trees such as oak, maple, beech, and chestnut, that lose their leaves in autumn. The climate on the mainland of Europe is more diverse than the climate of the British Isles. As a result, mainland Europe has a wider variety of plant life.

Farther inland, in Germany, Austria, and Switzerland, the drier climate as well as highlands and mountain ranges support other kinds of plants. Coniferous forests have become more common. **Coniferous** trees, such as fir and pine trees, have cones and needle-shaped leaves, and they keep their leaves during the winter. Many peaks in the Alps and Pyrenees lie above the tree line. Here, grasses and shrubs are the most common plants. Much of Western Europe was once covered in forests. Most of these forests were destroyed to make towns, cities, and roads.

Wildlife

Animals have had to adapt to these changes. Deer, wild boars, hare, and mice are common. Wildcats, lynx, and foxes roam the forests. Small populations of brown bears live in the Pyrenees. In the British Isles, the population of large mammals, such as wolves, reindeer, and boars, has decreased. The islands have more than 200 kinds of birds, many of which have adapted to life in towns and cities.

France is one of the largest producers of wheat in the world; it is the largest European producer.
▶ **CRITICAL THINKING**
Analyzing What makes the soils of the Northern European Plain good for growing crops?

 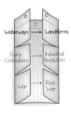
Include this lesson's information in your Foldable®.

✔ **READING PROGRESS CHECK**

Determining Central Ideas What effect did coal have on the Industrial Revolution?

Philippe Huguen/AFP/Getty Images

LESSON 1 REVIEW

Reviewing Vocabulary

1. How do the *Westerlies* affect the climate in Western Europe?

Answering the Guiding Questions

2. *Identifying* What landmasses make up the British Isles, and what countries do they form?

3. *Describing* Why is the North Sea important to Western Europe?

4. *Analyzing* How does the marine west coast climate in Western Europe differ from marine west coast climates in other parts of the world?

5. *Describing* Why has the use of coal dwindled in Western Europe? What discovery in the North Sea helped bring about that change?

6. *Informative/Explanatory Writing* Write a paragraph describing how Europe's rivers contribute to the region's industry.

Reading **HELP**DESK

Academic Vocabulary

- theory

Content Vocabulary

- smelting
- feudalism
- Middle Ages
- pilgrimage
- Parliament
- industrialized
- Holocaust

TAKING NOTES: *Key Ideas and Details*

Summarize Use a chart like this one to describe the different ways nationalism affected people in Britain, France, and Germany.

Britain	France	Germany

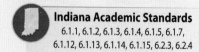

Indiana Academic Standards
6.1.1, 6.1.2, 6.1.3, 6.1.4, 6.1.5, 6.1.7, 6.1.12, 6.1.13, 6.1.14, 6.1.15, 6.2.3, 6.2.4

Lesson 2
History of Western Europe

ESSENTIAL QUESTION • *Why do civilizations rise and fall?*

IT MATTERS BECAUSE
Even though the nations of Western Europe are not geographically large, their culture and technology have had a worldwide impact.

History of the Region Through 1800

GUIDING QUESTION *How did Western Europe change from a land controlled by loose-knit tribes to a region of monarch-ruled nations?*

Western Europeans were not the first Europeans to begin farming, but the rich soil and moderate climate drew many early people to the region. As the forests were cleared for farmland, people began a long struggle to control the land.

Beginnings

Modern humans have lived in Europe for about 40,000 years. The early people were hunters and gatherers, but over time the practice of agriculture was introduced in Western Europe. Populations began to grow, and settlements became towns. People began to make tools from metal, especially bronze. To make bronze, people needed to know how to melt and fuse tin and copper, a process called **smelting**.

Roman Empire

Meanwhile, the Romans were spreading throughout Southern Europe and advancing into Western Europe. By A.D. 14, all of France and most of Germany were under Roman control. Within 100 years, Rome also controlled most of the island of Britain.

The Romans did not just conquer people and territories. They brought their beliefs, their language (Latin), and their technologies with them. They built concrete roads and bridges throughout the empire. They also built aqueducts, which carried water long distances, from remote areas to cities and towns. Some of these structures are still visible today.

Over the centuries, Rome's empire in Western Europe began to weaken. The Huns, a warrior people from Asia, invaded from the east, driving invading groups of Germanic peoples, such as the Visigoths, westward. Rome could no longer protect its colonies in Western Europe. Some Germanic groups settled there and created kingdoms. The Franks ruled what is now France, and the Angles and Saxons ruled what is today England.

Christianity and Western Europe

Christianity, which became Europe's major religion, began in the eastern Roman Empire. It gradually spread throughout the empire. Once Emperor Constantine converted to Christianity in A.D. 312, Christianity began to spread quickly. By the time Rome's western empire fell in A.D. 476, Christianity was common throughout most of Western Europe. The Christian Church played a key role in education and developed religious communities called monasteries. The Roman Catholic Church became a major force in Western European life.

The Middle Ages

As time went on, invaders threatened the region. No strong governments existed to help Western Europeans fight off invasion. To bring order, a system called **feudalism** arose. Under feudalism, kings gave land to nobles. The nobles in turn gave kings military service. Many nobles became knights, or warriors on horseback. Today, we call this period of time the **Middle Ages**, or the Medieval Age. This term describes a period of transition between ancient and modern times.

Conflicts also arose over religious beliefs. One of the most important rituals in medieval European society was the religious **pilgrimage**, a visit to lands that were important to the history of Christianity. Jerusalem was the most important destination for pilgrims, but in the late 1000s, Muslims controlled the city. Pope Urban II, leader of the Catholic Church, called for a crusade to conquer Jerusalem for Christianity. The kings and noblemen of Western Europe formed great armies to meet the pope's demand. They won Jerusalem in the First Crusade. More crusades followed, but they were not successful. Muslims regained control of Jerusalem, and Muslim power continued to grow.

Thinking Like a Geographer

The Channel in History

Because the English Channel is the most direct route from continental Europe to the island of Britain, it has served as the route of many invasions. When Julius Caesar first invaded Britain in 55 B.C., the Roman army crossed the English Channel. In World War II, American, British, and Canadian troops invaded from the opposite direction, from Britain to the beaches of Normandy in France.

This iron and bronze helmet dates from the 300s B.C.

▶ **CRITICAL THINKING**

Identifying What warrior people from Asia invaded Western Europe?

As the Crusades ended, monarchs began to lay the foundations of what became Europe's modern nation-states. A nation state is a political organization controlling a specific territory. It also includes a population that shares something in common, such as language, history, or ethnic background. In building new nations, Europe's monarchs strengthened control over territories as the power of local nobles declined. Promoting law and order, monarchs won the support of the rising merchant class, whose wealth provided the taxes needed for government services and whose writing and business skills were used in running new government offices.

Many of the crusaders returned to their European homes with changed ideas. They had seen a richer, more powerful, more modern world in the east. These ideas from the east began to spread across Northern Europe.

The economy in Western Europe was changing. Villages grew into towns. Traders and merchants began to play a bigger role in town life. Work became more specialized. People with important skills—metalsmiths, butchers, carpenters—began to organize into guilds. Guilds were not as powerful as the noblemen or the Church, but they helped the towns grow stronger.

Hundred Years' War

The threat of war between France and England flared throughout the 1200s and early 1300s. When war finally broke out between the two countries in 1337, the fighting lasted for a total of more than 100 years. England won important battles early on, gaining land in France. By the end of the Hundred Years' War, France had won all that land back. Several truces were agreed to during the war, some of them lasting many years. One of the most important developments of the war though was not a truce. It was the rapid spread of a terrible disease called a plague.

INFOGRAPHIC

BLACK DEATH

The Black Death, or bubonic plague, swept through Europe in the 1300s. It is estimated to have killed 25 million people, about a third of the region's population.

CONDITIONS / SANITATION
Garbage and human waste were often simply dumped onto city streets, attracting infected rats and further spreading the disease.

WHAT CAUSED IT?

BACTERIA: The bubonic plague is a bacterial infection caused by the bacteria *Yersinia pestis.*

FLEAS: These jumping pests carried and transmitted the bacteria with their bites.

RATS: Rats carried infected fleas. Rat bites could also transmit the infection.

TRAVEL/TRADE: Trading ships carried infected rats and fleas to ports across Europe.

KEY:

Spread of the Black Death
1347 1348
1349 1350

Location / percent of population lost to the plague

Areas totally or partially spared

IMPACT ON SOCIETY
As many as 90% of people in some villages died.

Norway

Hamburg
58%–68%

London
up to 50%

Mongols

Poland

Hungary

France

Portugal
Spain

Languedoc
40%–50%

Tuscany
60%–80%

Italy

Turkey

Morocco   Algeria

N

With the great loss of population, the production of food and goods also decreased. Workers demanded higher wages and farmers increased the price of their goods.

▶ **CRITICAL THINKING**
Analyzing Explain the effect of supply and demand on wages.

In 1347 the plague, called the Black Death, reached Western Europe, where it raged for four years. Whole towns were wiped out. Four more outbreaks struck Europe by the end of the century. Victims of the Black Death often stayed in monasteries and hospitals run by Roman Catholic officials.

Early Modern Europe

The Roman Catholic Church was wealthy and had power over numerous aspects of society. Many people wanted to reform, or change, some Church teachings and practices. For example, most people did not speak Latin. Yet, the Bible was largely available only in Latin. People began to demand translations of the Bible in their languages so that they could read and interpret it on their own.

People also began questioning the moneymaking practices of Church officials. This included the sale of indulgences, pardons from the Church for a person's sins. A German priest named Martin Luther protested this practice. He declared that only trust in God could save people from their sins. In 1517 Luther wrote the Ninety-Five Theses, a document that attacked the practice of selling indulgences. The Church expelled Luther for his beliefs, but his ideas spread quickly. His followers became known as Lutherans, and his efforts spurred a religious movement called the Protestant Reformation.

As the Catholic Church's power weakened, England's kings also were being forced to share power with a new government institution called **Parliament**. This lawmaking body was made up of two houses. The House of Lords represented the wealthy, powerful nobles. The "lower" house, or House of Commons, represented the common citizens, usually successful guild members and business owners.

The Enlightenment

A wave of discovery and scientific observation swept over Europe in the 1600s and 1700s. During this time, European explorers were traveling and mapping the world, and European astronomers were mapping the solar system.

In 1543 the Polish astronomer Nicolaus Copernicus proposed the **theory** that Earth and the other planets orbit the sun instead of the sun and other planets orbiting Earth, as was then believed. Philosophers began to consider ways of improving society. People began to use reason to observe and describe the world around them. Reason transformed the way people thought about how to answer questions about the natural world. This period is called the Enlightenment.

English philosophers John Locke and Thomas Hobbes used reason to study society itself. Locke believed that the best form of government was a contract, or agreement, between the ruler and the people. People began to question the authority of kings and of the Church.

Indiana CONNECTION

England's Political Heritage

During the Middle Ages and early modern period, important documents (the Magna Carta, the Petition of Right, and the English Bill of Rights) limited the power of English kings and recognized the rights of landowners—rights that eventually came to apply to all English people. As a result, Parliament, which came to be made up of the people's representatives, emerged as the most powerful government institution. England's system of parliamentary government shaped the political life of the United States and other countries that were for a time under English rule.

Academic Vocabulary

theory an explanation of why or how something happens

During the French Revolution in July 1789, a huge mob stormed the Bastille, an old fort used as a prison and a weapons armory.

▶ **CRITICAL THINKING**

Analyzing What was the result of the French Revolution?

Reform

In 1789 France was a powerful country, ruled by a king. Most of the people in France were peasants, living in poverty. But a growing, successful middle class resented not having a voice in government. In July of that year, a revolution limited the king's power and ended the privileges of nobles and church leaders. An important document was written: *The Declaration of the Rights of Man and of the Citizen.* It stated that government's power came from the people, not the king. A few years later, the king was removed and executed.

Not everyone in France supported the revolution. Violence raged in France for the next 10 years. Finally, in 1799 a young French general named Napoleon Bonaparte quickly took military and political control of the country. With a powerful army, he brought much of Europe under French control. By 1814, the combined might of France's enemies in conquered lands led to Napoleon's defeat and removal from power.

✓ **READING PROGRESS CHECK**

Identifying What roles did the Reformation and the Enlightenment play in changing the balance of power in Western Europe?

Change and Conflict

GUIDING QUESTION *How did the industrial system change life in Western Europe?*

During the 1800s, some Western European nations **industrialized**, or changed from an agricultural society to one based on industry. As a result, many people moved from the countryside to the city to

©Stefano Bianchetti/Corbis

find work in factories. The urban population grew, and the cities became powerful. At the same time, some Europeans began to feel strong loyalty to their country. A new, national spirit was rising.

The Industrial Revolution

A big change took place in Britain in the period from 1760 to 1830. People began to use steam-powered machines to perform work that had been done by humans or animals. For example, weavers in small villages once wove cloth on looms in their own homes, but new machines were invented to weave more cloth at greater speed for lower cost.

Machines of the Industrial Revolution did not affect only urban populations; they improved farm labor so much that fewer people were needed to work the land. People began to leave farms and villages for industrial cities where they could work in the factories.

As nations industrialized, loyalties shifted. Former enemies Great Britain and France grew closer as Germany gained military strength. As the possibility of war increased, alliances formed between countries.

World War I

Rivalries among European powers for new territory and economic power helped lead to World War I. Political changes also contributed as monarchies and empires were being replaced by modern nation-states.

A steam hammer in an English factory molds steel and iron into engine and machine parts.

▶ **CRITICAL THINKING**

Describing How did the Industrial Revolution affect rural populations?

Axis Control in World War II

NORWAY
SWEDEN
FINLAND
ESTONIA
DENMARK
LATVIA
IRELAND
GREAT BRITAIN
EAST PRUSSIA
LITHUANIA
USSR
NETHERLANDS
BELGIUM
GERMANY
POLAND
ATLANTIC OCEAN
FRANCE
SWITZERLAND
SLOVAKIA
HUNGARY
ITALY
ROMANIA
PORTUGAL
YUGOSLAVIA
Black Sea
SPAIN
BULGARIA
ALBANIA
GREECE
TURKEY
MOROCCO
ALGERIA
Mediterranean Sea
TUNISIA
LIBYA
EGYPT

60°N
50°N
40°N
30°N
0°
10°E
20°E
30°E
40°E

0 ____ 500 miles
0 ____ 500 kilometers
Lambert Azimuthal Equal-Area projection

N W E S

Maximum extent of territory under Axis control

MAP SKILLS

1 PLACES AND REGIONS Which country in central Europe did not fall under Axis control?

2 THE GEOGRAPHER'S WORLD Why would control of Southern Europe and northern Africa be important?

Fought between 1914 and 1918, World War I resulted in millions of deaths and great destruction. Germany lost the war, and the victorious countries—led by Great Britain, France, Italy, and the United States—demanded that Germany pay for damages.

The defeat nearly wrecked the German economy. Germans believed they were being punished too harshly for their role in the war. A political radical named Adolf Hitler used the people's anger to build a political party called the Nazi Party. By 1933, he was the dictator, or absolute ruler, of Germany. The Nazis believed the Germans were a superior race. They carried out the **Holocaust**, the government-sponsored murder of 6 million Jews. Other minorities also suffered at the hands of the Nazis. Hitler and his Nazi Party envisioned a new German empire.

World War II

War came when Hitler's armies began seizing other countries. World War II stretched far beyond Western Europe. Germany allied with Italy and Japan to form the Axis Powers. Great Britain, the United States, and the Soviet Union formed the Allied Powers. The war was fought in Western and Southern Europe, in Africa, and in the Pacific. A combination of American, British, and Canadian troops invaded

France in June 1944 and liberated it from the Germans.

After Hitler's death and Germany's surrender in May 1945, the war continued in East Asia and the Pacific for another three months. The fighting ended after the United States used atomic bombs on the cities of Hiroshima and Nagasaki in Japan. Worldwide, between 40 million and 60 million people died in World War II. More civilians died than military forces.

The Cold War

Before World War II, Britain, France, and Germany were among the most powerful nations in the world. However, World War II had weakened them. After the war, the United States and the Soviet Union emerged as the leading world powers. Both superpowers were interested in Europe's fate. The Soviet Union took control of most of Eastern Europe. The United States was a strong ally to nations in Western Europe. Germany was split in half, with Britain, the United States, and France occupying western Germany, and the Soviet Union controlling the eastern half.

For more than 40 years, the United States and the Soviet Union engaged in a cold war, a conflict that never erupted into war, but the threat of war always existed. Both sides stockpiled nuclear weapons. In the 1980s, Soviet influence began to weaken. Protest movements spread in European countries under Soviet control. The Cold War ended when the government of the Soviet Union collapsed in 1991.

☑ **READING PROGRESS CHECK**

Determining Central Ideas How were the causes of World War I and World War II similar? How were the causes different?

Hulton Archive/Getty Images

Nazi soldiers post a notice urging Germans not to buy from this establishment because it is owned by a Jew. Nazi actions against Jews became violent.

▶ CRITICAL THINKING
Describing What was the Holocaust?

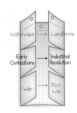

Include this lesson's information in your Foldable®.

LESSON 2 REVIEW

Reviewing Vocabulary
1. Describe how *feudalism* worked.

Answering the Guiding Questions
2. ***Identifying*** What was the economic result of the plague?

3. ***Determining Central Ideas*** How did thinking change during the historical period known as the Enlightenment?

4. ***Analyzing*** How did the Industrial Revolution change life in Western Europe?

5. ***Identifying*** What factors led to World War I?

6. ***Informative/Explanatory Writing*** Write a paragraph to discuss this statement: The printing press was one of the greatest inventions in history. Explain why you agree or disagree with the statement.

networks

There's More Online!

☑ **GRAPHIC ORGANIZER**

☑ **IMAGE** Dublin, Ireland

☑ **MAP** Empires: Western Europe

☑ **VIDEO**

Reading **HELP**DESK

Academic Vocabulary

- **cooperate**
- **regulate**
- **diverse**

Content Vocabulary

- **postindustrial**

TAKING NOTES: *Key Ideas and Details*

Identify Use a graphic organizer like the one here to identify three characteristics of major European cities.

Major European Cities

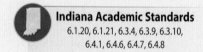

Indiana Academic Standards
6.1.20, 6.1.21, 6.3.4, 6.3.9, 6.3.10, 6.4.1, 6.4.6, 6.4.7, 6.4.8

Lesson 3
Life in Western Europe

ESSENTIAL QUESTION · *How do governments change?*

IT MATTERS BECAUSE
Western European nations recognize that in a global economy, they need to work together if their region is to prosper.

People, Places, and Cultures

GUIDING QUESTION *What are contributions of Western Europe to culture, education, and the arts?*

Western Europe's great cities are major population centers. They are also historical landmarks and tourist attractions. National capitals such as London, Dublin, Berlin, and Paris are among the world's most famous cities.

A Changing World

Political events in the 1900s threatened all of Europe. In order to survive and compete in a changing world, the nations of Western Europe needed to learn to work together.

When Western Europe began to rebuild after World War II, countries made efforts to **cooperate**. In April 1951, the Treaty of Paris called for an international agency to supervise the coal and steel industries in France, West Germany, Belgium, the Netherlands, Luxembourg, and Italy.

Those six nations then created the European Economic Community, or EEC, in 1958 to make trade among its member nations easier. The spirit of cooperation among these countries continued when they created the European Commission, or EC, in 1967. Two more Western European nations, the United Kingdom and Ireland, joined the EC in 1971. By the late 1980s, Denmark, Greece, Spain, and Portugal had also joined.

Forming the European Union

Those 12 nations formed the European Union, or EU, in 1993 with one goal being to strengthen trade among the countries of Europe. Member nations have control over their own political and economic decisions, but they also follow EU laws to **regulate** the use of natural resources and the release of pollutants. They also have agreements on law enforcement and security.

When the Soviets lost control of Eastern Europe in the late 1980s, those nations began forming their own governments. With the Soviet threat gone, East Germany and West Germany reunited. A united Germany became a strong voice in the EU.

The European Union now has 27 members. Eight of those nations lie in Western Europe: Austria, Belgium, France, Germany, Ireland, Luxembourg, the Netherlands, and the United Kingdom.

Ethnic and Language Groups

Celts, Saxons, Romans, Vikings, Visigoths, and others fought for dominance in ancient Western Europe. Those traditional ethnic divisions faded as the modern nations of Europe began to take shape. The people of a nation share a common language and a common history. Ethnic groups such as the French, the Germans, and the British rule entire countries. Their languages are the main languages of those nations.

Paul M O'Connell/Flickr/Getty Images

Academic Vocabulary

cooperate to act or work with others

regulate to adjust or control something according to rules or laws

O'Connell Street is Dublin, Ireland's, main thoroughfare. Dublin is among the world's most famous cities.

Employees prepare a food order at a fast-food restaurant that caters to the city's large Islamic population.

Identifying From what areas did many Muslims come to Western Europe?

Western Europe is home to significant numbers of other ethnic groups who are minorities in the country. Many are immigrants. They often speak the language of their homeland and continue their own culture and way of life.

Most people in the region speak one of the Indo-European languages. Indo-European is a family of related languages. It includes languages spoken in most of Europe, parts of the world that were colonized by Europeans, Persia, India, and some other parts of Asia.

Two major divisions of Indo-European languages spoken in Western Europe are Romance and Germanic. Romance languages are based on Latin, the language of the Roman Empire. The most common Romance language in Western Europe is French. The Germanic languages spoken in Western Europe include German, Dutch, and English, although about half of the English vocabulary comes from the Romance languages. Not all European languages are Indo-European, however. For example, Basque, a language spoken in the Pyrenees region of France and Spain, is unrelated to any other language spoken today. It is common for Western Europeans to speak more than one language—their native language in addition to English, French, or German.

Religion in Western Europe

Romans accepted Christianity, and Christian missionary-monks spread their religion during the Middle Ages. Christianity continues as Europe's major religion. Germany was the birthplace of the Protestant Reformation, and Western Europe was the first place that Protestantism took hold. Today, most Western European Christians are either Catholic or Protestant. The Roman Catholic faith is strongest in France, Ireland, and Belgium. Protestant churches are strongest in the United Kingdom and Germany.

Immigration from Africa and Asia has brought many Muslims to Western Europe, especially to the United Kingdom, France, Germany, Austria, the Netherlands, and Switzerland. Muslims follow the religion of Islam.

World War II and the Holocaust nearly wiped out Europe's Jewish population. Today, Europe's Jewish communities are growing in Western Europe, especially in France, the United Kingdom, and Germany.

Literature, Music, and the Arts

For centuries, Western Europe has been a world leader in the arts and culture. As European explorers spread European culture to other parts of the globe, the names of their greatest artists became known worldwide. England's William Shakespeare is one of the most famous playwrights in the world, nearly 400 years after his death. The music of German and Austrian composers such as Bach, Mozart, Beethoven, and Schubert is among the most important in all of classical music. The paintings of great artists from France, the Netherlands, and Belgium are among the most treasured in the world.

The arts are an important part of Western European culture. Museums and cultural institutions celebrate each nation's art and history, and national governments support the arts. The German government, for example, funds hundreds of theaters, and concerts and plays attract large audiences. Most important is the influence Western European culture has had on the rest of the world. German architects from the Bauhaus School influenced buildings in cities throughout the 1900s. British popular music and television have had an impact, especially on American culture.

MAP SKILLS

1 **THE GEOGRAPHER'S WORLD** The original members of the EU are from what part of Europe?

2 **HUMAN GEOGRAPHY** From a geographical standpoint, would you predict that the EU in the future will grow, stay the same, or decrease in members? Explain.

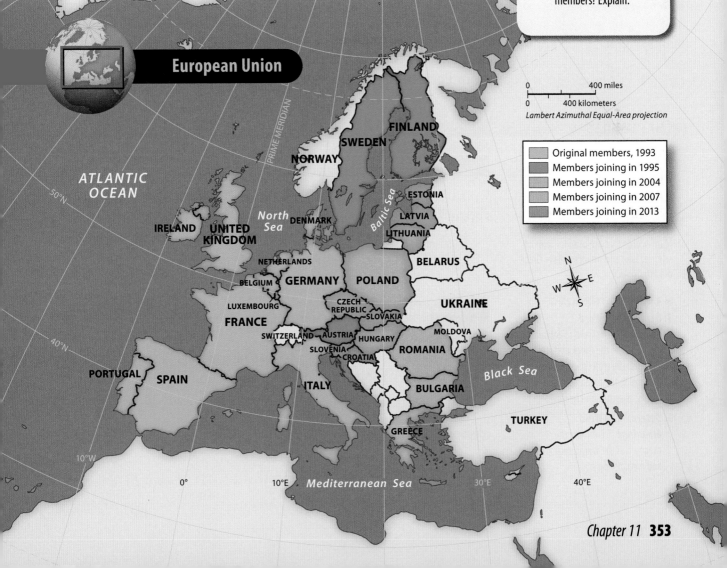

European Union

0 ——— 400 miles
0 ——— 400 kilometers
Lambert Azimuthal Equal-Area projection

Original members, 1993
Members joining in 1995
Members joining in 2004
Members joining in 2007
Members joining in 2013

Daily Life

The most popular team sport across Western Europe is football—what Americans call soccer. Professional leagues have formed throughout Western Europe. In the United Kingdom, cricket and rugby are popular team sports. Switzerland and Austria's rugged Alps and plentiful winter snow make mountain climbing, skating, downhill skiing, and cross-country skiing popular in both countries.

Because so much of the population of Western Europe lives in cities, roads are crowded. Automobile traffic and pollution are extensive in parts of the region. In Switzerland, traffic congestion has created serious air pollution in the Alpine valleys. To relieve congestion and address problems with pollution, much of Europe turned to high-speed rail travel.

In many areas, tradition is part of their everyday lives. The people of Scotland and Wales, for example, take pride in their ancient languages—Scottish Gaelic in Scotland and Welsh in Wales. These languages are taught in schools to keep the old cultures alive.

Railways and Highways

Europeans first began riding high-speed rail lines in France in 1981. France went on to build a high-speed line connecting all of its major cities. These trains travel at speeds of up to 185 miles (298 km) per hour. In the 1990s, the French high-speed rail lines began connecting to other high-speed rail lines: from Paris to London via a tunnel beneath the English Channel, from Paris to the Netherlands, and from Paris to Brussels, Belgium.

CHART SKILLS >

ENGLISH WORDS FROM OTHER LANGUAGES

Many of the words we use every day are derived from French or German. Many words that are used in fields of science, law, and medicine have Greek or Latin roots.

▶ **CRITICAL THINKING**

1. *Determining Word Meanings* The word *telephone* is derived from the Greek prefix *tele* (far) and the suffix *phone* (voice). Determine the meaning of the words: *telescope* and *telecast*.

2. *Determining Word Meanings* Determine the meaning of the words: *microphone* and *megaphone*.

French	German	Latin	Greek
ballet	kindergarten	a.m. /p.m.	atmosphere
denim	blitz	census	comedy
garage	poltergeist	millennium	democracy
infantry	noodle	lunar	geography
salon	hamster	solar	pediatrician
	pretzel		

Here are other English words derived from other languages:

African: banana, cola, jazz, zebra

Arabic: algebra, chemistry

Spanish: breeze, canyon, mesa

Japanese: anime, tycoon, tsunami

Norwegian: fjord, ski, slalom

The Louvre in Paris, France, is one of the world's great museums and the most-visited art museum in the world.

A well-developed highway system also links Europe's major cities. Germany's superhighways, called autobahns, are among Europe's best roads. Many European countries are participating in the Forever Open Roads project. By combining efforts to develop innovative technology, the planners are working to transform the way roads are designed, built, and maintained in the twenty-first century.

Education

Western Europe is one of the wealthiest, most urban, and well-educated regions in the world. In most of Western Europe, school is mandatory until students reach the age of 16, but many students then attend college.

Western Europe contains some of the oldest and most renowned universities. Oxford University in England and the University of Paris opened their doors to students before 1200. Many universities started at this time after Pope Gregory VII issued a ruling calling for the creation of schools of education for the clergy. Hundreds of secular colleges—those without religious affiliation—were established by the 1400s.

☑ **READING PROGRESS CHECK**

Describing In what ways do nations of Western Europe support art and culture?

Buena Vista Images/The Image Bank/Getty Images

Current Challenges

GUIDING QUESTION *Why is Western Europe considered a postindustrial region?*

France, Germany, and the United Kingdom are not the military giants they were in 1900, but they still have some of the biggest economies in the world. A global financial crisis, however, has hurt the entire region since the early 2000s.

Earning a Living

Since the Industrial Revolution, improvements in agriculture have made it possible for fewer people to cultivate larger areas of land. Today, more than half the population of Western Europe lives and works in cities. Even in France, Western Europe's leading agricultural nation, less than 4 percent of the workforce works in agriculture.

In the past few decades, the number of industrial workers has also declined. Only about 25 percent of Western Europeans work in the industrial, or secondary sector, of the economy. Many more people work in the tertiary sector, which is service industries. This sector includes government, education, health care, banking and financial services, retail, computing, and repair of mechanical equipment. The United Kingdom was the birthplace of modern industry. Yet today, only 18.2 percent of the workforce in the United Kingdom works in industry.

MAP SKILLS

1 **THE GEOGRAPHER'S WORLD** Which empire controlled the largest territory?

2 **THE GEOGRAPHER'S WORLD** What effect do you think world wars had on the British, French, and German empires?

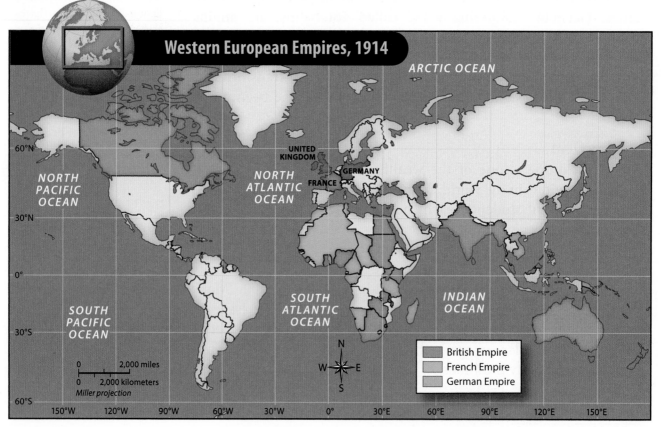

Western European Empires, 1914

Legend:
- British Empire
- French Empire
- German Empire

When the economy of a country depends more on services than it does on industry, that country is said to be **postindustrial**. Every nation in Western Europe has a postindustrial economy.

Challenges

For hundreds of years, the nations of Western Europe were among the most powerful in the world. In 1900 Great Britain, France, and Germany ruled over empires that extended beyond Europe to Asia, Africa, the Americas, and the Pacific Islands. The 1900s was hard on Western Europe. The two world wars did extensive damage to nearly the entire region. Then the Cold War kept Western Europe on the brink of war for more than 40 years.

Even so, Germany, France, and the United Kingdom have been economically strong for a long time and remain among the seven biggest economies in the world. The cooperation made possible by the European Union helps Western European nations compete with larger economies, such as the United States, China, and Japan. For that to continue, the economies of all the EU member nations must be healthy. The global financial crisis of 2008, however, had an impact on all of Europe. Governments of the EU disagreed about how to deal with ongoing financial problems.

Immigration Brings Changes

The population of Western Europe is changing. Most population growth in Western Europe is caused by immigration. Many people come to Western Europe from Africa, Asia, and Eastern Europe looking for job opportunities or trying to escape political oppression. When they immigrate, they bring parts of their culture with them, including their religions. Germany, France, the Netherlands, and the United Kingdom each have large Muslim populations. The mix of European and immigrant cultures creates a richer, more **diverse** culture, but it also creates racial and religious tensions. To avoid these problems, some countries have attempted to restrict immigration.

☑ **READING PROGRESS CHECK**

Determining Central Ideas What challenges do the nations of Western Europe face?

Academic Vocabulary

diverse different from each other

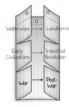
Include this lesson's information in your Foldable®.

LESSON 3 REVIEW

Reviewing Vocabulary
1. Define *postindustrial* as it relates to industry and services.

Answering the Guiding Questions
2. *Identifying* Name one advantage and one disadvantage resulting from the creation of the European Union.

3. *Analyzing* How did the collapse of the Soviet Union affect the European Union?

4. *Determining Central Ideas* Why is Western Europe considered a postindustrial region?

5. *Argument Writing* Write an open letter to the nations of Western Europe explaining the need to form the European Union.

What Do You **Think?**

Is the European Union an Effective Economic Union?

The European Union was formed to build peace among its member states, but it also aims to build prosperity. In the 1990s, member states agreed to do away with trade barriers, create a single market, and make the euro their common currency. But the global financial crisis of 2008 hurt the EU greatly. Greece, Ireland, and Portugal needed bailouts to cover their huge debts. Italy and Spain struggled economically, too. EU leaders disagreed over how to respond to the crisis, and tensions still exist.

No !

PRIMARY SOURCE

❝ The creation of the eurozone was presented as an unambiguous [clear] economic benefit to all the countries willing to give up their own currencies that had been in existence for decades or centuries. . . .[S]tudies promised that the euro would help accelerate economic growth and reduce inflation and stressed . . . that the member states of the eurozone would be protected against all kinds of unfavorable economic disruptions . . . It is absolutely clear that nothing of that sort has happened. After the establishment of the eurozone, the economic growth of its member states slowed down . . . Two distinct groups of countries have formed within the eurozone—one with a low inflation rate and one (Greece, Spain, Portugal, Ireland and some other countries) with a higher inflation rate. . . . [T]he economic performance of individual eurozone members diverged [went in different directions] and the negative effects of the 'straight-jacket' of a single currency over the individual member states have become visible. . . . [A]s a project that promised to be of considerable economic benefit to its members, the eurozone has failed." ❞

—Václav Klaus, president of the Czech Republic

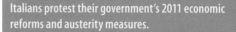

Italians protest their government's 2011 economic reforms and austerity measures.

TEXT: From "WHEN WILL THE EUROZONE COLLAPSE?" by Václav Klaus. Published May 26, 2010, Cato Institute Economic

Cargo containers at Felixstowe in the United Kingdom are prepared for transfer to cargo ships.

Yes!

PRIMARY SOURCE

" [T]he single market has transformed for the better many aspects of European life. . . . People move freely across most borders. . . . Going to work in another Member State is much easier. . . . Goods are no longer delayed for hours or days at borders by heavy paperwork: this makes delivery times shorter, allowing manufacturers to save money and reduce prices for customers. . . . Consumer choice is vast: the range of products on sale across the EU is wider than ever and in most cases prices are easily compared thanks to the euro. Manufacturers have to keep prices down because they are selling into one huge competitive market. . . . Capital—the investment that businesses need to start and to grow—flows easily within the single market, sustaining companies and generating jobs. "

—The European Commission

What Do You Think?　DBQ

① **Determining Central Ideas** How has the EU's single market benefited travelers, workers, consumers, and manufacturers?

② **Identifying Point of View** Why does Václav Klaus consider the EU's single currency to be a "straight-jacket"?

Critical Thinking

③ **Analyzing** Some people think the eurozone will dissolve. Others, including Václav Klaus, believe that it will survive, especially if new rules force member countries to behave responsibly. What factors might affect what happens in the future?

Directions: Write your answers on a separate piece of paper.

1 Use your FOLDABLES to explore the Essential Question.

INFORMATIVE/EXPLANATORY WRITING Write an essay explaining how the physical geography and the climate of Western Europe influenced the development of agriculture in the region.

2 21st Century Skills

INTEGRATING VISUAL INFORMATION Working in small groups, choose any country in Western Europe and create a slide show depicting the country's landforms, waterways, and natural resources. Include images of the country's capital city. Describe at least one cultural tradition or holiday that is not familiar to you or to most Americans.

3 Thinking Like a Geographer

INTEGRATING VISUAL INFORMATION Fill in this graphic organizer to help you remember the climates of Western European countries.

Climate	Characteristics

4 **GEOGRAPHY ACTIVITY**

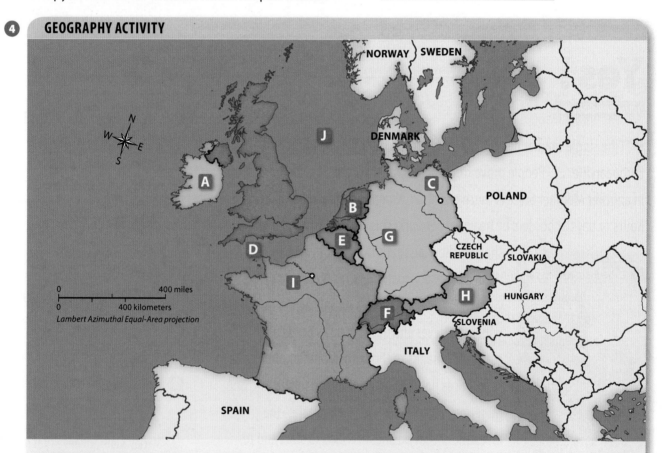

Locating Places

Match the letters on the map with the numbered places below.

1. Ireland   **4.** Switzerland   **7.** Austria   **10.** North Sea

2. English Channel   **5.** the Netherlands   **8.** Rhine River

3. Paris   **6.** Belgium   **9.** Berlin

REVIEW THE GUIDING QUESTIONS

Directions: Choose the best answer for each question.

1 People in the Netherlands build dikes and reclaim land from the sea that they call

  A. peat bogs.

  B. estuaries.

  C. polders.

  D. reserves.

2 Which two mountain ranges separate Western Europe from Southern Europe?

  F. Alps and Urals

  G. Pyrenees and Alps

  H. Middle Rhine Highlands and Carpathians

  I. Tian Shan and Apennines

3 In which Western European country did the Industrial Revolution begin?

  A. Germany

  B. France

  C. the Netherlands

  D. Great Britain

4 The system of land ownership and farming during the Middle Ages was called

  F. pilgrimage.

  G. channeling.

  H. feudalism.

  I. slavery.

5 The Industrial Revolution resulted in

  A. a shortage of workers on farms.

  B. a great migration of people from rural areas to cities.

  C. more time for people to spend with their families.

  D. handmade goods.

6 Many Europeans now work in the tertiary sector, which includes

  F. factory work.

  G. construction.

  H. retail sales.

  I. mining.

DBQ ANALYZING DOCUMENTS

7 DETERMINING CENTRAL IDEAS Read the following passage:

"*Intense green vegetation . . . covers most of [Ireland]. . . . Ireland owes its greenness to moderate temperatures and moist air. The Atlantic Ocean, particularly the warm currents in the North Atlantic Drift, gives the country a more temperate climate than most others at the same latitude.*"

—from NASA, Earth Observatory Web Site (2011)

Based on this passage, what is unusual about Ireland's climate?

A. warmer than expected given its latitude

B. colder than expected given its nearness to the ocean

C. wetter than expected given that it is an island

D. drier than expected given its northerly location

8 ANALYZING Which economic activities in Ireland most likely benefit most from this climate?

F. manufacturing and transportation

G. agriculture and tourism

H. finance and retailing

I. mining and fishing

SHORT RESPONSE

"*The European Union [EU] . . . was created in the aftermath of the Second World War. The first steps were to foster economic cooperation: the idea being that countries who trade with one another become economically interdependent and so more likely to avoid conflict.*"

—from European Union, "Basic Information" (2012)

9 IDENTIFYING What was the original purpose of the European Union?

10 CITING TEXT EVIDENCE Why did World War II convince European leaders to form the European Union?

EXTENDED RESPONSE

11 INFORMATIVE/EXPLANATORY WRITING Explain why more countries have joined the European Union since its founding and whether you think it has been successful.

Need Extra Help?

If You've Missed Question	**1**	**2**	**3**	**4**	**5**	**6**	**7**	**8**	**9**	**10**	**11**
Review Lesson	1	1	2	2	2	3	1	3	3	3	3

http://europa.eu ©European Union, 1995-2012

Northern and Southern Europe

ESSENTIAL QUESTIONS · *How do people adapt to their environment?*
· *Why do civilizations rise and fall?* · *How do new ideas change the way people live?*

Christophe Boisvieux/age fotostock

**Norwegian fisherman from a
Lofoten Island fishing village**

networks

There's More Online about Northern and
Southern Europe.

CHAPTER 12

Lesson 1
*Physical Geography of
the Regions*

Lesson 2
History of the Regions

Lesson 3
*Life in Northern and
Southern Europe*

The Story Matters...

**The people of Northern Europe
live in a land of cold, harsh
winters. In contrast, the people
of Southern Europe enjoy the
warm Mediterranean climate.
What do these people have in
common? The countries in this
subregion have long histories,
dating back to the Vikings in the
north and the ancient Greeks and
Romans in the south. These
ancient peoples influenced many
aspects of culture—from
exploration to philosophy to the
discovery of new ideas.**

FOLDABLES®
Study Organizer

Go to the Foldables® library in the back
of your book to make a Foldable® that
will help you take notes while reading
this chapter.

Northern and Southern Europe

Geography | History | Culture

NORTHERN AND SOUTHERN EUROPE

Despite the distance between Northern and Southern Europe, many of the regions' countries are members of the European Union. Their histories show conflict, but today they are unified as members of the European continent, with similar economic goals.

Step Into the Place

MAP FOCUS Use the map to answer the following questions.

1 **THE GEOGRAPHER'S WORLD** Northern and Southern Europe have several large peninsulas. Identify two peninsulas in Northern Europe and three in Southern Europe, and name the countries of each.

2 **PLACES AND REGIONS** With so many countries bordering water, such as the North Sea, the Atlantic Ocean, and the Mediterranean Sea, what activity would you expect to find in Northern and Southern Europe?

3 **PLACES AND REGIONS** Name the capital cities of Spain, Italy, Norway, and Sweden.

4 **CRITICAL THINKING**
Analyzing With the difference in latitude between Northern and Southern Europe, how would you expect their climates to differ?

SOUTHERN COASTLINE Warm, bright sunshine floods the ruins of an ancient temple that overlooks the Mediterranean Sea in Greece.

NORTHERN COASTLINE A red fox walks across rocks in a wilderness area of northern Finland. The rugged land in far northern areas of Europe was shaped by retreating glaciers at the end of the last Ice Age.

Step Into the Time

TIME LINE Choose an event from the time line and write a journal entry from the point of view of a teenager living during that time. Describe how that event has changed your perception of the world.

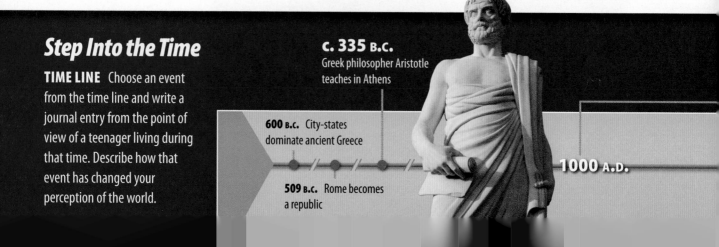

c. 335 B.C. Greek philosopher Aristotle teaches in Athens

600 B.C. City-states dominate ancient Greece

509 B.C. Rome becomes a republic

1000 A.D.

Northern and Southern Europe

ICELAND
Reykjavík

ARCTIC CIRCLE

PRIME MERIDIAN

60°N

Faeroe
Islands
(Denmark)

Norwegian
Sea

FINLAND

SWEDEN

NORWAY

Helsinki

N
W · E
S

50°N

Oslo · Stockholm

North
Sea

DENMARK

Baltic Sea

Copenhagen

ATLANTIC
OCEAN

B

WESTERN
RUSSIA

| ⊙ | National capital |
| • | City |

0 ____ 400 miles
0 ____ 400 kilometers
Lambert Azimuthal Equal-Area projection

40°N

WESTERN
EUROPE

EASTERN
EUROPE

PORTUGAL

Ebro R.

Andorra
la Vella

Milan

Po R.

Black Sea

Madrid

ANDORRA

Florence

SAN
MARINO

Tagus R.

Lisbon

SPAIN

ITALY

Rome

Istanbul

Balearic
Islands
(Spain)

Sardinia
(Italy)

Naples

TURKEY

ASIA

VATICAN CITY
(within Rome)

GREECE

Aegean Sea

Strait of
Gibraltar

10°W

A

Athens

Nicosia

Sicily
(Italy)

CYPRUS

30°N

AFRICA

MALTA ⊙ Valletta

Mediterranean Sea

Crete
(Greece)

CYPRUS

0°   10°E   30°E

---

c. 1000
Leif Eriksson reaches
North America

1436 Gutenberg
invents printing press

1492 Columbus sets
sail to the Americas

1995 Sweden joins the
growing European Union

1945 World War II ends

1400 — **1900** — **2000**

c. 1400 During Renaissance, astonishing
developments in arts and literature occur

1610 Italian astronomer Galileo
discovers moons orbiting Jupiter

1992 European Union forms

2008 Financial crisis leads
to worldwide recession

Reading **HELP**DESK

Academic Vocabulary

- **uniform**

Content Vocabulary

- **glaciation**
- **fjord**
- **tundra**
- **scrubland**
- **trawler**

TAKING NOTES: *Key Ideas and Details*

Identify Choose one of the countries in the region. List important resources of that country on a graphic organizer like the one below.

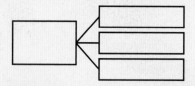

Indiana Academic Standards
6.3.7, 6.3.8, 6.3.12, 6.4.7

Lesson 1
Physical Geography of the Regions

ESSENTIAL QUESTION · *How do people adapt to their environment?*

IT MATTERS BECAUSE
Northern Europe and Southern Europe have different landforms, climates, and resources. The geography of the regions presents unique challenges to the people who live there.

Landforms and Waterways

GUIDING QUESTIONS *How are the landforms in Northern and Southern Europe similar? How are they different?*

Much of Northern Europe is a land of rugged mountains, rocky soils, and jagged coasts. A map of Northern Europe would show Denmark, Sweden, Norway, Finland, and Iceland. Together, these five far-northern lands are often called the Nordic countries.

The Mediterranean Sea dominates the coast of much of southern Europe, affecting the climate and the movement of people. Southern Europe is made up of Spain, Portugal, Italy, and Greece, as well as the tiny countries of Andorra, San Marino, and Vatican City. It also includes the island countries of Malta and Cyprus, the westernmost part of Turkey, and the tiny British territory of Gibraltar.

A Land of Peninsulas

A peninsula is an area of land surrounded on three sides by water. In the United States, Florida is a good example of a peninsula.

Two peninsulas make up most of Northern Europe. Jutland is a peninsula that extends northward from Germany and includes most of Denmark. The Scandinavian Peninsula is made up of Norway and Sweden. The large landmass east of Sweden is Finland. Much of the land of Norway, Sweden, and Finland lies north of 60° N latitude.

Northern Europe also includes many islands. Iceland is a large island in the northern Atlantic Ocean near the Arctic Circle. Denmark has about 400 islands. Its capital, Copenhagen, is on the largest of the islands.

Southern Europe also has several peninsulas. Spain and Portugal form the Iberian Peninsula. Most of Italy is the long, boot-shaped Italian peninsula. East of Italy, the larger Balkan Peninsula includes several Eastern European nations, with Greece at its southern tip.

In Southern Europe, the large Mediterranean islands of Sicily and Sardinia are part of Italy. The nearby islands of Malta form an independent country. The island of Crete is part of Greece. Farther east, the island of Cyprus contains the largely Greek but independent nation of Cyprus, as well as North Cyprus, a Turkish territory.

Mountains and Plains

The Scandinavian Peninsula has a spine of rugged mountains, formed when two tectonic plates collided. **Glaciation**, or the weathering and erosion caused by moving masses of ice called glaciers, carved the land into the mountains and plateaus we see today.

Iceland also has rugged terrain, but it formed from volcanic activity. The island is part of a mountain range, the Mid-Atlantic Ridge, which is mostly underwater. At Iceland, it rises above sea level. Iceland is home to more than 200 volcanoes and many hot springs, as well as Europe's largest glacier.

Part of the boundary between Western Europe and Southern Europe is formed by two mountain ranges, the Pyrenees and the Alps. The Pyrenees mark the boundary between southern France and the Iberian Peninsula. The Alps form the northern border of Italy and separate the Italian peninsula from the rest of Europe.

Hikers pass beneath steep rock formations in the Dolomites, a branch of the Alps in northeastern Italy.

▶ **CRITICAL THINKING**
Describing Describe the geographical relationship of the Alps to the Italian peninsula.

Maremagnum/Photographer's Choice/Getty Images

Some countries of Southern Europe are mountainous. The Apennines extend along the length of Italy. They are volcanic and subject to earthquakes. Greece also has rugged highlands. The tallest and most famous of its mountains is Mount Olympus, which legend says was the home of the gods of Greek mythology.

Most of Spain lies on a plateau called the Meseta Central. It is a harsh landscape. To the north and south of the Meseta Central are mountain ranges, but its western side slopes gently toward the Atlantic Ocean. Valencia, along Spain's eastern coast, is a coastal plain of rolling hills.

Portugal is divided geographically by the Tagus River. South of the river, the landscape is characterized by extensive rolling plains. Throughout the area are abundant trees and plants, including evergreen oak trees, olive trees, figs, and vineyards. The northeastern coast features a landscape of wide valleys and steep hills.

The Mediterranean Sea

The most important body of water in Southern Europe is the Mediterranean Sea. It stretches about 2,500 miles (4,023 km) from the southern coast of Spain in the west to the coasts of Greece, Turkey, and various countries of Southwest Asia. The Mediterranean is almost completely surrounded by land. In the west, it connects to the Atlantic Ocean through the Strait of Gibraltar. At the Strait's narrowest point, just 8 miles (13 km) separates the southern tip of Spain from Africa.

Waterways

Among the important rivers of Southern Europe is Italy's Po. Many smaller rivers drain into the Po as it travels from the Alps to the Adriatic Sea. The Ebro is the longest river that lies entirely in Spain. It originates in the Cantabrian Mountains in the north and ends at Spain's Mediterranean coast. The longest river on the Iberian Peninsula is the Tagus, which crosses Spain and Portugal on its way to the Atlantic Ocean.

Northern Europe has few important rivers, but it has a long coastline, indented by many seas and bays. Norway is surrounded on three sides by water, and much of Norway's west coast is dotted by narrow, water-filled valleys called **fjords**.

The Baltic Sea borders Sweden's southern and southeastern coasts. Sweden has major ports on the Baltic, but the sea is more likely to freeze than other bodies of salt water. This is because the Baltic is shallow and has a fairly low concentration of salt. Finland also borders the Baltic Sea. Its interior is better known for its many lakes. Glaciers have carved as many as 56,000 lakes in Finland.

✓ READING PROGRESS CHECK

Analyzing Which landforms best characterize Northern Europe?

Visual Vocabulary

Fjord A fjord is a deep, narrow, sea-filled valley at the base of steep cliffs.

©Doug Pearson/JAI/Corbis

Contrasting Climates

GUIDING QUESTION *How is the climate of Northern Europe different from the climate of Southern Europe?*

Northern Europe has a cool or cold climate. Yet, the conditions in some places are much harsher than in others. Landforms, distance from the sea, and ocean currents play a part in determining climate.

Chilly Northern Europe

The northern part of Norway is located as far north as Alaska, but it is not nearly as cold. The Norwegian Current, part of the Gulf Stream, flows past Norway carrying warm water from the tropics. As a result, western Norway has a marine climate, with mild winters and cool summers. The relatively mild climate does not extend far to the east. Mountains reduce the eastward flow of milder air. Eastern Norway thus has colder and snowier winters. Even so, the climate in eastern Norway and in Sweden is milder than in other parts of the world at the same latitude.

Finland's climate is considered continental because it receives little influence from the seas. It has cold winters and hot summers. Winters are harsh, especially north of the Arctic Circle. In the mountainous parts of northern Finland, the snow never melts. This land is mostly **tundra**, a region where subsoil is frozen and only plants such as lichens and mosses can survive.

The Gulf Stream brings warm water to the southern and western coasts of Iceland. This moderates the temperatures in these parts of the country. The effect does not reach northern Iceland, which just touches the Arctic Circle. There, drifting ice and fog are common in winter, and temperatures year-round are several degrees colder than they are in the southern and western parts of the country.

Warm Southern Europe

The most common climate in Southern Europe is Mediterranean. The climate features warm or hot summers and cool or mild winters. Spring and fall are rainy, but summers are dry. If you have ever been to the southern coast of California, you have experienced a Mediterranean climate.

Temperatures are not **uniform** across Southern Europe or even within countries. Northern Italy is nearer the Alps and has a cooler mountain climate, where winters are colder than in southern Italy and snow is heavy at higher elevations. The Meseta Central, a vast plateau in Spain, on the other hand, has a continental climate.

Academic Vocabulary

uniform the same or similar

Vacationers enjoy an attractive beach in Ibiza, one of the Balearic Islands. These scenic islands in the Mediterranean Sea are part of Spain.

▶ **CRITICAL THINKING**

Analyzing Why might many Northern Europeans vacation in Spain and other Mediterranean countries during the winter months?

Katja Kreder/age fotostock

The area's high elevation and mountain barriers cause dry winds and drought conditions year-round. The dryness causes temperature extremes, with cold winters and hot summers.

☑ **READING PROGRESS CHECK**

Identifying How do landforms and waterways affect the climates of Norway and Italy?

Natural Resources

GUIDING QUESTION *What natural resources are available to the people of Northern and Southern Europe?*

Northern and Southern Europe hold rich stores of resources. The sea also provides a variety of resources, from fish to oil and gas.

Vegetation

Plants need to be drought resistant in order to survive the dry summers in a Mediterranean climate. Because of the dry climate and poor soils in Southern Europe, many areas are **scrubland**, or places where short grasses and shrubs are the dominant plants. Trees such as olive, fig, and cypress are common. Two of the most important crops throughout Southern Europe are grapes and olives. Wine, which is made from grapes, is an important export for Spain, Portugal, Italy, and Greece. Italy and Greece are also major exporters of olive oil.

The Sautso Dam in Norway spans northern Europe's largest canyon. The water reservoir behind the dam is 11.8 miles (19 km) long.

▶ **CRITICAL THINKING**

Identifying What are Norway's main exports?

In widely forested Northern Europe, the main plant resource is wood. Forests cover nearly three-fourths of Finland, and wood from these forests is Finland's most important natural resource. Finland exports birch, spruce, and pine wood and paper products to Western Europe. Sweden's forests produce timber, paper, wood pulp, and furniture.

Northern Minerals and Energy

Norway might not be considered an energy powerhouse. But, thanks to the discovery of petroleum in the North Sea, Norway is now Europe's biggest exporter of oil and one of Europe's leading suppliers of natural gas.

Northern Europe also has rich mineral ore resources. With deposits of iron ore, copper, titanium, lead, nickel, and zinc, Norway remains one of the world's leading metal exporters. Sweden lacks fossil fuels, but it has rich mineral resources. These include iron ore, copper, gold, zinc, and lead.

Arnulf Husmo/Stone/Getty Images

Denmark has few natural mineral or energy resources. However, Denmark uses wind turbines to supply much of its electricity. Sweden and Norway get much of their electricity from hydroelectric power plants. Because of the volcanic nature of the island, Iceland has enormous reserves of geothermal energy. It provides energy for industries and all of the heating needs of Reykjavik, Iceland's major industrial area, as well as several other towns. Iceland's rivers also supply hydroelectric power.

Sea Resources

A rich variety of fish inhabit the North Sea, and the nations along its shores have long-standing fishing traditions. Norway has one of the biggest fishing industries in Europe. Some of the fish caught, such as sand eels and mackerel, are ground into fish meal, a powder that is used in animal feed and fertilizer. Today, fewer people work in the fishing industry. This is because ships that tow large nets behind them, called factory ships or **trawlers**, have increased catches.

At one time, whaling was an important industry for many countries by the sea, including Norway. Whales were hunted for their meat and oil, but by the mid-1900s, many of the largest species of whales were in danger of becoming extinct. Whaling is now limited to a few smaller species that are not believed to be endangered.

Spain and Portugal have extensive Atlantic Ocean coastlines, and fishing is an important industry in both countries. The cities of Vigo and La Coruña are Spain's biggest fishing ports. Spanish fishing fleets range far from their shores, however, leading to conflicts with other countries.

The Mediterranean Sea has long been an important fishing ground for the countries that border its shores, including Italy and Greece. However, overfishing and pollution have reduced fish populations. In addition, the Mediterranean Sea lacks the nutrients necessary to support large populations of fish. Commercial fisheries use fish hatcheries to cultivate and breed fish. After the fish mature, they are released into the sea. That way, they repopulate the fish population that has been overfished.

☑ **READING PROGRESS CHECK**

Analyzing How can a thriving fishing industry be a positive and a negative factor for a country?

FOLDABLES
Study Organizer

Include this lesson's information in your Foldable®.

Northern and Southern Europe
Geography | History | Culture

LESSON 1 REVIEW

Reviewing Vocabulary

1. How does *glaciation* affect the landscape?

Answering the Guiding Questions

2. *Describing* How are the landforms of Northern and Southern Europe alike and different?

3. *Describing* How would the climate of Norway be different without the Norwegian Current?

4. *Analyzing* What kinds of problems does the fishing industry in Southern Europe face?

5. *Informative/Explanatory* Describe what is unique about the island of Iceland.

networks

There's More Online!

- ☑ **MAP** Greek and Roman Empires
- ☑ **SLIDE SHOW** The Renaissance
- ☑ **VIDEO**

Reading **HELP**DESK

Academic Vocabulary

- achievement
- convert
- rational

Content Vocabulary

- city-state
- longship
- pagan
- Renaissance

TAKING NOTES: *Key Ideas and Details*

Summarizing Choose one of the countries in the lesson. As you read, use a graphic organizer like the one below to describe three or more important events in that country's history.

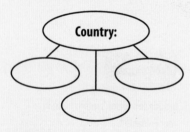

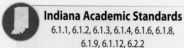

Indiana Academic Standards
6.1.1, 6.1.2, 6.1.3, 6.1.4, 6.1.6, 6.1.8,
6.1.9, 6.1.12, 6.2.2

Academic Vocabulary

achievement the result gained by a great or heroic deed

372

Lesson 2
History of the Regions

ESSENTIAL QUESTION • *Why do civilizations rise and fall?*

IT MATTERS BECAUSE
Some of the most influential civilizations in the world originated in these two regions. As you read, think about how the accomplishments of the past influence us today.

Early History of the Regions

GUIDING QUESTION *Why were early civilizations in Northern and Southern Europe important?*

Southern Europe produced two of the world's most influential civilizations: ancient Greece and ancient Rome. Greek and Roman, or classical, culture continue to affect our world. During the Middle Ages, the period between A.D. 500 and 1500, Christianity and other classical ideas helped build a new, orderly European civilization. But wars and invasions were still common. The Vikings of Northern Europe were seafarers and invaders. Their voyages changed the history of Western Europe and North America.

Ancient Greece

Greece's many mountains and seacoasts influenced the ancient Greeks to form separate communities called **city-states**. Each city-state was independent, but each one was linked to the other city-states by Greek language and culture. Powerful city-states, such as Athens and Sparta, were rivals, but they faced a common enemy to the east in mighty Persia. When the Persians invaded the Greek mainland in 490 B.C., the combined forces of Athens's navy and Sparta's army spent 40 years defeating them.

After the Persian Wars, Athens emerged as the most developed city-state. Its **achievements** were momentous. For

(l to r) Sigurgeir Jonasson/Nordic Photos/Getty Images; Salvator Barki/Flickr/Getty Images; Keystone-France/Gamma-Keystone/Getty Images

example, the ideas of philosophers such as Socrates, Plato, and Aristotle are still studied. Greek art set a standard for beauty that later influenced the Romans and inspires people to this day. Athens was also the first known democracy. The free citizens of Athens enjoyed a way of life that was unique in ancient times.

Wars weakened the Greek city-states, and Macedon, a kingdom north of Greece, took advantage. The Macedonian king, Alexander the Great, extended his rule over not only Greece, but also Asia Minor (now the Asian part of Turkey), Persia, and Egypt. Even though Alexander died at the age of 33, he spread Greek culture throughout an empire that lasted another 300 years.

Roman Empires

While the Greek city-states were at their height, another group was slowly gaining power to the west on the Italian peninsula. A series of small settlements built on hills along the Tiber River eventually merged into a single city that became Rome. Later, the Romans put government into the hands of consuls, who were elected to office annually. This was the birth of the Roman Republic.

The Romans had a talent for warfare, and they set out to conquer their neighbors. By 275 B.C., they controlled the Italian peninsula, inspiring Rome to add even more territory.

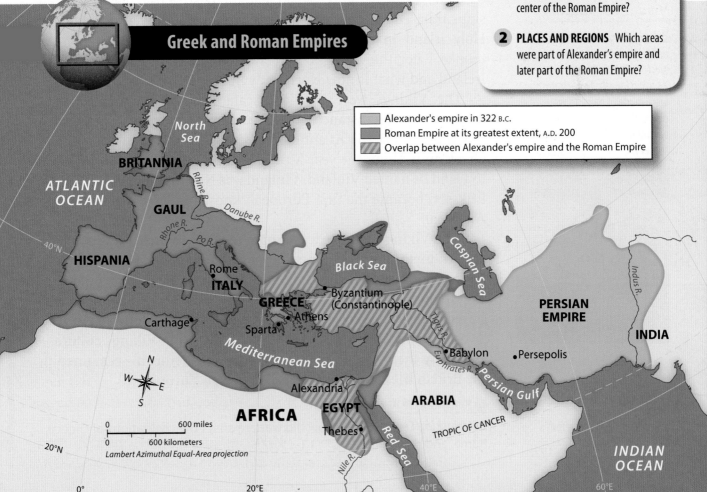

Greek and Roman Empires

Alexander's empire in 322 B.C.
Roman Empire at its greatest extent, A.D. 200
Overlap between Alexander's empire and the Roman Empire

Indiana CONNECTION

Democracy

Ancient Greek and Roman political practices live on today. Athens was a direct democracy in which citizens met to debate and vote on government matters. Today, many towns in New England hold meetings in which townspeople practice direct democracy. In the Roman Republic, citizens chose representatives to make laws and govern on their behalf. The United States today is a representative democracy, a republic in which citizens elect officials to government bodies such as the U.S. Senate, modeled in part on the ancient Roman Senate

MAP SKILLS

1 **PHYSICAL GEOGRAPHY** Why was Italy an ideal location to be the center of the Roman Empire?

2 **PLACES AND REGIONS** Which areas were part of Alexander's empire and later part of the Roman Empire?

"Father of Geography"

Eratosthenes was a Greek scholar who closely estimated the circumference of Earth by measuring shadow lengths at different latitudes. Educated people of this time understood that Earth was round. Eratosthenes, who was born around 276 B.C., was the first person to use the term *geography*. He is often referred to as the father of geography.

Spain, Sicily, Macedonia, Greece, and Asia Minor fell to Roman armies and were turned into Roman provinces. Eventually, a powerful military leader, Julius Caesar, seized control of Rome. After Caesar was assassinated, his great-nephew, Octavian, given the title Augustus, became the first of a series of emperors, and the Roman Republic was no more. The new Roman Empire expanded eastward to Egypt and westward to the British Isles.

In A.D. 330, Emperor Constantine moved the capital of the empire from Rome to the Greek city of Byzantium, in what is now Turkey. This location was closer to important trade routes to China and Southwest Asia. The new capital, renamed Constantinople, was also farther from the barbarians who were attacking the Roman Empire in the west. Repeated invasions continued to weaken the western empire. German invaders took control of Rome in A.D. 476. This was the end of the western Roman Empire. The eastern empire lasted for almost the next 1,000 years until it fell in 1453 to the Ottoman Turks. The Turks changed the name Constantinople to Istanbul.

The Viking Age

In the A.D. 700s, ships carrying warriors from Scandinavia began raiding the coasts of Western Europe. At home, these warriors were farmers or young men eager for adventure. At sea, they were pirates called Vikings, and they spread fear and destruction wherever their ships traveled. In A.D. 793, they raided and destroyed the abbey at Holy Island in northeastern England, killing and enslaving the monks. Later, the Vikings conquered other parts of Britain as well as Ireland and what is now Normandy in France.

The Vikings were excellent seafarers, and they sailed their **longships** great distances to explore and to trade. They sailed westward across the Atlantic, founding settlements in Iceland and Greenland. About the year A.D. 1000, Leif Eriksson led the Vikings to a land he named Vinland. Vinland was Newfoundland, in Canada. Eriksson became the first European known to have reached North America.

The Vikings followed a **pagan** religion, which was based on ancient myths and had a number of different gods. After about A.D. 1000, Viking groups throughout Scandinavia began to **convert** to Christianity. The Viking threat died out as more Scandinavians stayed home. They contributed to building the kingdoms of Norway, Sweden, and Denmark. However, traces of Viking culture—especially their epic tales of adventure and heroism—remained in the British Isles and other parts of Western Europe.

☑ **READING PROGRESS CHECK**

Determining Central Ideas How did warfare affect the civilizations of Greece, Rome, and the Vikings?

Visual Vocabulary

Longship The Vikings raided and explored on sturdy ships that were 45 feet to 75 feet (14 m to 23 m) in length. These longships were made of oak and were powered by wind and up to 60 oarsmen. The square sails were sometimes brightly colored.

Academic Vocabulary

convert to bring about a change in beliefs

Sigurgeir Jonasson/Nordic Photos/Getty Images

Discovery and "Rebirth"

GUIDING QUESTION *How did the Renaissance pave the way for voyages of discovery?*

During the long period known as the Middle Ages, many of the ancient achievements were forgotten. Important manuscripts were lost or destroyed. Many of the writings that survived ended up in the East, where scholars could still read classical Greek. Beginning in the 1300s, a curiosity for Greek and Roman learning took hold in the Italian city of Florence, where poets such as Dante and Petrarch were inspired by ancient literature. To them, these works were freer, more **rational**, and more joyous than the works of their world.

Renaissance

When the Byzantine Empire fell in 1453, many scholars traveled west with ancient Greek manuscripts. At the same time, a practical printing press was invented in Germany. Suddenly, it was possible to print many copies of manuscripts that until then had to be lettered by hand. People could now own and read books.

These breakthroughs resulted in a period of artistic and intellectual activity known as the **Renaissance**. The Italian city of Florence became a center of learning and culture. Architects drew inspiration from the ancients and created new architectural styles. Painters and sculptors, such as Leonardo da Vinci, looked to nature for inspiration.

Curiosity about the natural world also led to the birth of modern science. In 1609 Italian astronomer Galileo designed a telescope to observe the moon and the planets. Galileo's observations helped prove Copernicus's theory that the planets, including Earth, orbit the sun. The Renaissance began in Italy, but Galileo's work influenced scientists throughout Europe.

Empires and Exploration

By the 1400s, Europeans wanted to do more business with China and India, but overland routes were long and dangerous. Prince Henry of Portugal inspired sailors and navigators to find a sea route to Asia by sailing around Africa. In 1488 Portuguese sea captain Bartholomeu Dias reached the Cape of Good Hope at the southern tip of Africa.

Academic Vocabulary

rational based on reason or logic

Hagia Sophia (or "Holy Wisdom") in Istanbul, at first, was a Christian church. It later became a Muslim place of worship, and today it is a museum.
▶ **CRITICAL THINKING**
Identifying What was the name of the city of Istanbul when Hagia Sophia was built?

Salvator Barki/Flickr/Getty Images

Ten years later, Vasco de Gama rounded the Cape and sailed to India. The Portuguese established sea trade with South Asia.

Christopher Columbus, an Italian navigator, had a different idea: Why not reach Asia via a westward sea route? Spain agreed to finance the expedition, and Columbus left Spain with three ships on August 3, 1492. Columbus underestimated the size of Earth and overestimated the size of Asia. When he finally saw land, he assumed he had reached Asia. In fact, it was a Caribbean island. Columbus, like Leif Eriksson before him, had landed in the Americas.

Through its expeditions to the Americas, Spain became the most powerful country in Europe. The Spanish built an empire in Mexico, Central America, and South America. They conquered the Aztec of Mexico and the Inca Empire in South America, and they enslaved the native peoples of the Caribbean.

Contact between Europe and the Americas also resulted in the exchange of goods. Europeans brought wheat, olives, bananas, coffee, sugar, horses, sheep, pigs, and cattle to the Americas. In exchange, the Europeans received tomatoes, corn (maize), potatoes, squash, cacao (the source of chocolate), and hot peppers. This commerce is known as the Columbian Exchange.

Religion in the Regions

Meanwhile, Christianity had become identified with Europe. Rome considered itself the seat of Christianity as early as the A.D. 100s. When the Roman Empire was split into eastern and western empires, Christianity in the empire also split into eastern and western branches. The western branch evolved into the Roman Catholic Church, which was dominant in Italy and Spain and throughout Western Europe. The eastern branch became the Eastern Orthodox Church, centered in Greece and parts of Eastern Europe.

The rise of the religion of Islam threatened the power of the Christian churches. The Moors were Muslims, followers of Islam, who invaded Spain from Northern Africa in the A.D. 700s and ruled most of Spain for more than 700 years. The Byzantine Empire fell to the Ottoman Turks, another Muslim people, in 1453. Under Ottoman rule, Greek Christians were free to practice their religion, but their rights were limited compared to those of Muslims. Over the next few centuries, Greek Christians struggled to preserve their traditions and beliefs. Today, the vast majority of people in Greece still belong to the Greek Orthodox Church.

During the 1520s, the ideas of Martin Luther contributed to the spread of the Protestant Reformation. Kingdoms across Northern Europe broke away from the Roman Catholic religion. The countries adopted some form of Protestantism as their official state religion.

☑ **READING PROGRESS CHECK**

Determining Central Ideas Why did Christianity change as it took hold in Southern and Northern Europe?

History in the Modern Era

GUIDING QUESTION *What has been the relationship between Northern Europe and Southern Europe over the last 200 years?*

The 1800s brought sweeping changes to the regions. The Scandinavian countries saw their military glory vanish, but they became prosperous democracies. Spain and Portugal lost much of their overseas empires, followed by conflicts at home. Greece won freedom from the Turks. Italy's separate territories, except for San Marino and what later became Vatican City, united in 1870.

Conflict and War

During the 1900s, Northern and Southern Europe were involved in both world wars. After fighting a civil war in the 1930s, Spain stayed out of World War II. However, Italy, ruled by dictator Benito Mussolini, sided with Nazi Germany in that conflict. The Italians were defeated by allied U.S. and British forces in 1943, a year and a half before the war in Europe ended.

The period following World War II brought even more political changes. Italy became a democracy and began to rebuild its economy. Greece suffered a brutal civil war between Communists and opponents of communism. Spain, Portugal, and Greece joined Italy as democracies.

The Modern Era

Since 1945, Scandinavia has enjoyed a high standard of living as well as political and social freedoms. Its leaders worked for world peace and economic growth in the world's new nations. Beginning in the 1990s, the nations of Northern and Southern Europe developed closer ties with each other and with other European countries as members of the European Union (EU).

☑ **READING PROGRESS CHECK**

Identifying What country did Italy side with during World War II?

Swedish troops helped Finland in that country's conflict with the Soviet Union during World War II.

▶ **CRITICAL THINKING**
Describing How was Southern Europe involved in World War II?

Include this lesson's information in your Foldable®.

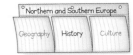

Keystone-France/Gamma-Keystone/Getty Images

LESSON 2 REVIEW

Reviewing Vocabulary
1. What changes did the *Renaissance* bring about in Europe?

Answering the Guiding Questions
2. *Identifying* Why were the ancient Greek and Roman civilizations important?

3. *Analyzing* How did the development of printing help promote voyages of discovery?

4. *Describing* How would you characterize the relationship among the countries of Northern Europe over the past 200 years?

5. *Argument Writing* Select a Renaissance thinker you believe was the most important. In a paragraph, explain why you believe this person's contributions were the most significant to the age and to people today.

There's More Online!

- ☑ **SLIDE SHOW** Recreation in Northern and Southern Europe
- ☑ **VIDEO**

Reading **HELP**DESK

Academic Vocabulary

- **contribution**

Content Vocabulary

- **homogeneous**
- **dialect**
- **welfare capitalism**
- **recession**

TAKING NOTES: *Key Ideas and Details*

Identify Use a graphic organizer like this one to list several artistic contributions of the countries of Northern and Southern Europe.

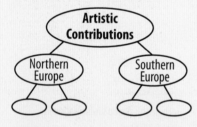

Artistic Contributions
Northern Europe
Southern Europe

Indiana Academic Standards
6.1.8, 6.1.9, 6.3.1, 6.3.4, 6.3.10

Lesson 3

Life in Northern and Southern Europe

ESSENTIAL QUESTION • *How do new ideas change the way people live?*

IT MATTERS BECAUSE

The countries of Europe are experiencing changes that will affect the lives of the people who live there. These changes will alter the relationships of the countries to each other and to the rest of the world.

People and Places

GUIDING QUESTION *What is the distribution of the populations in Northern Europe and in Southern Europe?*

Aging Populations

In most places in Northern and Southern Europe, improvements in the standard of living have helped reduce infant mortality—the number of babies who die in their first year of life. Improvements in the standard of living have also helped people live longer. You might think this would mean that populations in European countries are growing more quickly than in the past. That is not the case. Birthrates have declined as Europeans decided to have fewer children, and population growth has slowed in the past few decades. Older people have become a larger percentage of the population. Therefore, the population is aging.

Where People Live

As in other parts of Europe, the proportion of people who live in the rural areas of Southern Europe has decreased over the last 100 years. In 1900 Italy's three largest cities—Rome, Milan, and Naples—each had about 500,000 residents. The combined metropolitan area population of the three cities is now about 10 million. About 60 percent of the population of Greece live in cities. In fact, 25 percent of the population live in the capital city of Athens. More than 75 percent of Spain's population live in cities and towns, making rugged rural areas such as the Meseta Central region seem nearly empty in comparison. Madrid, Spain's capital, has a population of more than 5 million, and the population density is 1,750 persons per square mile (675 per sq. km). Compare this to the population density for the country as a whole: 220 persons per square mile (85 per sq. km).

In Northern Europe, the population is even more concentrated in cities. Most people live in the southern parts of the countries because of the milder climates. More than half of Iceland's population lives in the capital city, Reykjavik. The capital cities of Northern European countries—Copenhagen (Denmark), Oslo (Norway), Stockholm (Sweden), and Helsinki (Finland)—have by far the largest populations of any cities in their countries. They are also their countries' primary cultural centers.

✔ READING PROGRESS CHECK

Analyzing Why is the population of Northern and Southern Europe growing older?

Bikers cross a historic square in the central area of Copenhagen, the capital of Denmark.

▶ CRITICAL THINKING
Describing What are major characteristics of capital cities in Northern European countries?

The Euskera Language

The most unusual language in Spain is Euskera, an ancient language that is unrelated to other European languages. Euskera is spoken by the Basque people, who live near the Pyrenees mountains. The Basque are one of the oldest surviving ethnic groups in Europe. Today, most Basque people can also speak Spanish or French. Euskera is still widely spoken though, even in schools.

People and Cultures

GUIDING QUESTION *Why are most people in Northern Europe Protestant, whereas most people in Southern Europe are Catholic?*

Ethnic and Language Groups

The populations of the Northern European countries are relatively **homogeneous**, or alike, although they have some ethnic diversity due to immigration from Asia and Africa. The population of Norway is more than 90 percent Norwegian, and the population of Finland is more than 90 percent Finnish. Denmark and Sweden are similar. The original settlers of these lands were the Sami, or Lapps. Many of the Sami live by fishing and hunting, as they have for thousands of years. Most live in the northern parts of Sweden, Norway, and Finland. Iceland was first settled by Celts and later conquered by Norway. The bulk of the population in Iceland is a blend of the two ethnic groups.

The languages spoken in Denmark, Sweden, and Norway developed from a common German language base. Today, Danish, Swedish, and Norwegian are similar enough that the speakers can usually understand each other, even though each language is distinct. Finnish is unrelated to the other languages; it is closer to the Hungarian and Estonian languages.

The populations of the Southern European countries are relatively homogeneous but more diversified than it might seem at first. Spain is actually a nation of regions that constantly resist unifying pressures from the central government.

Greece also has small populations of Turks, Albanians, Macedonians, and Rom (gypsies). Most people speak Greek, which is closely related to the language spoken in ancient Greece. Most of Italy is ethnically Italian, but minority groups in northern Italy speak German, French, or Slovenian. The Italian language has many **dialects**, or regional variations, but most people speak and understand the Italian that originates around Florence and Rome. Most people in Spain speak a form of Spanish called Castilian. Other dialects are spoken in Spain as well. People in Catalonia, Valencia, and the Balearic Islands speak Catalan. Galicia is an area in northwestern Spain. People there speak Galician, which is closely related to Portuguese.

Religion and the Arts

The Protestant Reformation was successful in Northern Europe. To this day, more than three-quarters of all people living in Northern Europe belong to a Lutheran church. A small percentage are either Catholic or Muslim. In Finland, 15 percent of the people do not belong to any church.

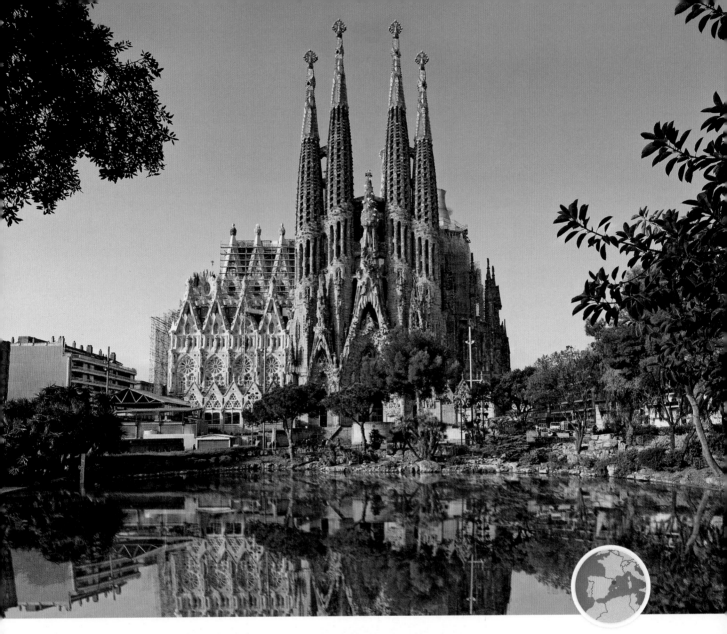

Sylvain Sonnet/Photographer's Choice/Getty Images

In Southern Europe, the Reformation had little lasting impact. Most people in Spain, Portugal, and Italy belong to the Roman Catholic Church. In Greece, nearly 98 percent of the population belongs to the Greek Orthodox Church. Immigrants from Muslim countries, however, have been slowly changing the religious makeup of Southern Europe.

As the birthplace of the Renaissance, Italy has made many contributions to painting, sculpture, and architecture. The Baroque artists who followed the Renaissance period also emerged from the Italian city-states. The genius of Renaissance and Baroque art can be seen in the palaces, churches, city squares, statues, and paintings of cities such as Rome and Florence. The art centers of Europe moved to Western Europe following the Baroque period—to countries such as France and the Netherlands. In the 1900s, the Italian futurist painters rebelled against traditional art. They painted scenes of modern life that celebrated the industrial age. Italy today is one of the world centers for architecture and fashion.

The Spanish Catalan architect Antoni Gaudí began to build the Sagrada Familia ("Holy Family") Church in Barcelona during the 1880s. This Roman Catholic place of worship, still being built, is due to be finished about 2026.

▶ **CRITICAL THINKING**

Analyzing Why are three different forms of Christianity practiced in Northern Europe and Southern Europe today?

A bin is filled with toy blocks at a factory in Billund, Denmark.

Academic Vocabulary

contribution an important part played by a person, bringing about a significant result or advancement

The musical legacies of Northern and Southern Europe are rich and deep. Italian operas are among the most popular in the world. Operas are stage presentations that use music to tell a story. Among the masterworks of opera are *The Barber of Seville* by Rossini and *Tosca* by Puccini.

Spanish music has been influential all over Europe and the Americas. It was the Spanish, more than any other people, who advanced and promoted the guitar as a serious musical instrument. Northern European composers such as Jean Sibelius (Finland), Edvard Grieg (Norway), and Carl Nielsen (Denmark) made important **contributions** to the classical music tradition. Northern European pop music, such as that by the 1970s Swedish group ABBA, has also had an impact around the world. Iceland has produced original and critically acclaimed popular musicians, including Björk and Sigur Rós.

Denmark has made a major contribution to children's toys: the LEGO™. Based in the town of Billund, LEGO™ (a play on the Danish words for "play well") has been creating plastic building-block toys since the late 1940s. LEGO™ sets were introduced in the United States in the 1960s, and their popularity has remained strong.

Daily Life

One of Northern Europe's most notable achievements is their literacy rate. In Norway, Sweden, and Denmark, the literacy rate is nearly 100 percent. The educational system in Northern European countries has strong support from the government, and most schooling is free. Northern European citizens pay relatively high taxes. In return, they receive a variety of public services and social welfare benefits. Every citizen is covered by health insurance.

Winter sports, such as skating and skiing, are popular in Northern Europe. Many people also enjoy skiing in the mountainous regions of Spain. Soccer (or, as it is known in Europe, football) is popular in Southern Europe. Spain, Portugal, Italy, and Greece have outstanding national soccer teams. Basketball is also common, especially in Spain and Greece. Bullfighting is still popular in Spain, although it is controversial because of the violence of the sport and cruelty to the bull.

Northern and Southern Europe are affected by the spread of popular culture, especially youth culture, from place to place. Young people throughout Europe are familiar with the same kinds of music. American fast-food chains, television programs, and films are popular throughout these regions.

✔ **READING PROGRESS CHECK**

Analyzing How is life in Southern Europe similar to life in the United States? How is it different?

Skiing is the most popular activity at a winter resort on Spain's side of the Pyrenees mountains.

▶ **CRITICAL THINKING**
Identifying Point of View Why does the sport of bullfighting arouse controversy?

Students relax in front of the Parliament building in Oslo, Norway's capital. The governments in Northern Europe and Southern Europe are all democracies. Some, such as Norway and Spain, are constitutional monarchies, while others like Finland and Italy are republics.

▶ **CRITICAL THINKING**
Determining Central Ideas What is the purpose of welfare capitalism?

Issues in Northern and Southern Europe

GUIDING QUESTION *How do the financial problems of one country in Southern Europe affect other European countries?*

The natural resources of the North Sea are key to many people's employment in Norway. The petroleum resources in the North Sea provide the single most important industry.

Earning a Living

Many Northern Europeans work in service industries. The standard of living is high in these countries, but so are taxes. Northern Europeans practice a form of **welfare capitalism**, where the government uses tax money to provide a variety of services, such as health care and education, to all citizens. The intention of welfare capitalism is to ensure that all people have access to those aspects of life that are considered essential, even those people who might not otherwise be able to afford them.

Following World War II, Italy had one of the weakest economies in Europe. In the decades since, the country has built a strong industrial base. Still, the economy is much stronger in northern Italy. The government has tried to stimulate the economy in the south, but most attempts have failed.

Making Connections

In 1992 a group of European countries signed a treaty creating the European Union (EU). The European Union is an attempt to provide common social, economic, and security policies throughout the continent. Today, 27 countries belong to the EU. Three Northern European countries are EU members: Denmark, Sweden, and Finland. In Southern Europe, Spain, Portugal, Italy, and Greece are members.

Meeting Challenges

Although the EU has helped ease conflicts among member countries, there have been many stumbling blocks to unifying the member nations. Immigration is one issue. Immigrants from many non-European countries bring their culture with them, creating a changing cultural landscape that can lead to conflict and violence. The EU promotes cooperation among member countries in easing conflicts, as well as preventing illegal immigration.

In 2008 a financial crisis swept the world. Europe experienced a **recession**, or a period of slow economic growth or decline. The three biggest banks in Iceland failed, forcing Iceland's prime minister to resign. The financial crisis hit Spain and Greece hard. Greece's inability to pay its skyrocketing debts became a threat to the economies of all the European countries.

Another issue surrounds two Northern European countries that are not members of the EU: Iceland and Norway. Whaling was an important industry in both countries. As populations of large whales began to shrink to near-extinction levels, a ban was placed on whaling. The ban has been lifted several times, but only to allow for hunting small, toothed whales. For now, the three remaining countries that hunt for whales—Norway, Iceland, and Japan—are under the watch of the International Whaling Commission.

Include this lesson's information in your Foldable®.

° Northern and Southern Europe °
Geography | History | Culture

☑ **READING PROGRESS CHECK**

Determining Central Ideas How does welfare capitalism work, and what advantages does it offer the people of Northern Europe?

LESSON 3 REVIEW

Reviewing Vocabulary
1. Why are the populations of Northern Europe considered *homogeneous*?

Answering the Guiding Questions
2. ***Determining Central Ideas*** Why do you think people in Northern and Southern Europe find living in cities more appealing than living in rural areas?

3. ***Analyzing*** Why do you think people in Norway, Denmark, and Sweden can easily understand each other's languages but find it more difficult to understand Icelandic?

4. ***Describing*** How has membership in the European Union changed the relationship among the countries of Northern and Southern Europe?

5. ***Informative/Explanatory*** Write a paragraph that compares the lives of people in Northern and Southern Europe.

Aging of Europe's
Population

An aging population is defined as a population in which the number of elderly (65 years old and older) is increasing at a faster rate than younger age groups. Europe is the region with the highest proportion of older persons.

Median Age The median age of a population is the age that divides the population into two equal groups. In the world as a whole, the median age is 28.4 years old. For Europe, the median age is 40 years old. The median age of Europeans has continued to rise and is expected to increase to 46 years by 2050.

Standard of Living Europeans enjoy a higher standard of living than people in some of the other regions of the world. In general, Europeans are living longer thanks to improvements in health, diet, and preventive health care.

> **"Some senior citizens continue working because they cannot afford to retire."**

Living Longer As people live longer, many experience age-related health problems. That puts an added strain on the health care system and health care costs. Some senior citizens continue working because they cannot afford to retire.

Smaller Families Over the past 40 years, families in Europe have grown smaller. As a result, fewer young workers are entering the labor force, while the number of senior citizens continues to grow.

Age of Retirement Some European nations are seeking ways to cut costs. Several European nations have raised the age of retirement. Adjusting the retirement age upward means more people in the workforce and

fewer people depending on government pensions.

The Workforce European countries are trying to increase their workforce by attracting skilled workers from other countries. Some are trying to attract workers only from other parts of Europe. Some European nations will have to change their immigration laws if they want to attract immigrants from regions outside of Europe.

Hikers make their way on a trek through ▶ the Alps.

THERE'S MORE ONLINE

HEAR how the EU is funded   •   *SEE* a slide show on European health care   •   *WATCH* a video on immigration

These numbers and statistics highlight the problems associated with an aging population.

population in 2060

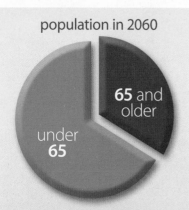

65 and older

under **65**

Up to
30%

The share of the population aged 65 and over will rise from 17% in 2010 to 30% in 2060. The percentage of people aged 80 or older will more than double: from 5% in 2010 to 12% in 2060.

Seventy Five
Years Old

Why is Europe's population growing older? People are living longer because of improved medical care and a better standard of living. In 1950 the average European lived to be 65. Today, the length of life is 75 years. At the same time, a decreasing birthrate means that there are fewer young people. In 2010 the birthrate—the number of births per 1,000 population—was half the birthrate in 1950.

65 **and Older**

Which countries have the highest percentage of residents who are 65 years old or older? Except for Japan, the world's 15 oldest countries are European countries. The five countries with the largest percentage of older citizens are: Japan and Italy, both at 19.5%, followed by Germany, 18.6%; Greece, 17.8%; and Sweden, 17.3%.

Life Expectancy at
FIFTY

What problems are caused by an aging population? As the aging population grows and more people retire, great strain is put on the health care and pension systems. Life expectancy varies greatly, depending on the country or region. For example, a 50-year-old female can expect 10.4 more healthy life years in Estonia, but 24.1 more in Denmark. The life expectancy for a 50-year-old male is slightly less.

14
or younger

European countries have many more senior citizens than most other nations—and far fewer young people. More than one-quarter—26.3%—of the world's population is 14 years old or younger. In Europe, only 15% are in that age group.

(t) Janine Wiedel Photolibrary/Alamy; (b) ©Alex Masi/Corbis

AN AGING POPULATION
Percentage of population age 60 and older

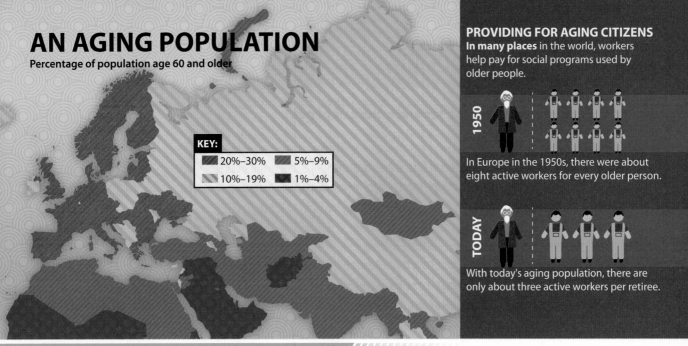

KEY:
- 20%–30%
- 10%–19%
- 5%–9%
- 1%–4%

PROVIDING FOR AGING CITIZENS
In many places in the world, workers help pay for social programs used by older people.

1950

In Europe in the 1950s, there were about eight active workers for every older person.

TODAY

With today's aging population, there are only about three active workers per retiree.

GLOBAL IMPACT

DEMOGRAPHIC TRENDS In many areas, the percentage of people who are older is increasing while the percentage of children is growing smaller.

The number of persons 60 years old and older has tripled between 1950 and 2000. By 2050, that number will more than triple again.

There is a higher proportion of older persons in Europe, but lower proportions in developing regions such as Africa. In 2050, 10 pecent of the population of Africa is projected to be 60 years old or older, up from 5 percent in 2000.

Europe, 2050

If projections hold, Europeans 60 years and older will comprise about 37 percent of the population by 2050. Persons younger than 15 will make up only about 14 percent of Europe's people.

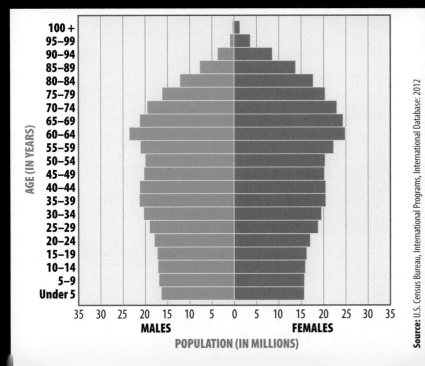

AGE (IN YEARS)

100 +
95–99
90–94
85–89
80–84
75–79
70–74
65–69
60–64
55–59
50–54
45–49
40–44
35–39
30–34
25–29
20–24
15–19
10–14
5–9
Under 5

35 30 25 20 15 10 5 0 5 10 15 20 25 30 35

MALES   **FEMALES**

POPULATION (IN MILLIONS)

Source: U.S. Census Bureau, International Programs, International Database: 2012

Thinking like a
Geographer

1. **Human Geography** Why is the percentage of elderly growing in Europe?

2. **Economic Geography** What kinds of businesses will grow in a place where the population is aging? What kinds of businesses will be in less demand?

3. **Human Geography** You and your classmates have been appointed by the mayor of a city to help companies find skilled workers from other countries. Prepare a PowerPoint presentation for the owners of those companies about ways to attract foreign workers.

Directions: Write your answers on a separate piece of paper.

1 Use your FOLDABLES to explore the Essential Question.

INFORMATIVE/EXPLANATORY Students in Europe participate in sports that might not be as well-known where you live. Identify one of those sports. Then, write at least two paragraphs to answer this question: Is the popularity of the sport related to the physical geography of the region?

2 **21st Century Skills**

ANALYZING The nations of Northern Europe have made progress in developing and using renewable energy sources. Work in small groups to choose a country in the region, and research that country's efforts to ensure that its people have adequate energy sources well into the future. Present your findings to the class in a slide show presentation.

3 **Thinking Like a Geographer**

DETERMINING CENTRAL ISSUES You have read that some locales in the regions are densely populated, but others are not. Identify the three major factors that affect population density.

4 **GEOGRAPHY ACTIVITY**

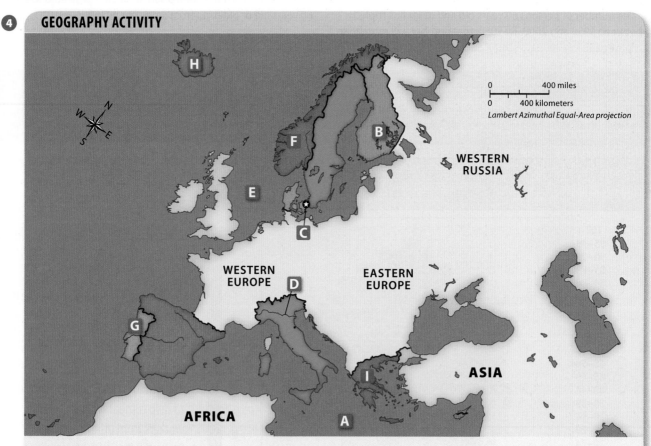

Locating Places
Match the letters on the map with the numbered places below.

1. Mediterranean Sea
2. Po River
3. Norway
4. Portugal
5. Greece
6. Copenhagen
7. Iceland
8. North Sea
9. Finland

REVIEW THE GUIDING QUESTIONS

Directions: Choose the best answer for each question.

1 What kind of energy heats all of the homes and most of the businesses in Reykjavík, Iceland?

A. nuclear

B. natural gas

C. geothermal

D. solar

2 Most of Southern Europe has a Mediterranean climate, which is much like the climate along the

F. southeastern coast of the United States.

G. Pacific northwest.

H. Chesapeake Bay.

I. coast of southern California.

3 What weakened Greece's city-states and eventually led to the fall of Greek civilization?

A. invasion by Vikings

B. years of war

C. Roman conquest

D. lack of trade routes

4 Who were the Scandinavian warriors and pirates who eventually became better known for exploring and trading?

F. Moors

G. Vikings

H. barbarians

I. Ottoman Turks

5 Most people in Northern and Southern Europe live in

A. rural areas.

B. small towns.

C. villages.

D. cities.

6 The Basques, one of the oldest minority groups in Europe, live in and around the Pyrenees mountains in

F. Spain and Portugal.

G. Italy and Greece.

H. France and Italy.

I. France and Portugal.

DBQ ANALYZING DOCUMENTS

7 **ANALYZING DOCUMENTS** Read the following passage about the Iberian Peninsula:

"*Its eastern seaboard forms part of the Mediterranean world, and in early times [it] was drawn successively [by turns] into the Carthaginian, Roman, and Muslim spheres. But much of the arid interior is drawn . . . towards the Atlantic.*"

—from Norman Davies, *Europe: A History* (1998)

What statement about Iberia it true based on the passage?

A. Iberians defeated the Carthaginians.

B. Iberia was isolated from all other regions.

C. Iberia was often under the control of other empires.

D. Iberians invaded Rome.

8 **IDENTIFYING** What impact did Iberia's nearness to the Atlantic have on its history?

F. The area was often invaded by the British.

G. Iberian nations led the way in exploring the Atlantic.

H. Navies from the region could not fight Mediterranean navies.

I. Iberian nations were cut off from trade with Asia.

SHORT RESPONSE

"*The expansion of most industry in Norway has largely been governed by private property rights and the private sector. Nevertheless, some industrial activities are owned or run by the state.*"

—from Aschehoug and Gyldendal's *Norwegian Encyclopedia*

9 **CITING TEXT EVIDENCE** Which economic sector, private or public, has been responsible for the expansion of most industry in Norway?

10 **ANALYZING** Why would economists label Norway's economy a "mixed economy"?

EXTENDED RESPONSE

11 **INFORMATIVE/EXPLANATORY WRITING** As you learned in your reading, the literacy rate in the Scandinavian countries is nearly 100 percent and education is free through college. How does that compare with literacy rates and the cost of education in Southern Europe and the United States? Research the issue, and report your findings.

Need Extra Help?

If You've Missed Question	**1**	**2**	**3**	**4**	**5**	**6**	**7**	**8**	**9**	**10**	**11**
Review Lesson	1	1	2	2	3	3	1	1	2	2	3

From EUROPE: A HISTORY, by Norman Davies, ©Norman Davies, 1996; Edited from Aschehoug and Gyldendal's Norwegian Encyclopedia, as seen on

EASTERN EUROPE AND WESTERN RUSSIA

ESSENTIAL QUESTIONS · *How does geography influence the way people live?* · *How do governments change?*

Dancers with Moscow's world-famous Bolshoi Ballet rehearse for an upcoming performance.

Dmitry Kostyukov/AFP/Getty Images

networks

There's More Online about Eastern Europe and Western Russia.

CHAPTER 13

Lesson 1
Physical Geography

Lesson 2
History of the Regions

Lesson 3
Life in Eastern Europe and Western Russia

The Story Matters...

The people of Eastern Europe and Western Russia share an agricultural background, which has been important in their history. Over the centuries, however, the people in this subregion have faced many political, ethnic, and economic challenges. The rise and fall of communism in Russia has had a tremendous impact on the countries of Eastern Europe.

FOLDABLES
Study Organizer

Go to the Foldables® library in the back of your book to make a Foldable® that will help you take notes while reading this chapter.

393

EASTERN EUROPE AND WESTERN RUSSIA

Rolling hills and fertile soil blanket much of Eastern Europe and Western Russia, two regions that merge just south of the Baltic Sea and north of the Black Sea and the Caspian Sea. As you study the map of Eastern Europe and Western Russia, look for geographic features that make this area unique.

Step Into the Place

MAP FOCUS Use the map to answer the following questions.

1 THE GEOGRAPHER'S WORLD
Which Eastern European country bordering Russia is located farthest north?

2 PLACES AND REGIONS Which river runs across the border of Russia and Eastern Europe?

3 THE GEOGRAPHER'S WORLD
Name four landlocked countries in Eastern Europe.

4 CRITICAL THINKING
Analyzing If you were a farmer in Belarus who needed to ship crops overseas, from which sea would you transport your crops? How would you get them to that sea?

A

CHURCH ARCHITECTURE St. Sophia Cathedral, with its 13 domes, is one of the oldest churches in Ukraine. It was built in 1037 to rival the church of the same name in Constantinople.

B

EUROPE'S LONGEST RIVER Boats line the Volga River, which flows about 2,300 miles (3,701 km) through west central Russia.

Step Into the Time

TIME LINE Using events on the time line, write a paragraph explaining how these events contributed to the rise of communism in Russia and Eastern Europe.

1700

1721 Peter the Great founds the Russian Empire

1762
Catherine the Great becomes empress of Russia

Eastern Europe and Western Russia

50°E   60°E   70°E

ARCTIC CIRCLE

- ⬦ National capital
- • City

60°N

**NORTHERN
EUROPE**

*North
Sea*

Baltic Sea

St. Petersburg

**WESTERN
RUSSIA**

*Europe/Asia
boundary*

ASIA

N. Dvina R.

Tallinn
ESTONIA

Riga
LATVIA

Kaliningrad
(RUS.)

LITHUANIA
Vilnius

Volga R.

Moscow

B

60°N

Minsk

POLAND

Warsaw

BELARUS

50°N

Prague

Oder R.

Vistula R.

A

Kiev

Ural R.

0   300 miles

0   300 kilometers

*Lambert Azimuthal
Equal-Area projection*

**CZECH
REPUBLIC**

**WESTERN
EUROPE**

SLOVAKIA
Bratislava

Dniester R.

UKRAINE

Dnieper R.

Don R.

Volgograd

SLOVENIA
Ljubljana

HUNGARY
Budapest

MOLDOVA
Chişinău

Zagreb
CROATIA

ROMANIA

*Sea of
Azov*

*Caspian
Sea*

**BOSNIA &
HERZEGOVINA**
Sarajevo

Belgrade

Danube R.

Bucharest

Adriatic

SERBIA

MONTENEGRO
Podgorica

KOSOVO
Priština

40°N

Sea

Tiranë
Sofia

Black Sea

**SOUTHERN
EUROPE**

Skopje
MACEDONIA

BULGARIA

ALBANIA

1945
Josip Broz Tito leads
postwar Yugoslavia

Tygodnik Rolników
SOLIDARNOŚĆ
REDAKCJA

1980
Lech Walesa founds
the labor union
Solidarity in Poland

1917 Czarist government
overthrown in Russia

1991 Baltic states declare
their independence

1900

2000

1920 The Communist Party gains
absolute power in Russia

1939 Germany invades Poland
and World War II begins

1987 Gorbachev introduces
reforms in USSR

1991 The Soviet Union breaks
up into 15 independent states

netw⌾rks

There's More Online!

☑ **IMAGE** The Ural Mountains

Physical Boundaries

☑ **VIDEO**

Reading **HELP**DESK

Academic Vocabulary

- **impact**
- **extract**

Content Vocabulary

- **upland**
- **steppe**
- **balkanization**
- **brackish**
- **reserves**

TAKING NOTES: *Key Ideas and Details*

Identify On a web diagram like this one, list at least four natural resources found in Eastern Europe and Western Russia.

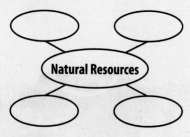

Natural Resources

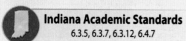

Indiana Academic Standards
6.3.5, 6.3.7, 6.3.12, 6.4.7

Lesson 1
Physical Geography

ESSENTIAL QUESTION • *How does geography influence the way people live?*

IT MATTERS BECAUSE

Both the rugged mountains and the gentle plains of Eastern Europe and Western Russia have shaped the cultures of the people living there.

Landforms and Waterways

GUIDING QUESTION *In what way have the landforms in the Balkan Peninsula shaped the cultures of that region?*

You can locate Eastern Europe and Western Russia on a map or a globe by identifying physical characteristics that border the regions. To the north are the Baltic and Barents Seas. The southern border is defined by the Caucasus Mountains and the Adriatic, Black, and Caspian Seas. The regions extend eastward to the Ural Mountains. Eastern Europe includes 10 countries in the north and 11 on the Balkan Peninsula. Russia is a huge country, extending through Europe and Asia and covering 11 time zones. Western Russia is the part of Russia that lies within Europe. Western Russia and Eastern Europe share characteristics of physical and human geography that unite them into a single region.

Vast Plains

Eastern Europe and Western Russia rest mostly on a group of plains. The largest plain is the Russian Plain, which begins in Belarus and Ukraine and stretches east about 1,000 miles (1,609 km) from Russia's western borders. In central European Russia, the Russian Plain rises to form the central Russian **upland**. An upland is an area of high elevation. To the east are the Ural Mountains, and beyond that, the west Siberian Plain.

(l to r) ©Serguei Fomine/Global Look/Corbis; ©Dallas and John Heaton/Corbis; Bloomberg/Getty Images

The Northern European Plain includes Poland in Eastern Europe, but it also extends into parts of Western Europe. South and southeast of the Northern European Plain is the Hungarian Plain, which includes parts of many different countries. Within Romania is the Transylvanian basin. A basin is an area of land that slopes gently downward from the surrounding land. Much of Ukraine is **steppe**, or vast, level areas of land that support only low-growing, vegetation-like grasses.

Bordering Mountains

To the south of the Russian Plain are two chains of mountains that make up the Greater and Lesser Caucasus Mountains. They extend from the northwest to the southeast with a valley between.

East of the Russian Plain, the Ural Mountains form a boundary between Europe and Asia. The Urals are up to 250 million years old. The northern mountains are covered in forests and some glaciers. Grasslands cover the southern Urals.

The Carpathian Mountains are much younger. On a map, the Carpathians appear almost as an eastward extension of the Alps. The Vienna basin in Austria separates the two mountain ranges.

The Balkan Peninsula is a mountainous region. In fact, *balkan* is a Turkish word for mountain. The Carpathian Mountains run through the peninsula's north and are linked to the Balkan Mountains. The region is so mountainous that human settlements are isolated from one another. This isolation results in cultural diversity among the people, but it is also the source of conflict among ethnic groups. Conflict among ethnic groups within a state, a country, or a region is known as **balkanization**.

Russia's Komi region borders the Urals and other mountain ranges. It is rich in coal, oil, natural gas, diamonds, gold, and timber.

▶ **CRITICAL THINKING**

Describing What characteristic of the Ural Mountains makes them unique?

Bulgaria, located in the Balkan Peninsula, has a mild climate and sandy beaches along its Black Sea coast.

▶ **CRITICAL THINKING**

Identifying What seas other than the Black Sea border the Balkan Peninsula?

Surrounding Seas

The Baltic Sea lies northwest of Russia and Eastern Europe. The Baltic is shallow and **brackish**, or somewhat salty, because it is seawater mixed with river water. In the southwest, the Adriatic, Ionian, and Black seas surround the Balkan Peninsula on three sides. The Black Sea borders the southern coast of Ukraine and southwestern Russia. The sea also separates Turkey from Ukraine and the Balkan Peninsula. At Europe's most southeastern point is the Caspian Sea. The Caspian Sea is the world's largest inland body of water, covering an area larger than Japan.

Rivers and Lakes

A vast number of rivers, canals, lakes, and reservoirs are found in Eastern Europe and Western Russia. The Volga River is the longest river in Europe and Russia's most important waterway. Originating northwest of Moscow, the Volga and its many tributaries carry more freight and passenger traffic than any other river in Russia. It provides hydroelectric power and water to many parts of Russia.

The Dnieper River also originates in Russia. It flows through Belarus and Ukraine before emptying into the Black Sea. Dams and reservoirs southeast of Kiev provide hydroelectric power. They also irrigate farmlands and help relieve water shortages in parts of Ukraine. Originating in the Carpathian Mountains and emptying into the Black Sea, the Dniester River is the second-longest river in Ukraine. The Dniester carries freight and passenger ships, and it serves as an important route to the Black Sea.

From its origins in southwestern Germany, the Danube flows toward the east through several countries before emptying into the Black Sea. The Danube provides transportation, hydroelectric power, fishing, and water for irrigation. Historically, the river transported traders as well as invading armies. Today, many cities are located along its banks, including three capital cities: Vienna, Austria; Budapest, Hungary; and Belgrade, Serbia. The Main River became connected to the Danube via the Main-Danube Canal, which linked the North Sea with the Black Sea.

✔ **READING PROGRESS CHECK**

Analyzing How is the location of the Black Sea strategic to the region?

Climates

GUIDING QUESTION *How does climate affect plants that are grown and harvested in Eastern Europe and Western Russia?*

Several different types of climate are found in Eastern Europe and Western Russia, from the hot summers and rainy winters in Albania to the cold, polar reaches of northern Russia.

Humid Continental Climates

Much of Eastern Europe and Western Russia have a humid continental climate. These areas experience mild or warm summers and long, cold winters. Farther south, in places such as Croatia, Serbia, and Bulgaria, summers are hotter, and winter weather is similar to that of areas farther north.

Albania and Macedonia experience a more Mediterranean climate, especially in the western areas. Summers tend to be hot and dry, and winters are mild to cool and rainy.

Russia's Far North

North of 60°N latitude, Western Russia has a subarctic climate. Winters are very cold, with temperatures as low as −40°F (−40°C). The summers are short and cool, though temperatures can range from 50°F (10°C) to 86°F (30°C).

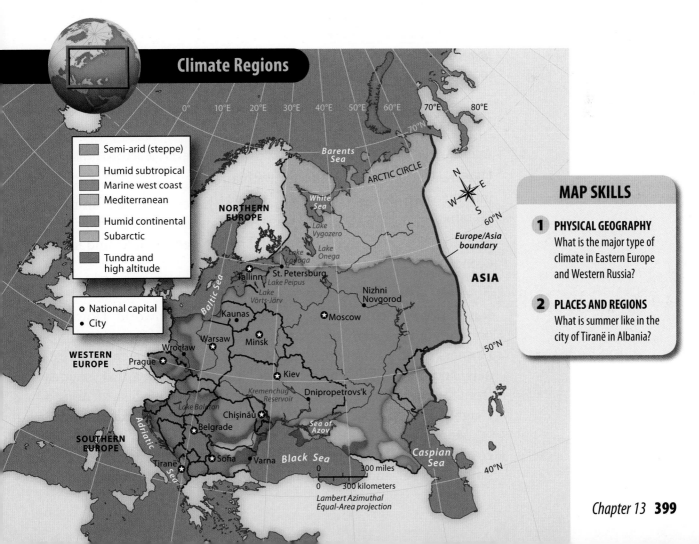

Climate Regions

Legend:
- Semi-arid (steppe)
- Humid subtropical
- Marine west coast
- Mediterranean
- Humid continental
- Subarctic
- Tundra and high altitude

- ⊕ National capital
- • City

MAP SKILLS

1 PHYSICAL GEOGRAPHY
What is the major type of climate in Eastern Europe and Western Russia?

2 PLACES AND REGIONS
What is summer like in the city of Tiranë in Albania?

A miner watches as the arm of a drill machine draws out coal from a mine in southern Poland.

▶ **CRITICAL THINKING**

Identifying What other minerals are important to Poland's economy?

Farther north is Novaya Zemlya, an archipelago consisting of two large islands and several small islands. The climate here is polar, and a large part of Novaya Zemlya is covered in ice year-round. Only the southern island is inhabited by a small number of the indigenous Nenets, who are herders and fishers.

☑ **READING PROGRESS CHECK**

Identifying What area in Eastern Europe or Western Russia has a climate most similar to where you live?

Natural Resources

GUIDING QUESTION *What are three important challenges to the development of resources in Eastern Europe and Western Russia?*

Eastern Europe and Western Russia have abundant mineral resources, as well as dense forests, fertile farmlands, and rich fishing grounds. These resources play a vital role in people's lives and in the economy of the countries in which they live.

Forests and Agriculture

Russia is a vast country—by far the largest in the world. However, only about one-sixth of Russia's land is suitable for agriculture. Farmers grow a number of crops, including grains such as wheat, oats, and barley. Most agricultural land is in an area that extends from the western shores of the Baltic Sea to the Black Sea, forming a roughly triangular shape. This area is known as the fertile triangle.

More than one-fifth of the world's forests are in Russia. They cover an area almost the size of the continental United States. Lumber, paper, and cardboard are important products of the forestry industry. The long, cold winters of Western Russia's continental climate, however, cause forests to grow slowly. The intense harvesting of forests and the slow rates of growth threaten the forests and the forestry industry.

In 2010 Russia experienced the hottest summer in 130 years, with drought conditions and temperatures reaching 104°F (40°C). That summer, wildfires destroyed 37 million acres (about 15 million ha) of forests, agricultural crops, and other vegetation. The **impact** of these fires was tremendous, taking lives, destroying homes, and damaging Russia's forestry and agricultural industries.

Academic Vocabulary

impact a dramatic or forceful effect or influence on something

Energy and Minerals

Most of Russia's vast coal, oil, and natural gas **reserves** are in Siberia. Reserves are the estimated total amount of a resource in a certain area. Russia's coal and rich deposits of iron ore fuel the country's steel industry. Machines made from steel are used to build Russia's automobiles, railroads, ships, and many consumer products.

Poland's important mineral resources include aluminum, coal, copper, lead, and zinc. Poland is one of the world's major sources of sulfur. Romania has rich deposits of coal, and it **extracts** oil from the Black Sea. Hydroelectric and thermal power plants also support Romania's energy needs. Other important mineral resources include copper and bauxite, the raw material for aluminum.

Academic Vocabulary

extract to draw or pull out

Fishing Industry

Russia's fishing industry is an important part of the country's economy. Salmon, cod, herring, and pollack are among the most important commercial fish in Russia. Many of Russia's lakes and rivers are also used for freshwater fishing.

Romania's fishing industry is concentrated in the southeastern area of that country. The Danube River and lakes and rivers near the Black Sea provide much of the fish. The European Union's restrictions on overfishing has hurt Romania somewhat, but fishing remains important to the country's economy.

✅ **READING PROGRESS CHECK**

Analyzing Why are Russia's mineral industries so important to its economy?

FOLDABLES®
Study Organizer

Include this lesson's information in your Foldable®.

LESSON 1 REVIEW

Reviewing Vocabulary
1. What caused the *balkanization* on the Balkan Peninsula?

Answering the Guiding Questions
2. *Describing* In what way have the landforms in the Balkan Peninsula shaped the cultures of the region?

3. *Analyzing* Why do few people live on the archipelago of Novaya Zemlya?

4. *Identifying* What are two important challenges to Russia's forestry industry?

5. *Informative/Explanatory Writing* Explain how the mineral resources in Russia are important to its industry.

Reading **HELP**DESK

Academic Vocabulary

- **strategy**
- **inevitable**

Content Vocabulary

- **czar**
- **serf**
- **genocide**
- **communism**
- **collectivization**

TAKING NOTES: *Key Ideas and Details*

Identify Use a graphic organizer like the one shown here to identify two ways that Joseph Stalin fashioned the Soviet Union into a communist dictatorship.

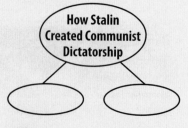

Indiana Academic Standards
6.1.15, 6.1.23, 6.2.5, 6.2.6, 6.2.7

Lesson 2
History of the Regions

ESSENTIAL QUESTION · *How do governments change?*

IT MATTERS BECAUSE
For most of the late 1900s, the USSR was one of the two most powerful countries in the world. The Soviets ruled Russia, nearly all of Eastern Europe, and much of Central Asia.

Early History

GUIDING QUESTION *How did Peter I and Catherine II change Russia?*

For the last 1,000 years, the people of Eastern Europe and Western Russia have been part of great empires that struggled against each other—and sometimes against their own people.

Early Slavic States

Many different ethnic groups settled in the regions of Eastern Europe and Western Russia long before modern national borders were set. Most of the people in the region are Slavs. Slavs are an ethnic group that includes Poles, Serbs, Ukrainians, and other Eastern Europeans.

Early Slavs migrated from Asia and settled in the area that now includes Ukraine and Poland. In the A.D. 400s and A.D. 500s, Slavs moved westward and southward, coming into contact with migrating Celtic and Germanic groups.

In the A.D. 800s, Slavic groups in the present-day Czech Republic formed Great Moravia, an empire covering much of central Europe. Other Slav people settled in the Balkans, eventually coming under the rule of the Ottoman Empire.

Another early Slav group settled in the forest and plains of present-day Ukraine and Belarus. The people of a settlement called Kiev organized the Slav communities into a union of city-states known as Kievan Rus. The leaders controlled the area's trade, using Russia's western rivers as a link between the

Baltic Sea and the Black Sea. Kievan Rus prospered from trade with the Mediterranean world and Western Europe. Later, non-Slavic people also settled in the region. Besides ethnic Russians who make up the majority of the population today, there are Hungary's Magyars, Romanians, Slavs, Ukrainians, and many others.

Throughout Russia's history, the Russian Slavs have dominated the country's politics and culture. Most Slavs practice Eastern Orthodoxy, a form of Christianity brought to Russia from the eastern Mediterranean area. By the A.D. 1000s, the ruler and people of Kievan Rus had accepted Eastern Orthodox Christianity. It remains Russia's largest religion today.

Imperial Russia

During the later 1200s, the warrior armies of the Mongols of Central Asia invaded Russia. For the next 250 years, they controlled most of Russia. Near the end of the Mongol reign, the princes of Muscovy (now the city of Moscow) rose to power.

MAP SKILLS

1 **HUMAN GEOGRAPHY**
Why did early Slav communities develop in the area of Western Russia?

2 **PLACES AND REGIONS**
In what time period did Russia gain control of the area around the Baltic Sea?

Expansion of Russia

Kievan Territory	Boundary of the Soviet Union in 1945
1360–1533	Present-day Russian boundary
1533–1689	
1689–1917	

0 1,000 miles
0 1,000 kilometers
Two-Point Equidistant projection

FEUDALISM IN EUROPE & RUSSIA

As in other parts of medieval Europe, Russia's feudal system depended on a large number of laborers. In exchange for a serf's labor, the lord or noble provided a place to live and protection.

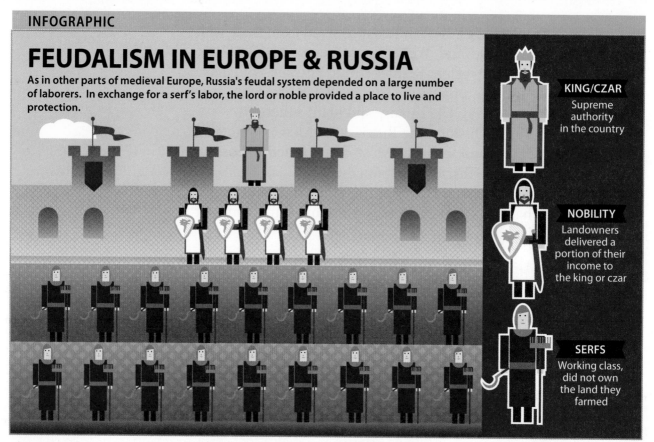

KING/CZAR
Supreme authority in the country

NOBILITY
Landowners delivered a portion of their income to the king or czar

SERFS
Working class, did not own the land they farmed

Under feudalism, monarchs gave land to nobles and lords in exchange for military protection. Serfs farmed the land of the nobles in exchange for the use of the land and protection.

▶ **CRITICAL THINKING**

Explaining How did constant warfare lead to the development of feudalism?

The most powerful of these princes, Ivan IV, defeated the Mongols and declared himself the **czar** of Russia. *Czar* is Russian for Caesar, or powerful ruler. The Russian nobility, dissatisfied with the czars who ruled after Ivan, looked for a young noble to lead the country. In 1613 they elected 16-year-old Michael Romanov as czar. The Romanovs ruled for the next 300 years.

Powerful Czars

Later, a czar now known as Peter the Great attempted to turn Russia into a major power. After Peter's death in 1725, Russia endured a string of weak czars. During the late 1700s, Empress Catherine the Great came to power. Catherine encouraged the development of Russian education, journalism, architecture, and theater. During her reign, Russia expanded its empire and took possession of the entire northern coast of the Black Sea.

Plight of the Serfs

The czars and nobles enjoyed rich, comfortable lives. At the bottom of society, however, were the great masses of people. Most were **serfs**, or farm laborers who could be bought and sold along with the land. These people lived hard lives. In 1861 Czar Alexander II abolished serfdom. The new law, however, did little to help the serfs.

They had no education and few ways to earn a living. Industrialization drew some serfs to cities, where they worked long hours for low wages.

☑ **READING PROGRESS CHECK**

Describing How did Catherine the Great expand the Russian Empire?

Conflict and Communism

GUIDING QUESTION *How did the Russian Communist Party plan to transform Russia into an industrial giant?*

In the early 1900s, discontent with the rule of the czars spilled into the streets. Strikes and demonstrations in 1905 nearly ended the reign of Czar Nicholas II. One event, called Bloody Sunday, began with workers marching toward the czar's palace in St. Petersburg to demand better working conditions. The march ended when soldiers fired into the marchers, killing nearly 1,000 people. Another much larger conflict was brewing—one that would involve millions of people, military and civilian.

Wars and Revolution

The threat of war in Europe had been brewing for many years. The major powers had already formed alliances. Austria-Hungary, Germany, and the Ottoman Empire made up the Central Powers. Great Britain, France, and Russia were called the Allies. An assassination triggered World War I. A Bosnian terrorist named Gavrilo Princip assassinated Archduke Francis Ferdinand of Austria-Hungary on June 28, 1914, in Bosnia. By August, nearly all of Europe was at war.

At first, the Russian people supported the war effort, but as military failures, high casualties, and food shortages began mounting, public opinion turned against the war and against the czar. Russia encouraged Armenians in Turkish-controlled lands to fight alongside them. The Turks responded by deporting 1.75 million Armenians to Syria and Mesopotamia. During this mass deportation, about 600,000 Armenians starved or were murdered by Turkish soldiers and police. The mass murder of vast numbers of an ethnic or cultural group is called **genocide**.

In 1917 food shortages in Russia triggered riots in the capital. Soldiers began deserting, joining civilians in their protests against the war. Even though the Allies won the war, Russia emerged as a weakened nation.

The killing of Archduke Ferdinand, heir to Austria-Hungary's throne, lit the fuse of World War I.

▶ **CRITICAL THINKING**
Determining Central Ideas Why did this killing lead to a war involving many nations?

DEA/A. DAGLI ORTI/De Agostini Picture Library/Getty Images

Czar Nicholas was forced to step down. A new government was installed, but it could not maintain power. By the end of 1917, a group of revolutionaries known as the Bolsheviks had seized control of the government.

Rise of Communism

The Bolsheviks had strong support all over Russia. Inspired by the writings of Karl Marx, they remade Russia into a communist state. **Communism** is an economic system built on the idea that all property should belong to the community or the state, not to private individuals. The Bolsheviks, who had become the Russian Communist Party, took control of all land and industry. Their leader, Vladimir Lenin, became the first premier of the new Union of Soviet Socialist Republics (also known as the Soviet Union).

When Lenin died in 1924, the secretary of the Central Committee of the Communist Party, Joseph Stalin, became leader of the Soviet Union. Stalin used terror and brute force to fashion the Soviet Union into a communist dictatorship. He forced the **collectivization** of all agriculture, so that all farmland was owned and controlled by the government. Peasants and landowners protested, especially in Ukraine. The clash between agricultural workers and the government resulted in a famine that killed millions.

May Day was an official holiday in the Soviet Union and Soviet satellite countries. Held on May 1, celebrations, like this one in Moscow, typically included military parades.

By the early 1930s, the Soviet Union was on its way to becoming one of the world's industrial giants. Stalin wanted something more, however. His vision was to spread a Soviet-style communist government throughout the world.

The USSR and Its Satellites

In 1941 Nazi Germany invaded the Soviet Union, drawing the country into World War II. During the conflict, the Soviets joined with Great Britain and the United States to defeat the Germans. At the end of World War II, the fate of Europe was left to the victors—the United States, Great Britain, and the USSR. The Soviet army already occupied Czechoslovakia, Poland, Romania, Hungary, and Bulgaria. Stalin agreed to allow elections in those countries but soon installed communist governments. Germany was split in two. The United States, Great Britain, and their allies set up West Germany as a democracy under their guidance, and East Germany became a communist state. Countries under Soviet rule came to be known as satellite countries, meaning they were under the economic and political domination of a more powerful country.

☑ **READING PROGRESS CHECK**

Analyzing How did the USSR come to control most of Eastern Europe?

The Regions in the Modern Era

GUIDING QUESTION *How is a "cold war" different from other kinds of war?*

After World War II, the Soviet Union shared superpower status with the United States. Both superpowers possessed weapons of unimaginable destructive force. Would they dare to use those weapons against each other?

The Cold War

The Cold War was the rivalry and conflict between the USSR and the United States and their allies. During the next four decades, the Soviet Union and the United States engaged in a struggle for world influence and power.

Although both superpowers built destructive weapons, they also used other **strategies**, such as the threat of force and providing military and financial aid to their allies. At times, however, an outcome of nuclear warfare seemed **inevitable**.

The United States and its allies created the North Atlantic Treaty Organization (NATO) in 1949. Any attack on a member country would be considered an attack on all of them, and NATO countries agreed to respond as a group. The original NATO countries included many of the non-Eastern European nations. When NATO admitted West Germany in 1955, the USSR responded by creating the Warsaw Pact. Member countries were Albania, Bulgaria, Czechoslovakia, East Germany, Hungary, Poland, Romania, and the USSR.

Academic Vocabulary

strategy a plan to solve a problem

inevitable sure to happen

Eastern Bloc

(1949) Date joined Soviet Union/Eastern Bloc
— Border of the Soviet Union
— Border of the Eastern Bloc states

ESTONIA *(1940)*
LATVIA *(1940)*
LITHUANIA *(1940)*
EAST GERMANY *(1949)*
POLAND *(1947)*
BELARUS *(1922)*
CZECHOSLOVAKIA *(1948)*
UKRAINE *(1922)*
HUNGARY *(1947)*
MOLDOVA *(1940)*
ROMANIA *(1947)*
YUGOSLAVIA *(1945)*
BULGARIA *(1946)*
ALBANIA *(1945)*
RUSSIA

Barents Sea
Baltic Sea
Black Sea
Caspian Sea

600 miles
600 kilometers
Lambert Azimuthal Equal-Area projection

MAP SKILLS

1 HUMAN GEOGRAPHY
What countries became tied to the Soviet Union during the early part of World War II?

2 PHYSICAL GEOGRAPHY
What about the location of Eastern Europe made this region important to the Soviet Union?

The two superpowers came close to war during the Cuban Missile Crisis. In October 1962, after learning that the Soviets were sending missiles to Cuba, the U.S. set up a naval blockade around Cuba to prevent the shipment of missiles. Both sides seemed prepared to go to war. As tensions grew, Soviet premier Nikita Khrushchev agreed to stop shipping the missiles to Cuba. Another crisis was brewing, however. The Soviet satellite countries in Eastern Europe began to rebel against Soviet control.

Unrest in the Soviet Satellites

In 1968 Czechoslovakia's leader Alexander Dubček announced sweeping reforms. He wanted to give the press more freedom and to guarantee citizens' civil rights. The Czech people welcomed the reforms, but the Soviets removed Dubček from power.

In 1980 dozens of Polish trade unions joined together to form Solidarity. Solidarity used strikes to put pressure on the government. The Polish government responded by declaring Solidarity illegal and putting its leaders in jail. Solidarity became an underground, or secret, organization.

Changes Under Gorbachev

Then in the 1980s, a new Soviet leader, Mikhail Gorbachev, came to power and implemented new policies. *Glasnost,* which means "openness," was an attempt to allow the people in the USSR and its satellite countries to have more social and political freedoms. *Perestroika,* which means "restructuring," was an attempt to reform the Soviet economy.

Change came quickly in Eastern Europe. Solidarity was legalized in Poland in 1989, then the Communists were voted out of power. By 1990, Hungary, Czechoslovakia, Bulgaria, and Romania had new governments. In Germany, the Berlin Wall that separated West and East Berlin was torn down. By the next year, East and West Germany were reunited. Soviet control of Eastern Europe was broken.

In 1991 a group of Soviet officials, who thought Gorbachev's policies meant the downfall of the Soviet Union, staged a coup and arrested Gorbachev. Gorbachev's allies resisted, the people protested, and the military turned against the coup leaders. Gorbachev was released, and the coup leaders were arrested. Communist control of the USSR was at an end. By the end of 1991, the Soviet Union was dissolved, and all the republics had become independent countries.

Divisions and Conflict

When Eastern Europe shook free of Soviet domination in 1989, ethnic tensions flared in the Balkan Peninsula. The former Yugoslav republics used to be one large country called Yugloslavia. In the early 1990s, disputes among ethnic groups tore the country apart. Croatia, Slovenia, Macedonia, and Bosnia and Herzegovina broke free of Yugoslavia and became separate countries. Serbia and Montenegro each became its own country in 2006.

Kosovo, which was considered part of Serbia, has a mostly Albanian Muslim population. When Yugoslavia broke apart, many people in Kosovo decided they wanted to break free of Serbian control. When the Kosovo Liberation Army began an armed rebellion in 1998, Serbs responded with military force. NATO intervened to end the bloodshed, and the United Nations began governing Kosovo. Finally, in 2008 Kosovo declared itself independent, though Serbia, Russia, and other countries refused to recognize this.

☑ **READING PROGRESS CHECK**

Determining Central Ideas How did glasnost and perestroika affect the USSR?

Soviet leader Mikhail Gorbachev worked to improve the Soviet Union's relations with the United States and other Western countries.

▶ **CRITICAL THINKING**
Citing Text Evidence How did Gorbachev's policy of glasnost affect Eastern Europe?

FOLDABLES Study Organizer

Include this lesson's information in your Foldable®.

Reviewing Vocabulary

1. Why did landowners protest *collectivization* of agriculture?

Answering the Guiding Questions

2. *Describing* How did Peter I and Catherine II change Russia?

3. *Determining Central Ideas* What were Stalin's main goals for the Soviet Union?

4. *Analyzing* How is a "cold war" different from other kinds of war?

5. *Argument Writing* Write a speech encouraging Czar Alexander II to abolish serfdom and to grant rights and liberties to all peasants.

©Peter Turnley/Corbis

networks

There's More Online!

☑ **CHART/GRAPH** Russia's
Population

☑ **MAP** Slavic Settlements

☑ **VIDEO**

Reading **HELP**DESK

Academic Vocabulary

- **decline**
- **unique**
- **factor**

Content Vocabulary

- **inflation**
- **oligarch**
- **devolution**

TAKING NOTES: *Key Ideas and Details*

Identify Use a graphic organizer like this one to identify three challenges faced by Russia and the countries of Eastern Europe.

Challenges for Russia and Eastern Europe Today

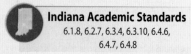

Indiana Academic Standards
6.1.8, 6.2.7, 6.3.4, 6.3.10, 6.4.6,
6.4.7, 6.4.8

Lesson 3
Life in Eastern Europe and Western Russia

ESSENTIAL QUESTION • *How does geography influence the way people live?*

IT MATTERS BECAUSE

It has been more than 20 years since the collapse of the USSR. The countries that the USSR once ruled in Eastern Europe have all moved forward, some more easily than others.

People and Places

GUIDING QUESTION *What were some of the challenges Russia faced after the fall of the Soviet Union?*

Life has changed in Eastern Europe and Western Russia. Some people have benefited from the changes; others have not. The attempt to change from a communist state to a free market economy has not been easy or particularly successful.

Economic Changes

Russia faced enormous challenges following the collapse of the USSR. The economy that was centrally controlled by the communist government had been in **decline** for years. Most of its industry had centered on military hardware and heavy industrial machinery. The country was not prepared to transform into a producer of consumer products that are the real engine of free market economies.

Inflation, or the rise in prices for goods and services, increased, while production slowed. The transfer of industry to private ownership was a great deal for wealthy individuals who had connections in government. However, these changes did not improve the living conditions for most Russians. People refer to the new owners of industrial Russia as **oligarchs**. An oligarch is one of a small group of people who control the government and use it to further their own goals.

Social and Political Changes

When the Soviet Union existed, the central government kept tight control over its many ethnic groups. Some groups wanted to form their own countries. Among them are the Chechens, who live in Chechnya near the Caspian Sea and Caucasus Mountains in southern Russia. The region has many oil reserves, and its oil pipelines transport fuel to major Russian cities. Russian troops fought Chechen rebels to keep Chechnya a part of Russia. When Russia finally pulled out in 1996, the Chechen rebellion was still not over. President Boris Yeltsin was widely blamed for being unable to solve these problems.

At the end of 1999, Yeltsin resigned and was replaced by Vladimir Putin, who was elected president in 2000. Putin, a former officer in the KGB, the country's secret police, was viewed as someone who wanted to keep a tight rein on government power. Putin launched reforms to reduce the power of the oligarchs and encouraged economic development. Although Putin helped stabilize the Russian economy, he dealt harshly with those who opposed him.

He was reelected for a second term in 2004. In 2008 Dmitry Medvedev was elected president, and he appointed Putin to be prime minister. In 2012 Putin ran for the presidency a third time and won. Soon after, new restrictive laws were passed strengthening penalties against demonstrators, blocking some Internet sites, and restricting free speech.

Where People Live

The two largest cities in Western Russia are Moscow and St. Petersburg. In addition to being the political capital of Russia, Moscow is the cultural, educational, and scientific capital. It has also been the spiritual home of the Russian Orthodox Church for more than 600 years. St. Petersburg was founded by Peter the Great in 1703.

The biggest population centers in Eastern Europe are the capital cities, such as Kiev, Ukraine; Minsk, Belarus; Budapest, Hungary; Warsaw, Poland; and Prague in the Czech Republic. Each of these cities is a center of national culture.

At one time, most Eastern Europeans lived in rural areas. Now more people live in urban areas. The urban population of Albania, Macedonia, and Croatia, for example, is above 50 percent. In some countries, urbanization is even higher. In Poland and Montenegro, the urban population is about 60 percent. In Hungary, the Czech Republic, and Belarus, urbanization is 70 percent or more.

Academic Vocabulary

decline a gradual deterioration into a weakened condition

A family sits in front of its farmhouse in Senj, a seaside town in Croatia. Although parts of Eastern Europe remain rural, most people now live in urban areas.

©Caro/Alamy

During the industrial age, people began moving from Eastern Europe to Western Europe and North America; that trend continues. Eastern Europeans have moved to escape political oppression and to seek better economic opportunities. Countries such as Romania have lost population since the lifting of Soviet travel restrictions.

☑ **READING PROGRESS CHECK**

Determining Central Ideas Why have so many Eastern Europeans emigrated to other parts of Europe or to the United States?

People and Cultures

GUIDING QUESTION *How did geographical barriers affect the development of Slavic culture in Eastern Europe?*

The history of Western Russia and Eastern Europe has created a rich mix of cultures and people. People take great pride in their folk and religious traditions, most of which were frowned upon by Soviet authorities.

Ethnic and Language Groups

At one time, a single Slavic language, understood by most Slavic people, was spoken. As Slavic people settled in different parts of Eastern Europe, geographical barriers separated and isolated them. These groups developed distinct languages and cultures.

Slavs generally belong to one of three categories. East Slavs are represented by the Slavic ethnic groups in Russia, Ukraine, and Belarus. West Slavs include ethnic Slavs in Poland, the Czech Republic, Slovakia, and parts of eastern Germany. The most diverse group are the South Slavs, who live in Bulgaria and other countries

GRAPH SKILLS >

POPULATION OF RUSSIA

The graph shows the changes in the population of Russia from 1940 to 2010.

▶ **CRITICAL THINKING**

1. Describing In what decades did the population of Russia increase?

2. Analyzing What is a possible explanation for Russia's sharp drop in population during the 1940s?

RUSSIAN POPULATION (IN MILLIONS)

YEAR

MAP SKILLS

1 HUMAN GEOGRAPHY
What three large divisions make up the Slavic people of Eastern Europe and Western Russia?

2 PLACES AND REGIONS
What countries in the region do not have majority Slavic populations?

of the Balkans. Each of these groups speaks its own language. Russia is made up of more than 120 ethnic groups, although almost 80 percent of the population is ethnic Russian.

The people of Albania are a distinct ethnic group that has been living in that region for about 4,000 years. Albanian is the last surviving language of an entire Indo-European language group. It is the ancestor of the language spoken by present-day Albanians, and it has survived thousands of years of conquest and cultural change.

Religion

For most of the 1900s, religious practice was strongly discouraged throughout Eastern Europe and Western Russia. In some countries in Eastern Europe, a sizable percentage of the population does not practice any religion. Less than half the population of the Czech Republic belongs to any church. In the Baltic region, nearly two-thirds of Latvians and one-third of Estonians are not affiliated with any church. In most of Eastern Europe, however, Soviet repression strengthened religious faith. The dominant religion in most of these countries is the Eastern Orthodox Church. Many different churches exist within the Orthodox faith.

Most of these churches are affiliated with a specific ethnic group or country. The majority of people in Belarus, Bulgaria, Moldova, Montenegro, Romania, Serbia, and Ukraine worship at an Eastern Orthodox Church. The majority of people in Croatia, Hungary, Lithuania, Poland, Slovakia, and Slovenia are Roman Catholics. Most of these countries also support minority populations of

Shoppers walk along the boulevard past a mall in Plovdiv, Bulgaria. Plovdiv is the country's second-largest city. Only the capital city of Sofia is larger.

Muslims, Roman Catholics, Eastern Orthodox, Protestants, and Jews. Nearly 70 percent of Albanians are Muslim, as are a sizable number of people in Bosnia and Herzegovina.

The Arts

In the 1800s and early 1900s, Russians produced some of the most important cultural works in all of Europe. The novels of Tolstoy and Dostoyevsky; the music of Mussorgsky, Tchaikovsky, and Rimsky-Korsakov; and the plays and short stories of Chekhov and Gogol are still considered among the world's finest.

People in the countries of Eastern Europe are proud of the great art produced by their people. In many cases, those works are an important symbol of their national character. Although a small amount of literary work was written in the Czech language, Czech literature did not became internationally important until Czechoslovakia became an independent country in 1918. Karel Čapek was a Czech writer who was famous for his plays and novels. His most well-known contribution to world literature is a word he coined—*robot*—in his 1921 play *R.U.R.* Eastern European composers, such as Béla Bartók and Zoltán Kodály, celebrated the traditional music of Hungary and Romania by using it in their compositions. Bulgaria also has a rich tradition in folk and choir music.

Western popular culture has had a huge impact on the art of Eastern Europe and Western Russia. Russian and Polish filmmakers can follow their national traditions, but they also can see how well-liked and influential American films and television are. Rock and pop music from the United States and Western Europe are extremely popular. Young artists in this part of the world are creating international popular culture while trying to bring something uniquely Eastern European to it.

Daily Life

For much of the 1900s, the people of Eastern Europe and Western Russia lived under communist governments that attempted to control their private lives. The collapse of the USSR brought about **devolution** in Russian government and in governments throughout Eastern Europe. Devolution occurs when a strong central government surrenders its powers to more local authorities.

One of the results of this change is the return of national traditions and identity. Most of these countries have strong cultural and religious traditions. These traditions were never really lost, but the lack of strong Soviet control has made it possible for people to live and speak more freely about their beliefs and interests. Such freedoms emphasize **unique** aspects of these countries and their people.

The other result of the loss of Soviet control is the rising influence of international popular culture. Soviet authorities did not trust the music, films, and television programs coming from capitalist countries such as the United States, but they could not effectively outlaw them. Young Russians and Eastern Europeans are now having the same cultural experiences that young people are having in the rest of Europe and in the Americas. This shared culture emphasizes those things that all these people have in common with each other.

One issue in Russia is the generation gap that exists between people who grew up in communist USSR or are old enough to remember it, and those people who have lived most or all of their lives in the post-Soviet era. One big question is how to teach the history of the USSR to young people who never experienced life under the Soviet system.

Bruce Yuanyue Bi/Lonely Planet Images/Getty Images

Academic Vocabulary

unique unusual or distinctive

Teenagers stroll on the main street of Minsk, the capital and largest city of Belarus.
▶ **CRITICAL THINKING**
Analyzing How has the end of Communist rule affected young people in Eastern Europe and Western Russia?

In the 1990s, historians took a critical look at the Russian Revolution, the leadership of Lenin and Stalin, and the excesses of the Soviet government. They took a more positive approach when looking back at the era of the Romanov czars.

☑ **READING PROGRESS CHECK**

Determining Central Ideas How has daily life changed in Russia since the fall of the USSR?

Academic Vocabulary

factor something that actively contributes to a result

Issues in Eastern Europe and Western Russia

GUIDING QUESTION *What are the economic advantages and disadvantages of Eastern Europe's location between continents?*

Eastern Europe and Western Russia are still in the process of change. They are trying to modernize their industries and governments during difficult economic times.

Earning a Living

Western Russia and the countries of Eastern Europe cover a vast area. The landforms, the soil, the mineral resources, the climate, the economies, and the national traditions are different throughout the region. These **factors** combine to determine how people earn a living in these places. Nearly half the working population in Albania is employed in agriculture. Romania, Serbia, and Bosnia and Herzegovina also have many people employed in agriculture.

Russia is one of the world's leading suppliers of oil and natural gas. Most of it comes from western Siberia or the region between the Volga River and the Ural Mountains. Pipelines link these regions to the rest of the country. Russia supplies oil as well as natural gas to European countries, especially the countries in Eastern Europe.

Russia is also a major supplier of iron ore and other metals. About 1 million people work in Russia's forestry industry. The majority of people in Russia and in most of Eastern Europe now work in the service industries, or businesses that provide services to individuals as well as other businesses.

Many of the industries in Eastern Europe fell on hard times after the collapse of the Soviet Union. Many industries suffered big losses, leading to high unemployment, especially in the

During Communist rule and shortly after its end, shoppers in Eastern Europe had to wait in long lines outside of stores. Food and other products often were in short supply and had to be rationed, or divided equally.

▶ **CRITICAL THINKING**

Describing What benefit has come to Eastern European workers in recent years?

©Shepard Sherbell/Corbis SABA

Balkans. One positive move is the number of Eastern European countries that have joined the European Union. Since 2004, the following countries have joined the EU: the Czech Republic, Hungary, Latvia, Lithuania, Poland, Slovakia, Slovenia, Bulgaria, and Romania. The EU has sought to protect employment, improve workers' living and working conditions, and create a strong European trading bloc that can compete effectively with the United States.

Connections: Europe and Asia

Developments in Europe and Asia have affected Russia. Russians lived under Mongol rulers for centuries and later were next door to the Ottoman Turks, one of the most powerful Islamic empires to exist. Russia was influenced by developments in Europe, and, in turn, made contributions to European culture. Russian culture has always been a mix of European and Asian influences.

Today, oil and natural gas extracted from the Siberian oil fields in Central Asia are delivered via pipeline to all of Russia, Eastern Europe, and as far west as Italy and Germany. Geographically and culturally, Russia plays a key role in the relationship between Europe and Asia.

Addressing Challenges

Russia and countries that were formerly part of the Soviet republic continue to discuss agreements on the borders of the countries. The countries involved are Estonia, Latvia, Lithuania, Ukraine, and Kazakhstan (in Central Asia). Even though fighting died down in Chechnya, occasional outbreaks of violence still occur in that republic, and rebels continue to call for independence.

The 2008 financial crisis hit Eastern Europe hard. The countries there struggled as they transformed into free market economies. Joining the EU should have been a great benefit to the economies of its new members, but all members of the EU have suffered as a result of the financial crisis.

Include this lesson's information in your Foldable®.

☑ **READING PROGRESS CHECK**

What have been two setbacks in the economies of Eastern Europe since the collapse of the USSR?

LESSON 3 REVIEW

Reviewing Vocabulary

1. Why were the *oligarchs* unpopular with some Russians?

Answering the Guiding Questions

2. *Analyzing* How did the rebellion in Chechnya affect the presidency of Boris Yeltsin?

3. *Describing* How did geographical barriers affect the development of Slavic culture in Eastern Europe?

4. *Determining Central Ideas* What are the economic advantages and disadvantages of Eastern Europe's location between continents?

5. *Narrative Writing* Imagine that you live in an Eastern European country and are writing a letter to your cousin who has lived in the United States for many years. Try to get your cousin to come visit you. Be sure to remind him or her of all the positive changes that have occurred since the collapse of the Soviet Union.

Chapter 13 ACTIVITIES

Directions: Write your answers on a separate piece of paper.

1 Use your **FOLDABLES** to explore the Essential Question.

INFORMATIVE/EXPLANATORY WRITING Explain the meaning of the word *balkanization* and how it acquired its meaning.

2 21st Century Skills

ANALYZING With a partner, research to find one primary source and one secondary source about the Cuban missile crisis. Then, discuss and answer these questions: Which source provided a clearer picture of the event? Why? Did either source seem to support or favor one side over the other?

3 Thinking Like a Geographer

IDENTIFYING On a chart like the one shown, list the most populous cities of Western Russia and Eastern Europe with their countries.

City	Country
Moscow, St. Petersburg	Russia
Kiev	Ukraine

4 GEOGRAPHY ACTIVITY

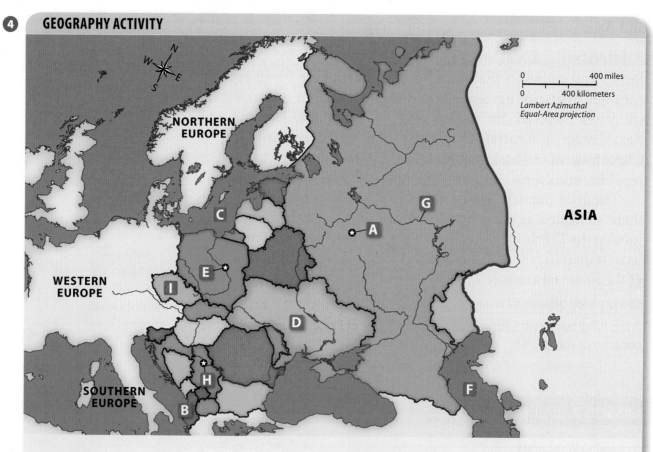

Identifying

Match the letters on the map with the numbered places listed below.

1. Baltic Sea
2. Moscow
3. Volga River
4. Caspian Sea
5. Ukraine
6. Belgrade
7. Albania
8. Warsaw
9. Czech Republic

REVIEW THE GUIDING QUESTIONS

Directions: Choose the best answer for each question.

1 Which mountains form a boundary between Europe and Asia?

A. the Carpathians

B. the Urals

C. the Balkans

D. the Alps

2 Although Russia is the largest country in the world, it has a relatively small percentage of land available for farming. Its main crop is

F. tobacco.

G. sunflowers.

H. grains.

I. lavender.

3 What did Czar Alexander II abolish in Russia in 1861?

A. the Russian Orthodox Church

B. diamond mining

C. the Russian army and navy

D. serfdom

4 The North Atlantic Treaty Organization (NATO), created in 1949, is an alliance of which countries?

F. the Balkan states

G. Western Russia and Eastern Europe

H. the United States and its allies

I. the USSR and Cuba

5 The dominant religion in most of Eastern Europe and Western Russia is

A. Islam.

B. Eastern Orthodox.

C. Protestantism.

D. Moldavan.

6 The difference in experience and viewpoints between those who grew up in the Communist USSR and Russians of the post-Soviet era is referred to as

F. a generation gap.

G. the diaspora.

H. cultural dissonance.

I. capitalism.

DBQ ANALYZING DOCUMENTS

7 **IDENTIFYING** Read the following passage about Russian culture:

"*As one looks at the history of Russian culture, it may be helpful to think of the forces rather than the forms behind it. Three in particular—the natural surroundings, the Christian heritage, and the Western contacts of Russia—hover bigger than life.*"

—from James H. Billington, *The Icon and the Axe* (1970)

Which theme of geography is represented by the influence of Western contacts on Russia?

A. human-environment interaction
B. location

C. movement
D. place

8 **ANALYZING** Which of these aspects of natural surroundings is most likely to have affected Russian culture?

F. sense of vast space

G. warm climate

H. fertile farmland

I. nearness to neighbors

SHORT RESPONSE

"*The southern half of Eastern Europe is referred to as the Balkans or Balkan Peninsula, after the name of a mountain range in Bulgaria. Balkanization [means] the recurrent division and fragmentation of this part of Eastern Europe, and it is now applied to any place where such processes take place.*"

—from H.J. de Blij and Peter O. Muller, *Geography* (2006)

9 **ANALYZING** What characteristics of the Balkans and the people who live there led to these frequent divisions?

10 **IDENTIFYING POINT OF VIEW** If you were a leader of one of the peoples of the Balkans, how would you try to bridge the divisions separating your group from others?

EXTENDED RESPONSE

11 **INFORMATIVE/EXPLANATORY WRITING** In the 1990s, Russia began to make a transition from a communist economy to a more capitalistic economy. Discuss how this transition has worked so far. What are some of the positive and negative factors in the change? You will want to examine some outside sources to explain this economic transition.

Need Extra Help?

If You've Missed Question	❶	❷	❸	❹	❺	❻	❼	❽	❾	❿	⓫
Review Lesson	1	1	2	2	3	3	3	1	3	3	3

Using **FOLDABLES** is a great way to organize notes, remember information, and prepare for tests. Follow these easy directions to create a Foldable® for the chapter you are studying.

CHAPTER 1: WHAT IS GEOGRAPHY?

Describing Make this Foldable and label the top *Geographer's View* and the bottom *Geographer's Tools*. Under the top fold, describe three ways you experience geography every day. Under the bottom fold, list and describe the tools of geography and explain how a map is a tool. In your mind, form an image of a map of the world. Sketch and label what you visualize on the back of your shutter fold.

Step 1
Bend a sheet of paper in half to find the midpoint.

Step 2
Fold the outer edges of the paper to meet at the midpoint.

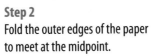

CHAPTER 2: EARTH'S PHYSICAL GEOGRAPHY

Identifying Make this Foldable and label the four tabs *Processes*, *Forces*, *Land*, and *Water*. Under *Processes*, identify and describe the processes that operate above and below Earth's surface. Include specific examples. Under *Forces*, give examples of how forces are changing Earth's surface where you live. Finally, under *Land* and *Water*, identify land and water features within 100 miles (161 km) of your community and explain how they influence your life.

Step 1
Fold the outer edges of the paper to meet at the midpoint. Crease well.

Step 2
Fold the paper in half from side to side.

Step 3
Open and cut along the inside fold lines to form four tabs.

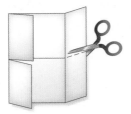

Step 4
Label the tabs as shown.

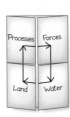

Processes | Forces
Land | Water

Foldables® Library

CHAPTER 3: EARTH'S PEOPLE

Analyzing Create this Foldable, and then label the tabs *Adaptations*, *Cultural Views*, and *Basic Needs*. Under *Adaptations*, describe how humans have adapted to life in two different geographic regions and describe population trends in each. Under *Cultural Views*, analyze what makes two different cultures unique. Finally, under *Basic Needs*, describe how your basic needs might be met in two different economic systems.

Step 1
Fold a sheet of paper in half, leaving a ½-inch tab along one edge.

Step 2
Then fold the paper into three equal sections.

Step 3
Cut along the folds on the top sheet of paper to create three tabs.

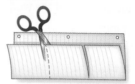

Step 4
Label your Foldable as shown.

CHAPTER 4: THE UNITED STATES EAST OF THE MISSISSIPPI

Organizing After you create the Foldable below, write the chapter title on the cover tab and label the three small tabs *East and West*, *Geographic Barriers*, and *Diversity*. Under *East and West*, sketch an outline of the United States and draw the Mississippi River. Then list facts about the region. Under *Geographic Barriers*, give examples of physical features that were barriers to westward expansion. Under *Diversity*, explain how cultural diversity makes the United States East of the Mississippi a unique region.

Step 1
Stack two sheets of paper so that the back sheet is 1 inch higher than the front sheet.

Step 2
Fold the paper to form four equal tabs.

Step 3
When all tabs are an equal distance apart, fold the papers and crease well.

Step 4
Open the papers, and then glue or staple them along the fold.

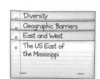

CHAPTER 5: THE UNITED STATES WEST OF THE MISSISSIPPI

Identifying Make a three-tab book with three columns. Label the columns *Geography*, *History*, and *Economy*. Under each column heading, write: *Know* and *Learned*. Use the book to record what you know and what you learn about the western region of the United States.

Step 1
Fold a sheet of paper in half, leaving a ½-inch tab along one edge.

Step 2
Then fold the paper into three equal sections.

Step 3
Cut along the folds on the top sheet of paper to create three tabs.

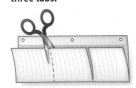

Step 4
Label your Foldable as shown.

CHAPTER 6: CANADA

Identifying Follow the steps below, and then label the four tabs *North*, *South*, *Past and Present*, and *World Relations*. Describe and give examples of the geography of the northern and southern regions of Canada under either the *North* or the *South* tab. Under *Past and Present*, identify important people, places, and events from Canada's history. Explain Canada's relations with other countries under the *World Relations* tab.

Step 1
Fold the outer edges of the paper to meet at the midpoint. Crease well.

Step 2
Fold the paper in half from side to side.

Step 3
Open and cut along the inside fold lines to form four tabs.

Step 4
Label the tabs as shown.

Foldables® Library

Foldables® Library

CHAPTER 7: MEXICO, CENTRAL AMERICA, AND THE CARIBBEAN ISLANDS

Analyzing Make the Foldable below. Write the chapter title on the cover tab, and label the three small tabs *Gulf of Mexico*, *Civilizations*, and *Trade and Commerce*. Under *Gulf of Mexico*, explain how the gulf has affected life in the region. Include information on weather, tourism, and the economy. Under *Civilizations*, sequence and describe the major civilizations that developed in this region and their cultural influences. Finally, under *Trade and Commerce*, compare and contrast trade events that are important to the economy of the region.

Step 1
Stack two sheets of paper so that the back sheet is 1 inch higher than the front sheet.

Step 2
Fold the paper to form four equal tabs.

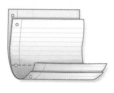

Step 3
When all tabs are an equal distance apart, fold the papers and crease well.

Step 4
Open the papers, and glue or staple them along the fold.

CHAPTER 8: BRAZIL

Organizing Create the Foldable below. On the back, write the chapter title and sketch a map of Brazil. Label the two front tabs *Valuable Natural Resources* and *Urban Population*. On your sketch, label Brazil's major geographic features. Under *Valuable Natural Resources*, outline when and where valuable natural resources were discovered and how the discoveries affected the native and colonial populations. Under *Urban Population*, discuss the impact of the population distribution.

Step 1
Bend a sheet of paper in half to find the midpoint.

Step 2
Fold the outer edges of the paper to meet at the midpoint.

CHAPTER 9: THE TROPICAL NORTH

Identifying Create the Foldable below. Label the cover *The Tropical North* and the layers *Geography*, *Foreign Influences and Resources*, and *Trade*. Under *Geography*, explain how geography and resources affect the countries and people of the region. Under *Foreign Influences and Resources*, explain how resources and foreign countries have impacted the Tropical North. Under *Trade*, explain how countries in the region are trying to expand trade and why.

Step 1
Stack two sheets of paper so that the back sheet is 1 inch higher than the front sheet.

Step 2
Fold the paper to form four equal tabs.

Step 3
When all tabs are an equal distance apart, fold the papers and crease well.

Step 4
Open the papers, and then glue or staple them along the fold.

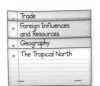

CHAPTER 10: ANDES AND MIDLATITUDE COUNTRIES

Describing Make the Foldable below, and then label the top of the sections *Geography*, *Culture*, and *Economy*. Under *Geography*, explain how the Andes Mountains affect the lives of the people who live near or around them. Under *Culture*, describe the rise and fall of the Inca Empire and what it tells about the history of the region. Finally, under *Economy*, explain how the terrain and the resources available affect the way people live.

Step 1
Fold a sheet of paper into thirds to form three equal columns.

Step 2
Label your Foldable as shown.

Foldables® Library

CHAPTER 11: WESTERN EUROPE

Analyzing Follow the steps below to create a Foldable. Then sketch an outline of Western Europe on the back. Label important geographic features in the region. On the front, label the tabs as illustrated. Under *Waterways—Landforms*, explain how waterways and landforms have influenced the development of Western Europe. Under *Early Civilizations—Industrial Revolution*, sequence the cultural and technological changes that occurred. Finally, under *War—Post-War*, summarize the effects of war on the region and explain why the EU was formed.

Step 1
Fold the outer edges of the paper to meet at the midpoint. Crease well.

Step 2
Open and cut three equal tabs from the outer edge to the crease on each side.

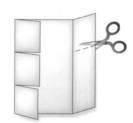

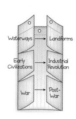

Step 3
Label the tabs as shown.

CHAPTER 12: NORTHERN AND SOUTHERN EUROPE

Describing Create the Foldable below, and label the anchor tab *Northern and Southern Europe*. Label the front of the tabs *Geography*, *History*, and *Culture*. Under *Geography*, describe how geography has influenced the lifestyles and economies in the region. Under *History*, explain the importance of the Silk Road and how trade with China influenced the people of the region. Finally, under *Culture*, identify different culture groups in each region.

Step 1
Fold a sheet of paper in half, leaving a ½-inch tab along one edge.

Step 2
Then fold the paper into three equal sections.

Step 3
Cut along the folds on the top sheet of paper to create three tabs.

Step 4
Label your Foldable as shown.

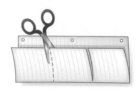

CHAPTER 13: EASTERN EUROPE AND WESTERN RUSSIA

Organizing Create this Foldable, and then sketch and label Eastern Europe, Western Russia, and the Ural Mountains on the back. On the front, label the top-left tab *Eastern Europe* and the top-right tab *Western Russia*. Under the tabs, describe landforms and natural resources found in each area. Label the two bottom tabs *Empires* and *Populations*. List the empires that once controlled this region and one important event from each. Finally, explain why populations are declining.

Step 1
Fold the outer edges of the paper to meet at the midpoint. Crease well.

Step 2
Fold the paper in half from side to side.

Step 3
Open and cut along the inside fold lines to form four tabs.

Step 4
Label the tabs as shown.

Eastern Europe *Western Russia*

Empires *Populations*

Foldables® Library

Gazetteer

A gazetteer (ga·zuh·TIHR) is a geographic index or dictionary. It shows latitude and longitude for cities and certain other places. Latitude and longitude are shown in this way: 48°N 2°E, or 48 degrees north latitude and two degrees east longitude. This Gazetteer lists many important geographic features and most of the world's largest independent countries and their capitals. The page numbers tell where each entry can be found on a map in this book. As an aid to pronunciation, most entries are spelled phonetically.

A

Abidjan [AH·BEE·JAHN] Capital of Côte d'Ivoire. 5°N 4°W (p. RA22)

Abu Dhabi [AH·BOO DAH·bee] Capital of the United Arab Emirates. 24°N 54°E (p. RA24)

Abuja [ah·BOO·jah] Capital of Nigeria. 8°N 9°E (p. RA22)

Accra [ah·KRUH] Capital of Ghana. 6°N 0° longitude (p. RA22)

Addis Ababa [AHD·dihs AH·bah·BAH] Capital of Ethiopia. 9°N 39°E (p. RA22)

Adriatic [AY·dree·A·tihk] **Sea** Arm of the Mediterranean Sea between the Balkan Peninsula and Italy. (p. RA20)

Afghanistan [af·GA·nuh·STAN] Central Asian country west of Pakistan. (p. RA25)

Albania [al·BAY·nee·uh] Country on the Adriatic Sea, south of Serbia. (p. RA18)

Algeria [al·JIHR·ee·uh] North African country east of Morocco. (p. RA22)

Algiers [al·JIHRZ] Capital of Algeria. 37°N 3°E (p. RA22)

Alps [ALPS] Mountain ranges extending through central Europe. (p. RA20)

Amazon [A·muh·ZAHN] **River** Largest river in the world by volume and second-largest in length. (p. RA17)

Amman [a·MAHN] Capital of Jordan. 32°N 36°E (p. RA24)

Amsterdam [AHM·stuhr·DAHM] Capital of the Netherlands. 52°N 5°E (p. RA18)

Andes [AN·DEEZ] Mountain system extending north and south along the western side of South America. (p. RA17)

Andorra [an·DAWR·uh] Small country in southern Europe between France and Spain. 43°N 2°E (p. RA18)

Angola [ang·GOH·luh] Southern African country north of Namibia. (p. RA22)

Ankara [AHNG·kuh·ruh] Capital of Turkey. 40°N 33°E (p. RA24)

Antananarivo [AHN·tah·NAH·nah·REE·voh] Capital of Madagascar. 19°S 48°E (p. RA22)

Arabian [uh·RAY·bee·uhn] **Peninsula** Large peninsula extending into the Arabian Sea. (p. RA25)

Argentina [AHR·juhn·TEE·nuh] South American country east of Chile. (p. RA16)

Armenia [ahr·MEE·nee·uh] European-Asian country between the Black and Caspian Seas. 40°N 45°E (p. RA26)

Ashkhabad [AHSH·gah·BAHD] Capital of Turkmenistan. 38°N 58°E (p. RA25)

Asmara [az·MAHR·uh] Capital of Eritrea. 16°N 39°E (p. RA22)

Astana Capital of Kazakhstan. 51°N 72°E (p. RA26)

Asunción [ah·SOON·see·OHN] Capital of Paraguay. 25°S 58°W (p. RA16)

Athens Capital of Greece. 38°N 24°E (p. RA19)

Atlas [AT·luhs] **Mountains** Mountain range on the northern edge of the Sahara. (p. RA23)

Australia [aw·STRAYL·yuh] Country and continent in Southern Hemisphere. (p. RA30)

Austria [AWS·tree·uh] Western European country east of Switzerland and south of Germany and the Czech Republic. (p. RA18)

Azerbaijan [A·zuhr·BY·JAHN] European-Asian country on the Caspian Sea. (p. RA25)

B

Baghdad Capital of Iraq. 33°N 44°E (p. RA25)

Bahamas [buh·HAH·muhz] Country made up of many islands between Cuba and the United States. (p. RA15)

Bahrain [bah·RAYN] Country located on the Persian Gulf. 26°N 51°E (p. RA25)

Baku [bah·KOO] Capital of Azerbaijan. 40°N 50°E (p. RA25)

Balkan [BAWL·kuhn] **Peninsula** Peninsula in southeastern Europe. (p. RA21)

Baltic [BAWL·tihk] **Sea** Sea in northern Europe that is connected to the North Sea. (p. RA20)

Bamako [BAH·mah·KOH] Capital of Mali. 13°N 8°W (p. RA22)

Bangkok [BANG·KAHK] Capital of Thailand. 14°N 100°E (p. RA27)

Bangladesh [BAHNG·gluh·DEHSH] South Asian country bordered by India and Myanmar. (p. RA27)

Bangui [BAHNG·GEE] Capital of the Central African Republic. 4°N 19°E (p. RA22)

Banjul [BAHN·JOOL] Capital of Gambia. 13°N 17°W (p. RA22)

Barbados [bahr·BAY·duhs] Island country between the Atlantic Ocean and the Caribbean Sea. 14°N 59°W (p. RA15)

Beijing [BAY·JIHNG] Capital of China. 40°N 116°E (p. RA27)

Beirut [bay·ROOT] Capital of Lebanon. 34°N 36°E (p. RA24)

Belarus [BEE•luh•ROOS] Eastern European country west of Russia. 54°N 28°E (p. RA19)

Belgium [BEHL•juhm] Western European country south of the Netherlands. (p. RA18)

Belgrade [BEHL•GRAYD] Capital of Serbia. 45°N 21°E (p. RA19)

Belize [buh•LEEZ] Central American country east of Guatemala. (p. RA14)

Belmopan [BEHL•moh•PAHN] Capital of Belize. 17°N 89°W (p. RA14)

Benin [buh•NEEN] West African country west of Nigeria. (p. RA22)

Berlin [behr•LEEN] Capital of Germany. 53°N 13°E (p. RA18)

Bern Capital of Switzerland. 47°N 7°E (p. RA18)

Bhutan [boo•TAHN] South Asian country northeast of India. (p. RA27)

Bishkek [bihsh•KEHK] Capital of Kyrgyzstan. 43°N 75°E (p. RA26)

Bissau [bihs•SOW] Capital of Guinea-Bissau. 12°N 16°W (p. RA22)

Black Sea Large sea between Europe and Asia. (p. RA21)

Bloemfontein [BLOOM•FAHN•TAYN] Judicial capital of South Africa. 26°E 29°S (p. RA22)

Bogotá [BOH•GOH•TAH] Capital of Colombia. 5°N 74°W (p. RA16)

Bolivia [buh•LIHV•ee•uh] Country in the central part of South America, north of Argentina. (p. RA16)

Bosnia and Herzegovina [BAHZ•nee•uh HEHRT•seh•GAW•vee•nuh] Southeastern European country bordered by Croatia, Serbia, and Montenegro. (p. RA18)

Botswana [bawt•SWAH•nah] Southern African country north of the Republic of South Africa. (p. RA22)

Brasília [brah•ZEEL•yuh] Capital of Brazil. 16°S 48°W (p. RA16)

Bratislava [BRAH•tih•SLAH•vuh] Capital of Slovakia. 48°N 17°E (p. RA18)

Brazil [bruh•ZIHL] Largest country in South America. (p. RA16)

Brazzaville [BRAH•zuh•VEEL] Capital of Congo. 4°S 15°E (p. RA22)

Brunei [bru•NY] Southeast Asian country on northern coast of the island of Borneo. (p. RA27)

Brussels [BRUH•suhlz] Capital of Belgium. 51°N 4°E (p. RA18)

Bucharest [BOO•kuh•REHST] Capital of Romania. 44°N 26°E (p. RA19)

Budapest [BOO•duh•PEHST] Capital of Hungary. 48°N 19°E (p. RA18)

Buenos Aires [BWAY•nuhs AR•eez] Capital of Argentina. 34°S 58°W (p. RA16)

Bujumbura [BOO•juhm•BUR•uh] Capital of Burundi. 3°S 29°E (p. RA22)

Bulgaria [BUHL•GAR•ee•uh] Southeastern European country south of Romania. (p. RA19)

Burkina Faso [bur•KEE•nuh FAH•soh] West African country south of Mali. (p. RA22)

Burundi [bu•ROON•dee] East African country at the northern end of Lake Tanganyika. 3°S 30°E (p. RA22)

C

Cairo [KY•roh] Capital of Egypt. 31°N 32°E (p. RA24)

Cambodia [kam•BOH•dee•uh] Southeast Asian country south of Thailand and Laos. (p. RA27)

Cameroon [KA•muh•ROON] Central African country on the northeast shore of the Gulf of Guinea. (p. RA22)

Canada [KA•nuh•duh] Northernmost country in North America. (p. RA6)

Canberra [KAN•BEHR•uh] Capital of Australia. 35°S 149°E (p. RA30)

Cape Town Legislative capital of the Republic of South Africa. 34°S 18°E (p. RA22)

Cape Verde [VUHRD] Island country off the coast of western Africa in the Atlantic Ocean. 15°N 24°W (p. RA22)

Caracas [kah•RAH•kahs] Capital of Venezuela. 11°N 67°W (p. RA16)

Caribbean [KAR•uh•BEE•uhn] **Islands** Islands in the Caribbean Sea between North America and South America, also known as West Indies. (p. RA15)

Caribbean Sea Part of the Atlantic Ocean bordered by the West Indies, South America, and Central America. (p. RA15)

Caspian [KAS•pee•uhn] **Sea** Salt lake between Europe and Asia that is the world's largest inland body of water. (p. RA21)

Caucasus [KAW•kuh•suhs] **Mountains** Mountain range between the Black and Caspian Seas. (p. RA21)

Central African Republic Central African country south of Chad. (p. RA22)

Chad [CHAD] Country west of Sudan in the African Sahel. (p. RA22)

Chang Jiang [CHAHNG jee•AHNG] Principal river of China that begins in Tibet and flows into the East China Sea near Shanghai; also known as the Yangtze River. (p. RA29)

Chile [CHEE•lay] South American country west of Argentina. (p. RA16)

China [CHY•nuh] Country in eastern and central Asia, known officially as the People's Republic of China. (p. RA27)

Chişinău [KEE•shee•NOW] Capital of Moldova. 47°N 29°E (p. RA19)

Colombia [kuh•LUHM•bee•uh] South American country west of Venezuela. (p. RA16)

Colombo [kuh•LUHM•boh] Capital of Sri Lanka. 7°N 80°E (p. RA26)

Comoros [KAH•muh•ROHZ] Small island country in Indian Ocean between the island of Madagascar and the southeast African mainland. 13°S 43°E (p. RA22)

Conakry [KAH•nuh•kree] Capital of Guinea. 10°N 14°W (p. RA22)

Congo [KAHNG•goh] Central African country east of the Democratic Republic of the Congo. 3°S 14°E (p. RA22)

Congo, Democratic Republic of the Central African country north of Zambia and Angola. 1°S 22°E (p. RA22)

Copenhagen [KOH•puhn•HAY•guhn] Capital of Denmark. 56°N 12°E (p. RA18)

Costa Rica [KAWS•tah REE•kah] Central American country south of Nicaragua. (p. RA15)

Côte d'Ivoire [KOHT dee•VWAHR] West African country south of Mali. (p. RA22)

Croatia [kroh•AY•shuh] Southeastern European country on the Adriatic Sea. (p. RA18)

Cuba [KYOO•buh] Island country in the Caribbean Sea. (p. RA15)

Cyprus [SY•pruhs] Island country in the eastern Mediterranean Sea, south of Turkey. (p. RA19)

Czech [CHEHK] **Republic** Eastern European country north of Austria. (p. RA18)

D

Dakar [dah•KAHR] Capital of Senegal. 15°N 17°W (p. RA22)

Damascus [duh•MAS•kuhs] Capital of Syria. 34°N 36°E (p. RA24)

Dar es Salaam [DAHR EHS sah•LAHM] Commercial capital of Tanzania. 7°S 39°E (p. RA22)

Denmark Northern European country between the Baltic and North Seas. (p. RA18)

Dhaka [DA•kuh] Capital of Bangladesh. 24°N 90°E (p. RA27)

Djibouti [jih•BOO•tee] East African country on the Gulf of Aden. 12°N 43°E (p. RA22)

Dodoma [doh•DOH•mah] Political capital of Tanzania. 6°S 36°E (p. RA22)

Doha [DOH•huh] Capital of Qatar. 25°N 51°E (p. RA25)

Dominican [duh•MIH•nih•kuhn] **Republic** Country in the Caribbean Sea on the eastern part of the island of Hispaniola. (p. RA15)

Dublin [DUH•blihn] Capital of Ireland. 53°N 6°W (p. RA18)

Dushanbe [doo•SHAM•buh] Capital of Tajikistan. 39°N 69°E (p. RA25)

E

East Timor [TEE•MOHR] Previous province of Indonesia, now under UN administration. 10°S 127°E (p. RA27)

Ecuador [EH•kwuh•dawr] South American country southwest of Colombia. (p. RA16)

Egypt [EE•jihpt] North African country on the Mediterranean Sea. (p. RA24)

El Salvador [ehl SAL•vuh•dawr] Central American country southwest of Honduras. (p. RA14)

Equatorial Guinea [EE•kwuh•TOHR•ee•uhl GIH•nee] Central African country south of Cameroon. (p. RA22)

Eritrea [EHR•uh•TREE•uh] East African country north of Ethiopia. (p. RA22)

Estonia [eh•STOH•nee•uh] Eastern European country on the Baltic Sea. (p. RA19)

Ethiopia [EE•thee•OH•pee•uh] East African country north of Somalia and Kenya. (p. RA22)

Euphrates [yu•FRAY•teez] **River** River in southwestern Asia that flows through Syria and Iraq and joins the Tigris River. (p. RA25)

F

Fiji [FEE•jee] **Islands** Country comprised of an island group in the southwest Pacific Ocean. 19°S 175°E (p. RA30)

Finland [FIHN•luhnd] Northern European country east of Sweden. (p. RA19)

France [FRANS] Western European country south of the United Kingdom. (p. RA18)

Freetown Capital of Sierra Leone. (p. RA22)

French Guiana [gee•A•nuh] French-owned territory in northern South America. (p. RA16)

G

Gabon [ga•BOHN] Central African country on the Atlantic Ocean. (p. RA22)

Gaborone [GAH•boh•ROH•nay] Capital of Botswana. (p. RA22)

Gambia [GAM•bee•uh] West African country along the Gambia River. (p. RA22)

Georgetown [JAWRJ•town] Capital of Guyana. 8°N 58°W (p. RA16)

Georgia [JAWR•juh] European-Asian country bordering the Black Sea south of Russia. (p. RA26)

Germany [JUHR•muh•nee] Western European country south of Denmark, officially called the Federal Republic of Germany. (p. RA18)

Ghana [GAH•nuh] West African country on the Gulf of Guinea. (p. RA22)

Great Plains The continental slope extending through the United States and Canada. (p. RA7)

Greece [GREES] Southern European country on the Balkan Peninsula. (p. RA19)

Greenland [GREEN•luhnd] Island in northwestern Atlantic Ocean and the largest island in the world. (p. RA6)

Guatemala [GWAH•tay•MAH•lah] Central American country south of Mexico. (p. RA14)

Guatemala Capital of Guatemala. 15°N 91°W (p. RA14)

Guinea [GIH•nee] West African country on the Atlantic coast. (p. RA22)

Guinea-Bissau [GIH•nee bih•SOW] West African country on the Atlantic coast. (p. RA22)

Gulf of Mexico Gulf on part of the southern coast of North America. (p. RA7)

Guyana [gy•AH•nuh] South American country between Venezuela and Suriname. (p. RA16)

H

Haiti [HAY•tee] Country in the Caribbean Sea on the western part of the island of Hispaniola. (p. RA15)

Hanoi [ha•NOY] Capital of Vietnam. 21°N 106°E (p. RA27)

Harare [hah•RAH•ray] Capital of Zimbabwe. 18°S 31°E (p. RA22)

Havana [huh•VA•nuh] Capital of Cuba. 23°N 82°W (p. RA15)

Helsinki [HEHL•SIHNG•kee] Capital of Finland. 60°N 24°E (p. RA19)

Himalaya [HI•muh•LAY•uh] Mountain ranges in southern Asia, bordering the Indian subcontinent on the north. (p. RA28)

Honduras [hahn•DUR•uhs] Central American country on the Caribbean Sea. (p. RA14)

Hong Kong Port and industrial center in southern China. 22°N 115°E (p. RA27)

Huang He [HWAHNG HUH] River in northern and eastern China, also known as the Yellow River. (p. RA29)

Hungary [HUHNG•guh•ree] Eastern European country south of Slovakia. (p. RA18)

I

Iberian [eye•BIHR•ee•uhn] Peninsula Peninsula in southwest Europe, occupied by Spain and Portugal. (p. RA20)

Iceland Island country between the North Atlantic and Arctic Oceans. (p. RA18)

India [IHN•dee•uh] South Asian country south of China and Nepal. (p. RA26)

Indonesia [IHN•duh•NEE•zhuh] Southeast Asian island country known as the Republic of Indonesia. (p. RA27)

Indus [IHN•duhs] River River in Asia that begins in Tibet and flows through Pakistan to the Arabian Sea. (p. RA28)

Iran [ih•RAN] Southwest Asian country that was formerly named Persia. (p. RA25)

Iraq [ih•RAHK] Southwest Asian country west of Iran. (p. RA25)

Ireland [EYER•luhnd] Island west of Great Britain occupied by the Republic of Ireland and Northern Ireland. (p. RA18)

Islamabad [ihs•LAH•muh•BAHD] Capital of Pakistan. 34°N 73°E (p. RA26)

Israel [IHZ•ree•uhl] Southwest Asian country south of Lebanon. (p. RA24)

Italy [IHT•uhl•ee] Southern European country south of Switzerland and east of France. (p. RA18)

J

Jakarta [juh•KAHR•tuh] Capital of Indonesia. 6°S 107°E (p. RA27)

Jamaica [juh•MAY•kuh] Island country in the Caribbean Sea. (p. RA15)

Japan [juh•PAN] East Asian country consisting of the four large islands of Hokkaido, Honshu, Shikoku, and Kyushu, plus thousands of small islands. (p. RA27)

Jerusalem [juh•ROO•suh•luhm] Capital of Israel and a holy city for Christians, Jews, and Muslims. 32°N 35°E (p. RA24)

Jordan [JAWRD•uhn] Southwest Asian country south of Syria. (p. RA24)

Juba [JU•buh] Capital of South Sudan. 5°N 31°E (p. RA22)

K

Kabul [KAH•buhl] Capital of Afghanistan. 35°N 69°E (p. RA25)

Kampala [kahm•PAH•lah] Capital of Uganda. 0° latitude 32°E (p. RA22)

Kathmandu [KAT•MAN•DOO] Capital of Nepal. 28°N 85°E (p. RA26)

Kazakhstan [kuh•ZAHK•STAHN] Large Asian country south of Russia and bordering the Caspian Sea. (p. RA26)

Kenya [KEHN•yuh] East African country south of Ethiopia. (p. RA22)

Khartoum [kahr•TOOM] Capital of Sudan. 16°N 33°E (p. RA22)

Kigali [kee•GAH•lee] Capital of Rwanda. 2°S 30°E (p. RA22)

Kingston [KIHNG•stuhn] Capital of Jamaica. 18°N 77°W (p. RA15)

Kinshasa [kihn•SHAH•suh] Capital of the Democratic Republic of the Congo. 4°S 15°E (p. RA22)

Kuala Lumpur [KWAH•luh LUM•PUR] Capital of Malaysia. 3°N 102°E (p. RA27)

Kuwait [ku•WAYT] Country on the Persian Gulf between Saudi Arabia and Iraq. (p. RA25)

Kyiv (Kiev) [KEE•ihf] Capital of Ukraine. 50°N 31°E (p. RA19)

Kyrgyzstan [s•gih•STAN] Central Asian country on China's western border. (p. RA26)

L

Laos [LOWS] Southeast Asian country south of China and west of Vietnam. (p. RA27)

La Paz [lah PAHS] Administrative capital of Bolivia, and the highest capital in the world. 17°S 68°W (p. RA16)

Latvia [LAT•vee•uh] Eastern European country west of Russia on the Baltic Sea. (p. RA19)

Lebanon [LEH•buh•nuhn] Country south of Syria on the Mediterranean Sea. (p. RA24)

Lesotho [luh•SOH•TOH] Southern African country within the borders of the Republic of South Africa. (p. RA22)

Liberia [ly•BIHR•ee•uh] West African country south of Guinea. (p. RA22)

Libreville [LEE•bruh•VIHL] Capital of Gabon. 1°N 9°E (p. RA22)

Libya [LIH•bee•uh] North African country west of Egypt on the Mediterranean Sea. (p. RA22)

Liechtenstein [LIHKT•uhn•SHTYN] Small country in central Europe between Switzerland and Austria. 47°N 10°E (p. RA18)

Lilongwe [lih•LAWNG•GWAY] Capital of Malawi. 14°S 34°E (p. RA22)

Lima [LEE•mah] Capital of Peru. 12°S 77°W (p. RA16)

Lisbon [LIHZ•buhn] Capital of Portugal. 39°N 9°W (p. RA18)

Lithuania [LIH•thuh•WAY•nee•uh] Eastern European country northwest of Belarus on the Baltic Sea. (p. RA21)

Ljubljana [lee•oo•blee•AH•nuh] Capital of Slovenia. 46°N 14°E (p. RA18)

Lomé [loh•MAY] Capital of Togo. 6°N 1°E (p. RA22)

London Capital of the United Kingdom, on the Thames River. 52°N 0° longitude (p. RA18)

Luanda [lu•AHN•duh] Capital of Angola. 9°S 13°E (p. RA22)

Lusaka [loo•SAH•kah] Capital of Zambia. 15°S 28°E (p. RA22)

Luxembourg [LUHK•suhm•BUHRG] Small European country bordered by France, Belgium, and Germany. 50°N 7°E (p. RA18)

M

Macao [muh•KOW] Port in southern China. 22°N 113°E (p. RA27)

Macedonia [ma•suh•DOH•nee•uh] Southeastern European country north of Greece. (p. RA19). Macedonia also refers to a geographic region covering northern Greece, the country Macedonia, and part of Bulgaria.

Madagascar [MA•duh•GAS•kuhr] Island in the Indian Ocean off the southeastern coast of Africa. (p. RA22)

Madrid Capital of Spain. 41°N 4°W (p. RA18)

Malabo [mah•LAH•boh] Capital of Equatorial Guinea. 4°N 9°E (p. RA22)

Malawi [mah•LAH•wee] Southern African country south of Tanzania and east of Zambia. (p. RA22)

Malaysia [muh•LAY•zhuh] Southeast Asian country with land on the Malay Peninsula and on the island of Borneo. (p. RA27)

Maldives [MAWL•DEEVZ] Island country southwest of India in the Indian Ocean. (p. RA26)

Mali [MAH•lee] West African country east of Mauritania. (p. RA22)

Managua [mah•NAH•gwah] Capital of Nicaragua. (p. RA15)

Manila [muh•NIH•luh] Capital of the Philippines. 15°N 121°E (p. RA27)

Maputo [mah•POO•toh] Capital of Mozambique. 26°S 33°E (p. RA22)

Maseru [MA•zuh•ROO] Capital of Lesotho. 29°S 27°E (p. RA22)

Masqat [MUHS•KAHT] Capital of Oman. 23°N 59°E (p. RA25)

Mauritania [MAWR•uh•TAY•nee•uh] West African country north of Senegal. (p. RA22)

Mauritius [maw•RIH•shuhs] Island country in the Indian Ocean east of Madagascar. 21°S 58°E (p. RA3)

Mbabane [uhm•bah•BAH•nay] Capital of Swaziland. 26°S 31°E (p. RA22)

Mediterranean [MEH•duh•tuh•RAY•nee•uhn] **Sea** Large inland sea surrounded by Europe, Asia, and Africa. (p. RA20)

Mekong [MAY•KAWNG] **River** River in southeastern Asia that begins in Tibet and empties into the South China Sea. (p. RA29)

Mexico [MEHK•sih•KOH] North American country south of the United States. (p. RA14)

Mexico City Capital of Mexico. 19°N 99°W (p. RA14)

Minsk [MIHNSK] Capital of Belarus. 54°N 28°E (p. RA19)

Mississippi [MIH•suh•SIH•pee] **River** Large river system in the central United States that flows southward into the Gulf of Mexico. (p. RA11)

Mogadishu [MOH•guh•DEE•shoo] Capital of Somalia. 2°N 45°E (p. RA22)

Moldova [mawl•DAW•vuh] Small European country between Ukraine and Romania. (p. RA19)

Monaco [MAH•nuh•KOH] Small country in southern Europe on the French Mediterranean coast. 44°N 8°E (p. RA18)

Mongolia [mahn•GOHL•yuh] Country in Asia between Russia and China. (p. RA23)

Monrovia [muhn•ROH•vee•uh] Capital of Liberia. 6°N 11°W (p. RA22)

Montenegro [MAHN•tuh•NEE•groh] Eastern European country. (p. RA18)

Montevideo [MAHN•tuh•vuh•DAY•oh] Capital of Uruguay. 35°S 56°W (p. RA16)

Morocco [muh•RAH•KOH] North African country on the Mediterranean Sea and the Atlantic Ocean. (p. RA22)

Moscow [MAHS•KOW] Capital of Russia. 56°N 38°E (p. RA19)

Mount Everest [EHV•ruhst] Highest mountain in the world, in the Himalaya between Nepal and Tibet. (p. RA28)

Mozambique [MOH•zahm•BEEK] Southern African country south of Tanzania. (p. RA22)

Myanmar [MYAHN•MAHR] Southeast Asian country south of China and India, formerly called Burma. (p. RA27)

N

Nairobi [ny•ROH•bee] Capital of Kenya. 1°S 37°E (p. RA22)

Namibia [nuh•MIH•bee•uh] Southern African country south of Angola on the Atlantic Ocean. 20°S 16°E (p. RA22)

Nassau [NA•saw] Capital of the Bahamas. 25°N 77°W (p. RA15)

N'Djamena [uhn•jah•MAY•nah] Capital of Chad. 12°N 15°E (p. RA22)

Nepal [NAY•PAHL] Mountain country between India and China. (p. RA26)

Netherlands [NEH•thuhr•lundz] Western European country north of Belgium. (p. RA18)

New Delhi [NOO DEH•lee] Capital of India. 29°N 77°E (p. RA26)

New Zealand [NOO ZEE•luhnd] Major island country southeast of Australia in the South Pacific. (p. RA30)

Niamey [nee•AHM•ay] Capital of Niger. 14°N 2°E (p. RA22)

Nicaragua [NIH•kuh•RAH•gwuh] Central American country south of Honduras. (p. RA15)

Nicosia [NIH•kuh•SEE•uh] Capital of Cyprus. 35°N 33°E (p. RA19)

Niger [NY•juhr] West African country north of Nigeria. (p. RA22)

Nigeria [ny•JIHR•ee•uh] West African country along the Gulf of Guinea. (p. RA22)

Nile [NYL] **River** Longest river in the world, flowing north through eastern Africa. (p. RA23)

North Korea [kuh•REE•uh] East Asian country in the northernmost part of the Korean Peninsula. (p. RA27)

Norway [NAWR•way] Northern European country on the Scandinavian Peninsula. (p. RA18)

Nouakchott [nu•AHK•SHAHT] Capital of Mauritania. 18°N 16°W (p. RA22)

O

Oman [oh•MAHN] Country on the Arabian Sea and the Gulf of Oman. (p. RA25)

Oslo [AHZ•loh] Capital of Norway. 60°N 11°E (p. RA18)

Ottawa [AH•tuh•wuh] Capital of Canada. 45°N 76°W (p. RA13)

Ouagadougou [WAH•gah•DOO•goo] Capital of Burkina Faso. 12°N 2°W (p. RA22)

P

Pakistan [PA•kih•STAN] South Asian country northwest of India on the Arabian Sea. (p. RA26)

Palau [puh•LOW) Island country in the Pacific Ocean. 7°N 135°E (p. RA30)

Panama [PA•nuh•MAH] Central American country on the Isthmus of Panama. (p. RA15)

Panama Capital of Panama. 9°N 79°W (p. RA15)

Papua New Guinea [PA•pyu•wuh NOO GIH•nee] Island country in the Pacific Ocean north of Australia. 7°S 142°E (p. RA30)

Paraguay [PAR•uh•GWY] South American country northeast of Argentina. (p. RA16)

Paramaribo [PAH•rah•MAH•ree•boh] Capital of Suriname. 6°N 55°W (p. RA16)

Paris Capital of France. 49°N 2°E (p. RA18)

Persian [PUHR•zhuhn] **Gulf** Arm of the Arabian Sea between Iran and Saudi Arabia. (p. RA25)

Peru [puh•ROO] South American country south of Ecuador and Colombia. (p. RA16)

Philippines [FIH•luh•PEENZ] Island country in the Pacific Ocean southeast of China. (p. RA27)

Phnom Penh [puh•NAWM PEHN] Capital of Cambodia. 12°N 106°E (p. RA27)

Poland [POH•luhnd] Eastern European country on the Baltic Sea. (p. RA18)

Port-au-Prince [POHRT•oh•PRIHNS] Capital of Haiti. 19°N 72°W (p. RA15)

Port Moresby [MOHRZ•bee] Capital of Papua New Guinea. 10°S 147°E (p. RA30)

Port-of-Spain [SPAYN] Capital of Trinidad and Tobago. 11°N 62°W (p. RA15)

Porto-Novo [POHR•toh•NOH•voh] Capital of Benin. 7°N 3°E (p. RA22)

Portugal [POHR•chih•guhl] Country west of Spain on the Iberian Peninsula. (p. RA18)

Prague [PRAHG] Capital of the Czech Republic. 51°N 15°E (p. RA18)

Puerto Rico [PWEHR•toh REE•koh] Island in the Caribbean Sea; U.S. Commonwealth. (p. RA15)

P'yŏngyang [pee•AWNG•YAHNG] Capital of North Korea. 39°N 126°E (p. RA27)

Q

Qatar [KAH•tuhr] Country on the southwestern shore of the Persian Gulf. (p. RA25)

Quito [KEE•toh] Capital of Ecuador. 0° latitude 79°W (p. RA16)

R

Rabat [ruh•BAHT] Capital of Morocco. 34°N 7°W (p. RA22)

Reykjavík [RAY•kyah•VEEK] Capital of Iceland. 64°N 22°W (p. RA18)

Rhine [RYN] **River** River in western Europe that flows into the North Sea. (p. RA20)

Riga [REE•guh] Capital of Latvia. 57°N 24°E (p. RA19)

Rio Grande [REE•oh GRAND] River that forms part of the boundary between the United States and Mexico. (p. RA10)

Riyadh [ree•YAHD] Capital of Saudi Arabia. 25°N 47°E (p. RA25)

Rocky Mountains Mountain system in western North America. (p. RA7)

Romania [ru•MAY•nee•uh] Eastern European country east of Hungary. (p. RA19)

Rome Capital of Italy. 42°N 13°E (p. RA18)

Russia [RUH•shuh] Largest country in the world, covering parts of Europe and Asia. (pp. RA19, RA27)

Rwanda [ruh•WAHN•duh] East African country south of Uganda. 2°S 30°E (p. RA22)

S

Sahara [suh•HAR•uh] Desert region in northern Africa that is the largest hot desert in the world. (p. RA23)

Saint Lawrence [LAWR•uhns] River River that flows from Lake Ontario to the Atlantic Ocean and forms part of the boundary between the United States and Canada. (p. RA13)

Sanaa [sahn•AH] Capital of Yemen. 15°N 44°E (p. RA25)

San José [SAN hoh•ZAY] Capital of Costa Rica. 10°N 84°W (p. RA15)

San Marino [SAN muh•REE•noh] Small European country located on the Italian Peninsula. 44°N 13°E (p. RA18)

San Salvador [SAN SAL•vuh•DAWR] Capital of El Salvador. 14°N 89°W (p. RA14)

Santiago [SAN•tee•AH•goh] Capital of Chile. 33°S 71°W (p. RA16)

Santo Domingo [SAN•toh duh•MIHNG•goh] Capital of the Dominican Republic. 19°N 70°W (p. RA15)

São Tomé and Príncipe [sow too•MAY PREEN•see•pee] Small island country in the Gulf of Guinea off the coast of central Africa. 1°N 7°E (p. RA22)

Sarajevo [SAR•uh•YAY•voh] Capital of Bosnia and Herzegovina. 43°N 18°E (p. RA18)

Saudi Arabia [SOW•dee uh•RAY•bee•uh] Country on the Arabian Peninsula. (p. RA25)

Senegal [SEH•nih•GAWL] West African country on the Atlantic coast. (p. RA22)

Seoul [SOHL] Capital of South Korea. 38°N 127°E (p. RA27)

Serbia [SUHR•bee•uh] Eastern European country south of Hungary. (p. RA18)

Seychelles [say•SHEHL] Small island country in the Indian Ocean off eastern Africa. 6°S 56°E (p. RA22)

Sierra Leone [see•EHR•uh lee•OHN] West African country south of Guinea. (p. RA22)

Singapore [SIHNG•uh•POHR] Southeast Asian island country near tip of the Malay Peninsula. (p. RA27)

Skopje [SKAW•pyay] Capital of the country of Macedonia. 42°N 21°E (p. RA19)

Slovakia [sloh•VAH•kee•uh] Eastern European country south of Poland. (p. RA18)

Slovenia [sloh•VEE•nee•uh] Southeastern European country south of Austria on the Adriatic Sea. (p. RA18)

Sofia [SOH•fee•uh] Capital of Bulgaria. 43°N 23°E (p. RA19)

Solomon [SAH•luh•muhn] Islands Island country in the Pacific Ocean northeast of Australia. (p. RA30)

Somalia [soh•MAH•lee•uh] East African country on the Gulf of Aden and the Indian Ocean. (p. RA22)

South Africa [A•frih•kuh] Country at the southern tip of Africa, officially the Republic of South Africa. (p. RA22)

South Korea [kuh•REE•uh] East Asian country on the Korean Peninsula between the Yellow Sea and the Sea of Japan. (p. RA27)

South Sudan [soo•DAN] East African country south of Sudan. (p. RA22)

Spain [SPAYN] Southern European country on the Iberian Peninsula. (p. RA18)

Sri Lanka [SREE LAHNG•kuh] Country in the Indian Ocean south of India, formerly called Ceylon. (p. RA26)

Stockholm [STAHK•HOHLM] Capital of Sweden. 59°N 18°E (p. RA18)

Sucre [SOO•kray] Constitutional capital of Bolivia. 19°S 65°W (p. RA16)

Sudan [soo•DAN] East African country south of Egypt. (p. RA22)

Suriname [SUR•uh•NAH•muh] South American country between Guyana and French Guiana. (p. RA16)

Suva [SOO•vah] Capital of the Fiji Islands. 18°S 177°E (p. RA30)

Swaziland [SWAH•zee•land] Southern African country west of Mozambique, almost entirely within the Republic of South Africa. (p. RA22)

Sweden Northern European country on the eastern side of the Scandinavian Peninsula. (p. RA18)

Switzerland [SWIHT•suhr•luhnd] European country in the Alps south of Germany. (p. RA18)

Syria [SIHR•ee•uh] Southwest Asian country on the east side of the Mediterranean Sea. (p. RA24)

T

Taipei [TY•PAY] Capital of Taiwan. 25°N 122°E (p. RA27)

Taiwan [TY•WAHN] Island country off the southeast coast of China; the seat of the Chinese Nationalist government. (p. RA27)

Tajikistan [tah•JIH•kih•STAN] Central Asian country east of Turkmenistan. (p. RA26)

Tallinn [TA•luhn] Capital of Estonia. 59°N 25°E (p. RA19)

Tanzania [TAN•zuh•NEE•uh] East African country south of Kenya. (p. RA22)

Tashkent [tash•KEHNT] Capital of Uzbekistan. 41°N 69°E (p. RA26)

Tbilisi [tuh•bih•LEE•see] Capital of the Republic of Georgia. 42°N 45°E (p. RA26)

Tegucigalpa [tay•GOO•see•GAHL•pah] Capital of Honduras. 14°N 87°W (p. RA14)

Tehran [TAY•uh•RAN] Capital of Iran. 36°N 52°E (p. RA25)

Thailand [TY•LAND] Southeast Asian country east of Myanmar. 17°N 101°E (p. RA27)

Thimphu [thihm•POO] Capital of Bhutan. 28°N 90°E (p. RA27)

Tigris [TY•gruhs] **River** River in southeastern Turkey and Iraq that merges with the Euphrates River. (p. RA25)

Tiranë [tih•RAH•nuh] Capital of Albania. 42°N 20°E (p. RA18)

Togo [TOH•goh] West African country between Benin and Ghana on the Gulf of Guinea. (p. RA22)

Tokyo [TOH•kee•OH] Capital of Japan. 36°N 140°E (p. RA27)

Trinidad and Tobago [TRIH•nuh•DAD tuh•BAY•goh] Island country near Venezuela between the Atlantic Ocean and the Caribbean Sea. (p. RA15)

Tripoli [TRIH•puh•lee] Capital of Libya. 33°N 13°E (p. RA22)

Tshwane [ch•WAH•nay] Executive capital of South Africa. 26°S 28°E (p. RA22)

Tunis [TOO•nuhs] Capital of Tunisia. 37°N 10°E (p. RA22)

Tunisia [too•NEE•zhuh] North African country on the Mediterranean Sea between Libya and Algeria. (p. RA22)

Turkey [TUHR•kee] Country in southeastern Europe and western Asia. (p. RA24)

Turkmenistan [tuhrk•MEH•nuh•STAN] Central Asian country on the Caspian Sea. (p. RA25)

U

Uganda [yoo•GAHN•dah] East African country south of Sudan. (p. RA22)

Ukraine [yoo•KRAYN] Eastern European country west of Russia on the Black Sea. (p. RA25)

Ulaanbaatar [oo•LAHN•BAH•TAWR] Capital of Mongolia. 48°N 107°E (p. RA27)

United Arab Emirates [EH•muh•ruhts] Country made up of seven states on the eastern side of the Arabian Peninsula. (p. RA25)

United Kingdom Western European island country made up of England, Scotland, Wales, and Northern Ireland. (p. RA18)

United States of America Country in North America made up of 50 states, mostly between Canada and Mexico. (p. RA8)

Uruguay [YUR•uh•GWAY] South American country south of Brazil on the Atlantic Ocean. (p. RA16)

Uzbekistan [uz•BEH•kih•STAN] Central Asian country south of Kazakhstan. (p. RA25)

V

Vanuatu [VAN•WAH•TOO] Country made up of islands in the Pacific Ocean east of Australia. (p. RA30)

Vatican [VA•tih•kuhn] City Headquarters of the Roman Catholic Church, located in the city of Rome in Italy. 42°N 13°E (p. RA18)

Venezuela [VEH•nuh•ZWAY•luh] South American country on the Caribbean Sea between Colombia and Guyana. (p. RA16)

Vienna [vee•EH•nuh] Capital of Austria. 48°N 16°E (p. RA18)

Vientiane [vyehn•TYAHN] Capital of Laos. 18°N 103°E (p. RA27)

Vietnam [vee•EHT•NAHM] Southeast Asian country east of Laos and Cambodia. (p. RA27)

Vilnius [VIL•nee•uhs] Capital of Lithuania. 55°N 25°E (p. RA19)

W

Warsaw Capital of Poland. 52°N 21°E (p. RA19)

Washington, D.C. Capital of the United States, in the District of Columbia. 39°N 77°W (p. RA8)

Wellington [WEH•lihng•tuhn] Capital of New Zealand. 41°S 175°E (p. RA30)

West Indies Caribbean islands between North America and South America. (p. RA15)

Windhoek [VIHNT•HUK] Capital of Namibia. 22°S 17°E (p. RA22)

Y

Yamoussoukro [YAH•MOO•SOO•kroh] Second capital of Côte d'Ivoire. 7°N 6°W (p. RA22)

Yangon [YAHNG•GOHN] City in Myanmar; formerly called Rangoon. 17°N 96°E (p. RA27)

Yaoundé [yown•DAY] Capital of Cameroon. 4°N 12°E (p. RA22)

Yemen [YEH•muhn] Country south of Saudi Arabia on the Arabian Peninsula. (p. RA25)

Yerevan [YEHR•uh•VAHN] Capital of Armenia. 40°N 44°E (p. RA25)

Z

Zagreb [ZAH•GREHB] Capital of Croatia. 46°N 16°E (p. RA18)

Zambia [ZAM•bee•uh] Southern African country north of Zimbabwe. (p. RA22)

Zimbabwe [zihm•BAH•bway] Southern African country northeast of Botswana. (p. RA22)

- Content vocabulary words are words that relate to world geography content.
- Words that have an asterisk (*) are academic vocabulary. They help you understand your school subjects.
- All vocabulary words are **boldfaced** or highlighted in yellow in your textbook.

aboriginal • autonomy

ENGLISH	A	ESPAÑOL

aboriginal a native people (p. 184)

absolute location the exact location of something (p. 21)

*****access** a way to reach a distant area (p. 183)

*****accurate** without mistakes or errors (p. 44)

*****achievement** a great accomplishment due to hard work (pp. 372–73)

acid rain rain that contains harmful amounts of poisons due to pollution (p. 65)

*****adapt** to change a trait in order to survive (p. 336)

aerospace the industry that makes vehicles that travel in the air and in outer space (p. 170)

agribusiness an industry based on huge farms that rely on machines and mass-production methods (p. 170)

agriculture the practice of growing crops and raising livestock (p. 129)

altiplano the high plains (p. 299)

altitude the height above sea level (p. 301)

annex to declare ownership of an area (p. 159)

*****annual** yearly or each year (p. 165)

archipelago a group of islands (p. 181)

*****area** a geographic location (p. 241)

atmosphere the layer of gases surrounding Earth (p. 44)

autonomy having independence from another country (p. 194)

aborigen persona nativa (pág. 184)

localización absoluta ubicación exacta de algo (pág. 21)

*****acceso** vía para llegar a un lugar distante (pág. 183)

*****exacto** sin faltas o errores (pág. 44)

*****logro** consecución importante que resulta de un trabajo arduo (págs. 372–73)

lluvia ácida lluvia que contiene cantidades nocivas de venenos debido a la polución (pág. 65)

*****adaptar** cambiar un rasgo para sobrevivir (pág. 336)

aeroespacial industria que construye vehículos que viajan por el aire y el espacio exterior (pág. 170)

agronegocio industria basada en granjas extensas que dependen de máquinas y técnicas de producción masiva (pág. 170)

agricultura actividad que consiste en cultivar la tierra y criar ganado (pág. 129)

altiplano meseta elevada (pág. 299)

altitud altura sobre el nivel del mar (pág. 301)

anexionar declarar la propiedad de un territorio (pág. 159)

*****anual** cada año (pág. 165)

archipiélago grupo de islas (pág. 181)

*****área** territorio geográfico (pág. 241)

atmósfera capa de gases que rodea la Tierra (pág. 44)

soberanía independencia respecto de otro país (pág. 194)

axis an imaginary line that runs through Earth's center from the North Pole to the South Pole (p. 42)

eje línea imaginaria que atraviesa el centro de la Tierra desde el Polo Norte hasta el Polo Sur (pág. 42)

B

balkanization to break a country up into smaller units that are often hostile to one another (p. 397)

balcanizar fragmentar un país en partes más pequeñas, con frecuencia hostiles entre sí (pág. 397)

basin an area of land that is drained by a river and its tributaries (p. 241)

cuenca área de terreno drenada por un río y sus afluentes (pág. 241)

bauxite the mineral that is used to make aluminum (p. 206)

bauxita mineral metalífero que se utiliza para producir aluminio (pág. 206)

*behalf in the interest of (p. 87)

*a favor de en beneficio de (pág. 87)

*benefit an advantage (p. 203)

*beneficio ventaja (pág. 203)

bilingual able to use two languages (p. 191)

bilingüe que habla dos idiomas (pág. 191)

birthrate the number of babies born compared to the total number of people in a population at a given time (p. 72)

tasa de natalidad número de nacimientos comparado con el número total de habitantes de una población en un tiempo determinado (pág. 72)

brackish water that is somewhat salty (p. 398)

salobre agua algo salada (pág. 398)

C

canopy the umbrella-like covering formed by the tops of trees in a rain forest (p. 241)

manto cubierta en forma de sombrilla formada por las copas de los árboles en una selva tropical (pág. 241)

cash crop a farm product grown for sale (p. 213)

cultivo comercial producto agrícola que se cultiva para la venta (pág. 213)

caudillo a person who often ruled a Latin American country as a dictator and was generally a high-ranking military officer or a rich man (p. 213)

caudillo persona que gobernaba un país latinoamericano como dictador; por lo general, era un oficial de alto rango o un hombre pudiente (pág. 213)

central city the densely populated center of a metropolitan area (p. 257)

ciudad central centro densamente poblado de un área metropolitana (pág. 257)

chinook a dry wind that sometimes blows over the Great Plains in winter (p. 154)

chinook viento seco que sopla a veces sobre las Grandes Llanuras en invierno (pág. 154)

*circumstances conditions (p. 217)

*circunstancias condiciones (pág. 217)

city-state an independent political unit that includes a city and the surrounding area (p. 372)

ciudad-Estado unidad política independiente que incluye una ciudad y el área circundante (pág. 372)

civil rights the basic rights that belong to all citizens (p. 137)

derechos civiles los derechos fundamentales de todos los ciudadanos (pág. 137)

Glossary/Glosario

climate the average weather in an area over a long period of time (p. 23)

clima tiempo atmosférico promedio en una zona durante un periodo largo (pág. 23)

coastal plain the flat, lowland area along a coast (p. 121)

llanura litoral planicie de baja altitud que bordea la costa (pág. 121)

collectivization a system in which small farms were combined into huge, state-run enterprises with work done by mechanized techniques in the hopes of making farming more efficient and reducing the need for farmworkers (p. 406)

colectivización sistema en el cual pequeños granjeros se integran a empresas gigantescas administradas por el Estado, en las que el trabajo se realiza mediante métodos técnicos con la esperanza de hacer más eficiente la agricultura y reducir la demanda de trabajadores agrícolas (pág. 406)

colonialism a policy based on control of one country by another (p. 212)

colonialismo política que se basa en el control o dominio de un país sobre otro (pág. 212)

colonist a person sent to live in a new place and claim land for his or her home country (p. 126)

colonizador persona enviada a establecerse en un nuevo lugar y reclamar territorios para su país de origen (pág. 126)

Columbian Exchange the transfer of plants, animals, and people between Europe, Asia, and Africa on one side and the Americas on the other (p. 214)

intercambio colombino traslado de plantas, animales y personas entre Europa, Asia y África, de un lado, y América, del otro (pág. 214)

command economy an economy in which the means of production are publicly owned (p. 96)

economía planificada sistema económico en el que los medios de producción son de propiedad pública (pág. 96)

communism a system of government in which the government controls the ways of producing goods (p. 406)

comunismo forma de gobierno en la que el gobierno controla los modos de producción de los bienes (pág. 406)

compass rose the feature on a map that shows direction (p. 28)

rosa de los vientos convención de un mapa que señala la dirección (pág. 28)

*component a part of something (p. 23)

*componente parte de algo (pág. 23)

*comprise to make up (p. 179)

*incluir integrar (pág. 179)

compulsory mandatory; enforced (p. 255)

compulsivo obligatorio; forzoso (pág. 255)

condensation the result of water vapor changing to a liquid or a solid state (p. 64)

condensación cambio del vapor de agua a un estado líquido o sólido (pág. 64)

*conflict a serious disagreement (p. 283)

*conflicto desacuerdo grave (pág. 283)

coniferous describing evergreen trees that produce cones to hold seeds and that have needles instead of leaves (p. 179)

coníferas árboles perennes que producen conos para contener las semillas y tienen agujas en vez de hojas (pág. 179)

conquistador a Spanish explorer of the early Americas (p. 211)

conquistador explorador español de América en sus inicios (pág. 211)

*contemporary of the present time; modern (p. 315)

*contemporáneo perteneciente al tiempo presente; moderno (pág. 315)

contiguous joined together inside a common boundary (p. 149)

contiguo unido dentro de un límite común (pág. 149)

continent a large, unbroken mass of land (p. 52)

continente extensión de tierra grande e ininterrumpida (pág. 52)

Continental Divide an imaginary line through the Rocky Mountains that separates rivers that flow west from rivers that flow east (p. 152)

divisoria continental línea imaginaria que atraviesa las montañas Rocosas para separar los ríos que fluyen hacia el oeste de los que fluyen hacia el este (pág. 152)

continental shelf the part of a continent that extends into the ocean in a plateau, then drops sharply to the ocean floor (p. 60)

plataforma continental parte de un continente que se adentra en el océano en forma de meseta y luego desciende abruptamente hasta el fondo oceánico (pág. 60)

*contribution something that is given (p. 382)

*contribución algo que se entrega (pág. 382)

*convert to change from one thing to another (p. 27)

*convertir cambiar de una cosa a otra (pág. 27)

*cooperate to work together (pp. 350–51)

*cooperar trabajar en unión (págs. 350–51)

cordillera a region of parallel mountain chains (pp. 149, 299)

cordillera región de cadenas montañosas paralelas (págs. 149, 299)

coup an action in which a group of individuals seize control of a government (p. 311)

golpe (de Estado) acción mediante la cual un grupo de individuos se apodera del control de un gobierno (pág. 311)

*create to make (p. 151)

*crear hacer (pág. 151)

Creole a group of languages developed by enslaved people on colonial plantations that is a mixture of French, Spanish, and African (p. 289)

Criollo grupo de lenguas desarrollado por las personas esclavizadas en las plantaciones colonials, que consiste en una mezcla de francés, espanol y africano (pág. 289)

cultural region a geographic area in which people have certain traits in common (p. 86)

región cultural área geográfica donde las personas tienen ciertos rasgos comunes (pág. 86)

culture the set of beliefs, behaviors, and traits shared by a group of people (p. 82)

cultura conjunto de creencias, comportamientos y rasgos compartidos por un grupo de personas (pág. 82)

*currency the paper money and coins in circulation (p. 101)

*moneda dinero en billetes y monedas en circulación (pág. 101)

czar the title given to an emperor of Russia's past (p. 403)

zar título dado a un emperador de Rusia en el pasado (pág. 403)

D

*data information (p. 158)

*dato información (pág. 158)

death rate the number of deaths compared to the total number of people in a population at a given time (p. 72)

tasa de mortalidad número de defunciones comparado con el número total de habitantes de una población en un tiempo determinado (pág. 72)

deciduous describing trees that shed their leaves in the autumn (p. 179)

caducifolios árboles que pierden sus hojas en el otoño (pág. 179)

*decline to reduce in number (p. 170)

*declinar reducirse en número (pág. 170)

delta an area where sand, silt, clay, or gravel is dropped at the mouth of a river (p. 62)

delta área donde se deposita arena, sedimento, lodo o gravilla en la desembocadura de un río (pág. 62)

democracy a type of government run by the people (p. 86)

democracia tipo de gobierno dirigido por el pueblo (pág. 86)

Glossary/Glosario

dependence too much reliance (p. 218)

dependencia confianza excesiva (pág. 218)

desalinization a process that makes salt water safe to drink (p. 61)

desalinización proceso que elimina la sal del agua para hacerla potable (pág. 61)

***despite** in spite of (p. 276)

***a pesar de** no obstante (pág. 276)

devolution the process by which a large, centralized government gives power away to smaller, local governments (p. 415)

autonomía proceso mediante el cual un gran gobierno centralizado cede poder a gobiernos locales menores (pág. 415)

dialect a regional variety of a language with unique features, such as vocabulary, grammar, or pronunciation (p. 83)

dialecto variedad regional de una lengua con características únicas, como vocabulario, gramática o pronunciación (pág. 83)

dictatorship a form of government in which one person has absolute power to rule and control the government, the people, and the economy (p. 87)

dictadura forma de gobierno en la que una persona detenta el poder absoluto para mandar y controlar al gobierno, el pueblo y la economía (pág. 87)

dike a large barrier built to keep out water (p. 336)

dique barrera grande construida para no dejar pasar el agua (pág. 336)

***distort** to change something so it is no longer accurate (p. 27)

***distorsionar** cambiar algo de modo que ya no es correcto (pág. 27)

***diverse** composed of many distinct and different parts (p. 256)

***diverso** compuesto de muchas partes distintivas y diferentes (pág. 256)

dormant still capable of erupting but showing no signs of activity (p. 208)

inactivo que aún es capaz de entrar en erupción pero no muestra señales de actividad (pág. 208)

doubling time the number of years it takes a population to double in size based on its current growth rate (p. 73)

tiempo de duplicación número de años que le toma a una población doblar su tamaño con base en la tasa de crecimiento actual (pág. 73)

Dust Bowl the southern Great Plains during the severe drought of the 1930s (p. 167)

Dust Bowl las Grandes Llanuras meridionales durante la fuerte sequía de la década de 1930 (pág. 167)

***dynamic** always changing (p. 20)

***dinámico** en permanente cambio (pág. 20)

E

earthquake an event in which the ground shakes or trembles, brought about by the collision of tectonic plates (p. 54)

terremoto suceso en el cual el suelo se agita o tiembla como consecuencia de la colisión de placas tectónicas (pág. 54)

economic system how a society decides on the ownership and distribution of its economic resources (p. 96)

sistema económico la forma en que una sociedad decide la propiedad y distribución de sus recursos económicos (pág. 96)

elevation the measurement of how much above or below sea level a place is (p. 29)

elevación medida de cuánto más alto o más bajo está un lugar respecto del nivel del mar (pág. 29)

emancipate to make free (p. 254)

emancipar liberar (pág. 254)

emigrate to leave one's home to live in another place (p. 78)

emigrar abandonar el hogar propio para vivir en otro lugar (pág. 78)

encomienda the Spanish system of enslaving Native Americans and making them practice Christianity (p. 281)

encomienda sistema español de esclavizar a los indígenas americanos y obligarlos a profesar el cristianismo (pág. 281)

environment the natural surroundings of a place (p. 23)

medioambiente entorno natural de un lugar (pág. 23)

Equator a line of latitude that runs around the middle of Earth (p. 21)

ecuador línea de latitud que atraviesa la mitad de la Tierra (pág. 21)

equinox one of two days each year when the sun is directly overhead at the Equator (p. 46)

equinoccio uno de dos días al año cuando el sol se halla situado directamente sobre el ecuador (pág. 46)

erosion the process by which weathered bits of rock are moved elsewhere by water, wind, or ice (p. 55)

erosión proceso por el cual fragmentos desgastados de rocas son llevados a otra parte por acción del agua, el viento o el hielo (pág. 55)

escarpment a steep cliff at the edge of a plateau with a lowland area below (p. 242)

escarpado acantilado pendiente, al borde de una meseta, que tiene debajo un área de tierras bajas (pág. 242)

*establish** to start (p. 157)

*establecer** comenzar (pág. 157)

estuary an area where river currents and the ocean tide meet (p. 301)

estuario área donde convergen corrientes fluviales y la marea oceánica (pág. 301)

ethanol a liquid fuel made in part from plants (p. 155)

etanol combustible líquido que se fabrica a partir de vegetales (pág. 155)

ethnic group a group of people with a common racial, national, tribal, religious, or cultural background (p. 83)

grupo étnico grupo de personas con un antecedente racial, nacional, tribal, religioso o cultural común (pág. 83)

evaporation the change of liquid water to water vapor (p. 63)

evaporación cambio del agua en estado líquido a vapor (pág. 63)

*eventually** at a later time (p. 185)

*finalmente** en un tiempo posterior (pág. 185)

*exceed** to go beyond a limit (pp. 274–75)

*exceder** traspasar un límite (págs. 274–75)

export to send a product produced in one country to another country (p. 99)

exportar enviar un bien producido en un país a otro país (pág. 99)

extinct describing a particular kind of plant or animal that has disappeared completely from Earth (p. 160); describing a volcano that is no longer able to erupt (p. 208)

extinto espécimen específico de una planta o un animal que ha desaparecido por completo de la Tierra (pág. 160); volcán que ya no puede entrar en erupción (pág. 208)

*extract** to remove or take out (p. 249)

*extraer** remover o sacar (pág. 249)

F

*factor** a cause (p. 416)

*factor** causa (pág. 416)

fall line the area where waterfalls flow from higher to lower ground (p. 122)

fault a place where two tectonic plates grind against each other (p. 54)

favela an overcrowded city slum in Brazil (p. 257)

*feature a noteworthy characteristic (pp. 210–11)

feudalism the political and social system in which kings gave land to nobles in exchange for the nobles' promise to serve them; those nobles provided military service as knights for the king (p. 343)

fishery an area where fish come to feed in huge numbers (p. 182)

fjord a narrow, U-shaped coastal valley with steep sides formed by the action of glaciers (p. 368)

free trade arrangement whereby a group of countries decides to set little or no tariffs or quotas (p. 100)

free-trade zone an area where trade barriers between countries are relaxed or lowered (p. 219)

frontier a region just beyond the edge of a settled area (p. 158)

línea de descenso área donde las cascadas fluyen de un terreno más alto a uno más bajo (pág. 122)

falla lugar donde dos placas tectónicas chocan entre sí (pág. 54)

favela tugurio urbano superpoblado de Brasil (pág. 257)

*rasgo característica notable (págs. 210–11)

feudalismo sistema social y político en el cual los reyes entregaban tierras a los nobles, que a cambio prometían servirles; estos nobles proveían de servicio militar al rey como caballeros (pág. 343)

pesquería zona donde los peces llegan a alimentarse en gran número (pág. 182)

fiordo valle costero estrecho, en forma de U con laderas escarpadas, formado por la acción de glaciares (pág. 368)

libre comercio acuerdo por el cual un grupo de países decide imponer aranceles bajos a las cuotas o no fija ningún arancel (pág. 100)

zona de libre comercio área donde las barreras comerciales entre los países se distienden o reducen (pág. 219)

frontera región inmediatamente posterior al borde de un área poblada (pág. 158)

G

genocide the mass murder of people from a particular ethnic group (p. 405)

geography the study of Earth and its people, places, and environments (p. 18)

glaciation the process of becoming covered by glaciers (p. 367)

glacier a large body of ice that moves slowly across land (p. 56)

globalization the process by which nations, cultures, and economies become mixed (p. 89)

granary a building used to store harvested grain (p. 188)

genocidio asesinato masivo de personas de un grupo étnico específico (pág. 405)

geografía estudio de la Tierra y de sus gentes, lugares y entornos (pág. 18)

glaciación proceso en el cual los glaciares cubren zonas amplias del planeta (pág. 367)

glaciar masa de hielo enorme que se mueve lentamente sobre la tierra (pág. 56)

globalización proceso mediante el cual naciones, culturas y economías se integran (pág. 89)

granero edificación en la cual se almacena el grano cosechado (pág. 188)

gross domestic product (GDP) the total dollar value of all final goods and services produced in a country during a single year (p. 98)

producto interno bruto (PIB) valor total en dólares de todos los bienes y servicios finales producidos en un país durante un año (pág. 98)

groundwater the water contained inside Earth's crust (p. 61)

agua subterránea agua contenida en el interior de la corteza terrestre (pág. 61)

guerrilla a member of a small, defensive force of irregular soldiers (p. 309)

guerrillero miembro de una fuerza pequeña y defensiva de soldados irregulares (pág. 309)

H

hacienda a large estate (p. 281)

hacienda gran propiedad rural (pág. 281)

hemisphere each half of Earth (p. 26)

hemisferio cada mitad de la Tierra (pág. 26)

*hierarchy a ruling body arranged by rank or class (p. 307)

*jerarquía cuerpo de gobierno organizado por rango o clase (pág. 307)

hinterland an inland area that is remote from the urban areas of a country (p. 256)

hinterland zona interior distante de las áreas urbanas de un país (pág. 256)

Holocaust the mass killing of 6 million European Jews by Germany's Nazi leaders during World War II (p. 348)

Holocausto exterminio masivo de 6 millones de judíos europeos ejecutado por los líderes nazis de Alemania durante la Segunda Guerra Mundial (pág. 348)

homogeneous made up of many things that are the same (p. 380)

homogéneo compuesto por muchas cosas iguales (pág. 380)

human rights the rights belonging to all individuals (p. 87)

derechos humanos los derechos que tienen todos los individuos (pág. 87)

hurricane a storm with strong winds and heavy rains (p. 122)

huracán tormenta con vientos fuertes y lluvias copiosas (pág. 122)

I

immigrate to enter and live in a new country (p. 78)

inmigrar entrar a un nuevo país y vivir allí (pág. 78)

immunity the ability to resist infection by a particular disease (p. 281)

inmunidad capacidad de resistir la infección provocada por una enfermedad específica (pág. 281)

*impact the effect or influence (pp. 312–13)

*impacto efecto o influencia (págs. 312–13)

import when a country brings in a product from another country (p. 99)

importación cuando un país ingresa un producto de otro país (pág. 99)

indigenous living or existing naturally in a particular place (p. 125)

nativo que vive o existe de modo natural en un lugar específico (pág. 125)

Glossary/Glosario

industrialized describing a country in which manufacturing is a primary economic activity (p. 346)

industrializado país en el cual la manufactura es la principal actividad económica (pág. 346)

industry the manufacturing and making of products to sell (p. 130)

industria manufactura y fabricación de bienes para la venta (pág. 130)

*inevitable sure to happen (p. 407)

*inevitable que sucederá con certeza (pág. 407)

inflation a sharp increase in the price of goods, sometimes caused by a shortage of goods (p. 410)

inflación incremento drástico en el precio de los bienes, a veces ocasionado por su escasez (pág. 410)

*initiate to begin (p. 218)

*iniciar comenzar (pág. 218)

*intense strong (p. 54)

*intenso poderoso (pág. 54)

irrigation the process of collecting water and using it to water crops (p. 153)

irrigación proceso de recolección del agua para regar los cultivos (pág. 153)

*isolate to make separate from others (p. 125)

*aislar separar de otros (pág. 125)

isthmus a narrow strip of land that connects two larger land areas (p. 59)

istmo franja estrecha de tierra que conecta dos áreas de tierra más grandes (pág. 59)

K

key the feature on a map that explains the symbols, colors, and lines used on the map (p. 28)

clave elemento de un mapa que explica los símbolos, colores y líneas usados en este (pág. 28)

L

landform a natural feature found on land (p. 23)

accidente geográfico formación natural que se encuentra sobre la tierra (pág. 23)

landscape the portions of Earth's surface that can be viewed at one time from a location (p. 19)

paisaje partes de la superficie terrestre que se pueden observar a un mismo tiempo desde una ubicación (pág. 19)

latitude the lines on a map that run east to west (p. 21)

latitud líneas sobre un mapa que van de este a oeste (pág. 21)

levee a raised riverbank used to control flooding (p. 120)

dique ribera elevada que sirve para controlar las inundaciones (pág. 120)

lock a gated passageway used to raise or lower boats in a waterway (p. 119)

esclusa compartimento con puertas que se utiliza para subir o bajar los barcos en un canal (pág. 119)

longitude the lines on a map that run north to south (p. 21)

longitud líneas sobre un mapa que van de norte a sur (pág. 21)

longship a ship with oars and a sail used by the Vikings (p. 374)

drakkar barco con remos y velas que utilizaban los vikingos (pág. 374)

M

Manifest Destiny the idea that it was the right of Americans to expand westward to the Pacific Ocean (p. 158)

Destino Manifiesto ideología según la cual los estadounidenses tenían derecho a expandirse al oeste hacia el océano Pacífico (pág. 158)

map projection one of several systems used to represent the round Earth on a flat map (p. 28)

proyección cartográfica uno de los varios sistemas que se usan para representar la esfera terrestre en un mapa plano (pág. 28)

maquiladora a foreign-owned factory where workers assemble parts (p. 216)

maquiladora fábrica de propiedad extranjera donde los obreros ensamblan partes (pág. 216)

market economy an economy in which most of the means of production are privately owned (p. 96)

economía de mercado economía en la cual la mayoría de los medios de producción son de propiedad privada (pág. 96)

*mature fully grown and developed as an adult; also refers to older adults (p. 72)

*maduro adulto plenamente crecido y desarrollado; también se refiere a los adultos mayores (pág. 72)

megalopolis a huge city or cluster of cities with an extremely large population (p. 80)

megalópolis ciudad enorme o cúmulo de ciudades que tienen una población extremadamente grande (pág. 80)

Métis the child of a French person and a native person (p. 187)

métis hijo de un francés y una indígena (pág. 187)

metropolitan having to do with a large city (p. 132); an area that includes a city and its surrounding suburbs (pp. 190; 257)

metropolitano(a) relativo a una ciudad grande (pág. 132); área que incluye una ciudad y los suburbios que la rodean (págs. 190; 257)

Middle Ages the period in European history from about A.D. 500 to about 1450 (p. 343)

Edad Media periodo de la historia europea que abarca aproximadamente del año 500 al 1450 (pág. 343)

*migrate to move to an area to settle (pp. 185; 288)

*migrar trasladarse a un lugar para establecerse allí (págs. 185; 288)

mission a Catholic-based community in the west (p. 157)

misión comunidad católica del Oeste (pág. 157)

mixed economy an economy in which parts of the economy are privately owned and parts are owned by the government (p. 96)

economía mixta economía en la cual unos sectores son de propiedad privada y otros son de propiedad del gobierno (pág. 96)

monarchy the system of government in which a country is ruled by a king or queen (p. 87)

monarquía sistema de gobierno en el que un rey o una reina gobiernan un país (pág. 87)

Mormon a member of the Church of Jesus Christ of Latter Day Saints (p. 167)

mormón miembro de la Iglesia de Jesucristo de los Santos de los Últimos Días (pág. 167)

multinational a company that has locations in more than one country (p. 310)

multinacional compañía que tiene oficinas en más de un país (pág. 310)

mural a large painting on a wall (p. 217)

mural pintura de gran tamaño hecha sobre un muro (pág. 217)

N

national park a park that has been set aside for the public to enjoy for its great natural beauty (p. 155)

nomadic describes a way of life in which a person or group lives by moving from place to place (p. 156)

nonrenewable resources the resources that cannot be totally replaced (p. 95)

parque nacional parque destinado al público para que disfrute sus grandes bellezas naturales (pág. 155)

nómada forma de vida en la que una persona o grupo vive trasladándose de un lugar a otro (pág. 156)

recursos no renovables recursos que no se pueden reponer por completo (pág. 95)

O

*occupy to settle in a place (p. 184)

*occur to happen or take place (p. 244)

oligarch a member of a small ruling group that holds great power (p. 410)

orbit to circle around something (p. 42)

*ocupar establecerse en un lugar (pág. 184)

*ocurrir suceder o acontecer (pág. 244)

oligarca miembro de un pequeño grupo gobernante que detenta gran poder (pág. 410)

orbitar moverse en círculo alrededor de algo (pág. 42)

P

pagan someone who believes in more than one god or someone who has little or no religious belief (p. 374)

pampas the treeless grassland of Argentina and Uruguay (p. 242)

*parallel running side by side with something; following the same general course and direction (p. 121)

Parliament the national legislature of England (now the United Kingdom), consisting of the House of Lords and the House of Commons (p. 345)

peacekeeping sending trained members of the military to crisis spots to maintain peace and order (p. 193)

pilgrimage a journey to a sacred place (p. 343)

plain a large expanse of land that can be flat or have a gentle roll (p. 58)

plantation a large farm (p. 213)

plateau a flat area that rises above the surrounding land (p. 58)

pagano persona que cree en más de un dios o cuya creencia religiosa es escasa o nula (pág. 374)

pampas praderas sin árboles de Argentina y Uruguay (pág. 242)

*paralelo que corre lado a lado con algo; que sigue el mismo curso y dirección generales (pág. 121)

Parlamento asamblea legislativa nacional de Inglaterra (hoy Reino Unido) integrada por la Cámara de los Lores y la Cámara de los Comunes (pág. 345)

pacificación envío de miembros entrenados de las fuerzas armadas a sitios críticos para mantener la paz y el orden (pág. 193)

peregrinación viaje a un lugar sagrado (pág. 343)

llanura gran extensión de tierra plana o con ligeras ondulaciones (pág. 58)

plantación granja grande (pág. 213)

meseta área plana que se eleva por encima del terreno circundante (pág. 58)

polder the land reclaimed from building dikes and then draining the water from the land (p. 336)

population density the average number of people living within a square mile or a square kilometer (p. 76)

population distribution the geographic pattern of where people live (p. 76)

postindustrial describing an economy that is based on providing services rather than manufacturing (p. 357)

precipitation the water that falls on the ground as rain, snow, sleet, hail, or mist (p. 48)

Prime Meridian the starting point for measuring longitude (p. 21)

productivity the measurement of what is produced and what is required to produce it (p. 98)

province an administrative unit similar to a state (p. 178)

pueblo a town built by the Pueblo people in the American Southwest (p. 156)

pueblo jóven shantytown with poor housing and little or no infrastructure built outside a large metropolitan area (p. 313)

pólder terreno ganado al mar a partir de la construcción de diques y el posterior desecado de la tierra (pág. 336)

densidad de población número promedio de personas que habitan en una milla cuadrada o un kilómetro cuadrado (pág. 76)

distribución de la población patrón geográfico que muestra dónde habita la gente (pág. 76)

posindustrial economía que se basa en la prestación de servicios, no en la manufacturación (pág. 357)

precipitación agua que cae al suelo en forma de lluvia, nieve, aguanieve, granizo o rocío (pág. 48)

primer meridiano punto de partida para medir la longitud (pág. 21)

productividad medición de lo que se produce y lo que se requiere para producirlo (pág. 98)

provincia unidad administrativa similar a un estado (pág. 178)

pueblo poblado construido por las tribus pueblo del sudeste estadounidense (pág. 156)

pueblo jóven barrio marginal con viviendas precarias y poca o ninguna infraestructura, construido en las afueras de una gran área metropolitana (pág. 313)

R

rain forest a dense stand of trees and other vegetation that receives a great deal of precipitation each year (p. 241)

rain shadow an area that receives reduced rainfall because it is on the side of a mountain facing away from the ocean (p. 49)

*****ratio** the relationship in amount or size between two or more things (p. 286)

*****rational** reasonable (p. 375)

recession a time when many businesses close and people lose their jobs (p. 385)

refugee a person who flees a country because of violence, war, persecution, or disaster (p. 78)

reggae a traditional Jamaican style of music that uses complex drum rhythms (p. 221)

selva tropical formación densa de árboles y otra vegetación que recibe una gran cantidad de precipitación todos los años (pág. 241)

sombra pluviométrica zona que recibe pocas precipitaciones porque se halla en la ladera de una montaña que está en el lado contrario al océano (pág. 49)

*****ratio** relación en cantidad o tamaño entre dos o más cosas (pág. 286)

*****racional** razonable (pág. 375)

recesión época en que muchos negocios cierran y las personas pierden sus empleos (pág. 385)

refugiado persona que huye de un país por la violencia, una guerra, una persecución o un desastre (pág. 78)

reggae género musical tradicional de Jamaica que utiliza complejos ritmos de tambor (pág. 221)

Glossary/Glosario

region a group of places that are close to one another and that share some characteristics (p. 22)

región agrupación de lugares cercanos que comparten algunas características (pág. 22)

*regulate to control something (p. 351)

*regular controlar algo (pág. 351)

relative location the location of one place compared to another place (p. 20)

localización relativa la ubicación de un lugar comparada con la de otro (pág. 20)

relief the difference between the elevation of one feature and the elevation of another feature near it (p. 29)

relieve diferencia entre la elevación de una formación y la de otra formación cercana (pág. 29)

remittance the money sent back to the homeland by people who have gone somewhere else to work (p. 221)

remesa dinero enviado al país de origen por personas que se han ido a trabajar a otro lugar (pág. 221)

remote sensing the method of getting information from far away, such as deep below the ground (p. 32)

detección remota método para obtener información muy lejana, como de las profundidades del subsuelo (pág. 32)

Renaissance the period in Europe that began in Italy in the 1300s and lasted into the 1600s, during which art and learning flourished (p. 375)

Renacimiento periodo de Europa que comenzó en Italia en el siglo XII y finalizó en el siglo XV, durante el cual florecieron el arte y la cultura (pág. 375)

renewable resources a resource that can be totally replaced or is always available naturally (p. 95)

recursos renovables recursos que pueden reponerse totalmente o siempre se encuentran disponibles en la naturaleza (pág. 95)

representative democracy a form of democracy in which citizens elect government leaders to represent the people (p. 87)

democracia representativa forma de democracia en la que los ciudadanos eligen líderes de gobierno para que representen al pueblo (pág. 87)

reservation an area of land that has been set aside for Native Americans (p. 161)

reservación territorio que ha sido destinado a los indígenas americanos (pág. 161)

reserves a large amount of a resource that has not yet been tapped (p. 401)

reservas gran cantidad de un recurso que aún no ha sido explotada (pág. 401)

resource a material that can be used to produce crops or other products (p. 23)

recurso materia prima que se puede utilizar para obtener cultivos u otros productos (pág. 23)

*revenue the income generated by a business (p. 133)

*renta ingresos generados por un negocio (pág. 133)

revolution a complete trip of Earth around the sun (p. 42); a period of violent and sweeping change (p. 212)

revolución recorrido completo de la Tierra alrededor del Sol (pág. 42); periodo de cambio violento y radical (pág. 212)

Ring of Fire a long, narrow band of volcanoes surrounding the Pacific Ocean (p. 54)

Cinturón de Fuego banda larga y estrecha de volcanes que rodean el océano Pacífico (pág. 54)

rural describes an area that is lightly populated (p. 77)

rural área poco poblada (pág. 77)

Rust Belt the area of the Midwest, Mid-Atlantic, and New England where many factories closed during the 1980s (p. 139)

Rust Belt zona del Medio Oeste, Atlántico Medio y Nueva Inglaterra donde muchas fábricas se cerraron durante la década de 1980 (pág. 139)

S

scale the relationship between distances on the map and on Earth (p. 29)

escala relación entre distancias en un mapa y en la Tierra (pág. 29)

scale bar the feature on a map that tells how a measured space on the map relates to the actual distance on Earth (p. 28)

escala numérica elemento cartográfico que muestra la relación entre un espacio medido sobre el mapa y la distancia real sobre la Tierra (pág. 28)

scrubland land that is dry and hot in the summer and cool and wet in the winter (p. 370)

chaparral territorio seco y caliente en el verano, y frío y húmedo en el invierno (pág. 370)

separatists a group that wants to break away from control by a dominant group (p. 194)

separatista grupo que quiere sustraerse del control de un grupo dominante (pág. 194)

serf a farm laborer who could be bought and sold along with the land (p. 404)

siervo trabajador agrícola que podía comprarse y venderse junto con la tierra (pág. 404)

service industry a type of business that provides services rather than products (p. 139)

industria de servicios tipo de negocio que provee servicios en vez de productos (pág. 139)

shield a large area of relatively flat land made up of ancient, hard rock (p. 179)

escudo extensa área de terreno relativamente plano formado por rocas duras y antiguas (pág. 179)

*significant important (p. 151)

*significativo importante (pág. 151)

*similar having qualities in common (p. 203)

*similar que tiene cualidades en común (pág. 203)

slash-and-burn agriculture a method of farming that involves cutting down trees and underbrush and burning the area to create a field for crops (p. 249)

agricultura de tala y quema método agrícola que consiste en talar árboles y rastrojos y quemar el área despejada para crear un campo de cultivo (pág. 249)

smallpox an often-fatal disease that causes a rash and leaves marks on the skin (p. 308)

viruela enfermedad, por lo general mortal, que causa sarpullido y deja marcas en la piel (pág. 308)

smelting the process of refining ore to create metal (p. 342)

fundición proceso mediante el cual se refinan minerales para producir metales (pág. 342)

solstice one of two days of the year when the sun reaches its northernmost or southernmost point (p. 45)

solsticio uno de dos días al año cuando el sol alcanza su máxima declinación norte o sur (pág. 45)

spatial Earth's features in terms of their places, shapes, and relationships to one another (p. 18)

espaciales características de la Tierra en cuanto a sus lugares, formas y relaciones entre sí (pág. 18)

*sphere a round shape like a ball (p. 26)

*esfera figura redonda como una pelota (pág. 26)

*stable staying in the same condition; not likely to change or fail (p. 285)

*estable que permanece en la misma condición; algo que es improbable que cambie o decaiga (pág. 285)

standard of living the level at which a person, group, or nation lives as measured by the extent to which it meets its needs (p. 98)

estándar de vida nivel en que vive una persona, grupo o nación, medido según la capacidad de satisfacer sus necesidades (pág. 98)

staple a food that is eaten regularly (p. 210)

alimento básico alimento que se consume habitualmente (pág. 210)

steppe a partly dry grassland often found on the edge of a desert (p. 397)

estepa pradera parcialmente seca que se encuentra con frecuencia al borde de un desierto (pág. 397)

*strategy a plan to solve a problem (p. 407)

*estrategia plan para resolver un problema (pág. 407)

subregion a smaller part of a region (p. 116)

subregión parte más pequeña de una región (pág. 116)

surplus extra; more than needed (p. 210)

excedente sobrante; más de lo que se necesita (pág. 210)

sustainability the economic principle by which a country works to create conditions where all the natural resources for meeting the needs of society are available (p. 101)

sostenibilidad principio económico según el cual un país crea condiciones para que estén disponibles todos los recursos naturales que satisfacen las necesidades de la sociedad (pág. 101)

T

tariff a tax added to the price of goods that are imported (p. 290)

arancel impuesto añadido al precio de los productos importados (pág. 290)

technology any way that scientific discoveries are applied to practical use (p. 30)

tecnología cualquier forma en que los descubrimientos científicos se aplican para un uso práctico (pág. 30)

tectonic plate one of the 16 pieces of Earth's crust (p. 53)

placa tectónica uno de las 16 partes de la corteza terrestre (pág. 53)

territory the land administered by the national government (p. 178)

territorio tierra administrada por el gobierno nacional (pág. 178)

thematic map a map that shows specialized information (p. 30)

mapa temático mapa que muestra información especializada (pág. 30)

*theory an explanation of why or how something happens (p. 345)

*teoría explicación de por qué o cómo ocurre algo (pág. 345)

tierra caliente the warmest climate zone, located at lower elevations (p. 205)

tierra caliente la zona climática más cálida, ubicada en elevaciones bajas (pág. 205)

tierra fría a colder climate zone, located at higher elevations (p. 205)

tierra fría zona climática fría, ubicada entre elevaciones altas (pág. 205)

tierra templada a temperate climate zone, located at mid-level elevations (p. 205)

tierra templada zona de clima templado, ubicada entre elevaciones medias (pág. 205)

timberline the elevation above which it is too cold for trees to grow (p. 149)

límite forestal elevación por encima de la cual hace demasiado frío para que los árboles prosperen (pág. 149)

topsoil the fertile soil that crops depend on to grow (p. 168)

mantillo suelo fértil del cual dependen las plantas para crecer (pág. 168)

tourism the industry that provides services to people who are traveling for enjoyment (p. 133)

turismo industria que presta servicios a las personas que viajan por placer (pág. 133)

Glossary/Glosario

trade winds the winds that blow regularly in the Tropics (p. 277)

vientos alisios vientos que soplan regularmente en los trópicos (pág. 277)

traditional economy an economy where resources are distributed mainly through families (p. 96)

economía tradicional economía en la que los recursos se distribuyen principalmente entre las familias (pág. 96)

transcontinental describing something that crosses a continent (p. 187)

transcontinental que atraviesa un continente (pág. 187)

*transform to change something completely (p. 64)

*transformar cambiar algo por completo (pág. 64)

trawler a large fishing boat (p. 371)

trainera barco pesquero grande (pág. 371)

trench a long, narrow, steep-sided cut on the ocean floor (p. 60)

fosa depresión larga, estrecha y profunda del fondo oceánico (pág. 60)

tributary a small river that flows into a larger river (p. 119)

tributario río pequeño que desemboca en uno más grande (pág. 119)

Tropics an area between the Tropic of Cancer and the Tropic of Capricorn that has generally warm temperatures because it receives the direct rays of the sun for much of the year (p. 243)

trópicos zona entre el trópico de Cáncer y el trópico de Capricornio que generalmente tiene temperaturas cálidas porque recibe los rayos directos del sol la mayor parte del año (pág. 243)

tsunami a giant ocean wave caused by volcanic eruptions or movement of the earth under the ocean floor (p. 54)

tsunami gigantesca ola oceánica provocada por erupciones volcánicas o movimientos de la tierra bajo el lecho oceánico (pág. 54)

tundra a flat, treeless plain with permanently frozen ground (pp. 181; 369)

tundra llanura plana y sin vegetación cuyo suelo permanece helado (págs. 181; 369)

U

*uniform not varying across several parts (p. 369)

*uniforme que no cambia en varias partes (pág. 369)

*unique unusual (p. 259)

*único inusual (pág. 259)

upland the high land away from the coast of a country (p. 396)

tierra alta tierra elevada de un país, alejada de la costa (pág. 396)

urban describes an area that is densely populated (p. 77)

urbana área densamente poblada (pág. 77)

urbanization when a city grows larger and spreads into nearby areas (p. 80)

urbanización cuando una ciudad crece y se expande hacia las áreas adyacentes (pág. 80)

V

*vary to show differences between things (p. 191)

*variar mostrar diferencias entre cosas (pág. 191)

*via on the way through (pp. 190–91)

*vía en el camino hacia (págs. 190–91)

Glossary/Glosario

W

water cycle the process in which water is used and reused on Earth, including precipitation, collection, evaporation, and condensation (p. 63)

ciclo del agua proceso en el cual el agua se usa y reutiliza en la Tierra; incluye la precipitación, recolección, evaporación y condensación (pág. 63)

weathering the process by which Earth's surface is worn away by natural forces (p. 55)

meteorización proceso mediante el cual la superficie terrestre se deteriora por la acción de fuerzas naturales (pág. 55)

welfare capitalism a system in which the government is the main provider of support for the sick, the needy, and the retired (p. 384)

capitalismo de bienestar sistema en el cual el gobierno es el principal proveedor de ayuda para los enfermos, necesitados y jubilados (pág. 384)

Westerlies strong winds that blow from west to east (p. 338)

Vientos del oeste vientos fuertes que soplan de oeste a este (pág. 338)

Glossary/Glosario

The following abbreviations are used in the index: *m=map, c=chart, d=diagram, i= infographic, p=photograph or picture, g=graph, crt=cartoon, ptg=painting, q=quote*

Aboriginal: culture of, 85. *See also* First Nations people; Native Americans

absolute location, 21

acid rain, 65, 195

Adriatic Sea, 368, 396, 398

aerospace industry, 170, *p170*

Africa: Cape of Good Hope, 375; Congo Basin in, 264; as continent, 52; emigration from, 78; population growth in, 73, 75; urbanization in, 81. *See also* North Africa; West Africa *and entries for individual regions*

African Americans, 138

African enslaved people: in Brazil, 252, 256; in Peru, 312. *See also* slave trade

aging population: demographic trends, *i389*; statistics about, 388

agribusinesses, 170

agriculture: Brazil, 247; Canada, 188; defined, 129; eastern United States, 138–39; industry, 97; Mexico, 216; modern, United States west of the Mississippi River, 169–70; in Paraguay, 81; Tropical North, 279; U.S. west of the Mississippi, early, 162. *See also* farming

Alabama: Montgomery bus boycott, 115; in Southeast subregion, 117, *m117*. *See also* United States, east of the Mississippi River

Alaska: climate in, 154; energy resources in, 155; *Exxon Valdez* disaster, 168; landforms of, 150; purchase by U.S. from Russia, 161

Albania, *m395*; agriculture jobs in, 416; climate in, 399; ethnic group in, 413; Muslim population in, 409; religion in, 414; urban population of, 411; in Warsaw Pact, 407. *See also* Eastern Europe

Alberta, Canada, 180, 187, 191; Jasper national Park in, *p180*

Alexander (Macedonian king), 373

Alexander II (Russian Czar), 404

Allegheny River, 121

Allende, Isabel, 315

Allende, Salvador (President), 311, 315

alpaca, 305

Alps mountain range, 335, 367, *p367*

altiplano, 299

altitude, effects of, *i302*

Amazon Basin, 240–41, 264; Ecuador in, 285; logging in, 246; Tropical North in, 275

Amazon rain forest: future of, 263; highway through, *p263*; resources from, 264–67; statistics about, 266. *See also* rain forest; tropical rain forest

Amazon River, 240, *p244*; Mississippi River vs., 240, *p241*

America, early, 124–31

American Petroleum Institute, *q141*

American Revolution, 127, 186

Amman, Jordan, 81

***Ancient Cuzco: Heartland of the Inca* (Bauer),** *q322*

Ancient Egypt. *See* **Egypt**

Ancient Greece, 372–73, *m373*

Andean music, pan flute, *p290*

Andean region, Native American societies in, 306

Anderson, Mark T, *q174*

Andes and midlatitude countries, 295–322, *m297*; Andes mountains in, 299; border disputes, 317; climate diversity in, 301–4; daily life in, 315; economic challenges, 310; economies of, 316–17; environmental issues, 317; ethnic and language groups, 314; history of, 306–11; Inca Empire, 306–7; large cities and communities in, 313; minerals and metals in, 305; overthrow of Spanish rule in, 309; physical geography of, 298–305; plains and deserts in, 299–300; political instability in, 310–11; popular sports in, 315; population density and distribution, 312–13; "population rim" in, 312; power and governance in, 309–10; waterways in, 300–1; wildlife in, 305

Andes Mountains: culture of people, 290; early people, 280; formation of, 299; in Tropical North, 274; as world's longest mountain system, 296

Andorra, 366. *See also* Southern Europe

Angel Falls, 275, *p276*, 277

Angles and Saxons, 343

animals. See **wildlife**

annexed, 159

Antarctica: as continent, 52; freshwater in ice caps of, 61; ice sheets covering, 56; surrounded by water, 59

Antigua, *p207*

Apennines mountain range, 368

Appalachian Mountains, 121–22; Great Smoky Mountains, *p114*

Arab Spring: social media and, 90, *p91*

Arawak people, 249

Arcadia National Park, *p137*

archipelago, 181

Arctic Circle, 369

Arctic climate. See **polar climate**

Arctic Ocean: bordering Canada, 181–83; freshwater in ice caps of, 61; as smallest ocean, 62

Argentina: climate in, 302; Falkland Islands, 311; Patagonia Desert in, 300; plateaus and valleys in, 299; Spanish colonies in, 308. *See also* Andes and midlatitude countries

Aristotle, 373

Arizona, *m147*; Treaty of Guadalupe Hidalgo and, 159. *See also* United States, west of the Mississippi River

Arkansas, *m147*. *See also* United States, west of the Mississippi River

Arkansas River, 152

Armstrong, Neil, *p17*

arts/culture, 85; Andean and midlatitude countries, 314–15; Brazil, 259–60; Caribbean islands, 221; Eastern Europe and Western Russia, 414–15; Mexico, 217; New Orleans, 135; New York City, 133–34; Northern and Southern Europe, 380–82; Tropical North, 289–90; Western Europe, 353. *See also* culture

ASEAN. *See* **Association of Southeast Asian Nations (ASEAN)**

Ashaninka family, in Amazon rain forest, *p249*

Asia: as continent, 52; early migration to Canada, 184; emigration from, 78; population growth in, 73, 75; Russian connections with, 417; U.S. ports vital to trade with, 164

Asia Minor: Alexander's rule over, 373; Roman conquest of, 374

Asian Americans, 138

Aspen, Colorado: Maroon Bells near, *p150*

Association of Southeast Asian Nations (ASEAN), 101

Asunción, Paraguay, 81

Atacama Desert, 300, 301

Atahualpa (Incan emperor), 308

Athabasca Tar Sands, 195

Athens, Greece, 372, 379

Atlanta, Georgia, 134

Atlantic Archipelago, 336

Atlantic coastal plain, 121

Atlantic lowlands, Brazil, 242–43, 247

Atlantic Ocean: Canadian coast on, 182; East Coast of the United States, 118; Europe, 368; Mexican coast on, 204; Orinoco River flows to, 275; relative size of, 62; Río de la Plata flows into, 301; Spain, 368; Spanish fishing ports along, 371; St. Lawrence Seaway to, 190; Tropical North coast on, 276; Western European coast on, 335–36

Atlantic Provinces, Canada, 179

Atlas of Canada, q198

atmosphere, 44; pollution of Earth's, 74

Augustus Caesar, 374

Aussies. *See* **Australia**

Austin, Texas, 165

Australia: as continent, 52; surrounded by water, 59

Austria: in Central Powers (WWI), 405; plant life in, 341; Vienna basin in, 397; winter sports in, 354. *See also* Western Europe

Austro-Hungarian Empire, 405

autonomy, 194

axis, 42

Aymara people, 306

Azores High, 339

Aztec civilization, 211; Spanish conquest of, 211, 376

Bach, Johannes, 353

Bachelet, Michelle (President of Chile), 311, *p297*

Baffin Island, 181

Bahamas, *m201*; as Caribbean islands, 208; tourism in, 221. *See also* Caribbean Islands

Baja, California, 203, 204

Balearic Islands, 380, *p369*

balkan, 397

balkanization, 397

Balkan Peninsula: Eastern European nations on, 367, 396–97; ethnic tensions in, 409; as mountainous region, 397; seas surrounding, 398

Ballet Folklorico, 217

Baltic Sea: fertile triangle and, 400; location of, 396, 398; Swedish coastline of, 368; in Western Europe, 335

Baltic states, independence of, 395

Barber of Seville, The **(Rossini),** 382

Barcelona, Spain, Sagrada Familia Church in, *p381*

Barents Sea, 396

Baroque art, 381

Bartók, Béla, 414

basin, 150, 241

Basque language, 352, 380

Bastille: storming of, *p346*

Battle of Quebec, *p186*

Bauer, Brian S, *q322*

Bauhaus School, 353

bauxite, 206

bay, 62

Beaumont, Texas, 165

Beethoven, Ludwig van, 353

behalf, 87

Belarus, *m395*; Dnieper River in, 398; Kievan Rus in, 402; religion in, 413; Russian Plain in, 396;

Slavic ethnic groups in, 412; teenagers in Minsk, *p415*. *See also* Eastern Europe

Belgium: forest in southern, *p339*; on Northern European Plain, 335. *See also* Western Europe

Belgrade, Serbia, 398

Belize, 210, 213, *m201*. *See also* Central America

Benelux countries, 334

Berlin, Germany, 350; Brandenburg Gate in, *p332*

Berlin Wall, 333, 409

bilingual, 191

Billington, James H, *q420*

Billund, Denmark, 382

Bing Streetside, 34

biome, 50

biosphere, 44

birthrate, 72–73, 75; European, 378

bison, 149, *p149*

Black Death, *i344*, 345

Black Hills, 149; gold in South Dakota's, 160

Black Sea: Bulgarian coast of, *p398*; fertile triangle and, 400; location of, 396, 398; Main-Danube Canal and, 337; oil from, 401

Blue Mountains, Jamaica, 208

Blue Ridge Mountains, 121

Bogotá, Colombia: Iglesia de San Francisco, *p287*; population of, 287; Spanish rule in, 281, *p281*

Boise, Idaho, 165

Bolivar, Simón, 282–83, *p282*, 309

Bolivia: as Andean country, 298; Illimani mountains, *p300*; Native American populations in, 312; natural gas reserves in, 316; Tarabuco, woman from, *p295*; territorial disputes, 317; two capitals of, 309. *See also* Andes and midlatitude countries

Bolsheviks, 406

Bolshoi Ballet, *p393*

Bonaparte, Napoleon, 253, 333, 346

Boone, Daniel, *p128*

Borges, Jorge Luis, 315

Bosnia, 409; agriculture jobs in, 416; urban population of, 411; WWI triggered in, 405

Bosnia Herzegovina, *m395*. *See also* Eastern Europe

Boston, Massachusetts, 132–33

brackish, 398

Brandenburg Gate, *p332*

Brasília, Brazil, 242, *p238*

Brazil, 237–70, *m239*; agriculture in, 247; Amazon Basin, 240–41; Amazon River, 240; Ashaninka family in rain forest, *p249*; Atlantic Lowlands, 242–43; Brazilian Highlands, 241–42; coalition governments in, 255; colonies in, *m252*; connections and challenges, 262–63; crowded cities in, 257; diverse population in, 256–57; economy of, 261; education and earning a living, 261–62; environmental concerns in, 263;

ethnic and language groups in, 259; Europeans arrive in, 249–50; favelas in, 257–58; history of, 248–55; independence and monarchy, 253–54; indigenous populations, 248–49; life in, 256–63; military rule in, 255; natural resources of, 245–47; people and cultures in, 259–60; physical geography of, 240–47; Portuguese colonial rule in, 250–52; public school in Amazon area, *p262*; religion and arts in, 259–60; as republic, 254–55; rural life in, 260; soccer in, 260, *p261*; sugarcane plantation workers in, *p251*; timeline, *c238–39*; tropical climate of, 243–45; urban life in, 260; voting for president in, *p254*; waterways and landforms, 240–43; worker on coffee plantation in, *p246*

Brazilian Highlands, 241–42; agriculture in, 247; climate of, 244; Japanese in, 259; mineral resources in, 246; Nambicuara people of, 249

"Brazilian Way," 259

Britain, 336, *m336*; Canada's relationship with, 193; Cold War and, 349; Industrial Revolution in, 347, *p347*. *See also* England; Great Britain; United Kingdom

British Columbia, settlers in, 187

British Guiana, European colonization of, 282

British Isles, 336, *m336*; plants in, 341; wildlife in, 341

Bronx, the, 133

Brooklyn, 133

Bryce Canyon National Park, 55; Thor's Hammer in, *p55*

Budapest, Hungary, 398, 411

Buddhism, *c84*; in the U.S., 138

Buenos Aires, Argentine: shantytowns in, 313, *p313*

buffalo, 149

Bulgaria, *m395*; Black Sea coast of, *p398*; boulevard in Plovdiv, *p414*; climate in, 399; as EU member, 417; new government in, 409; religion in, 413; Slavic ethnic groups in, 412; in Warsaw Pact, 407. *See also* Eastern Europe

buttes, in Monument Valley Navajo Tribal Park, *p146*

Byzantine Empire, 374, 375, 376

Cabral, Pedro, 248

Caesar, Julius, 374

CAFTA-DR. *See* **Central America Free Trade Agreement (CAFTA-DR); Dominican Republic-Central America Free Trade Agreement**

Cairo, Egypt: as primate city, 81

Calgary, Alberta, Canada, 191

California, *m147*; Baja, 203, 204; early oil industry in, 163; early Spanish settlements in, 157; energy resources in, 155; Gold Rush, 160; information science companies in, 171; Los Angeles, 165, *p166*, 169; modern agriculture in, 170; movie industry in, 163; port cities in, 164;

Silicon Valley, 171; Treaty of Guadalupe Hidalgo and, 159. *See also* United States, west of the Mississippi River

California Central Valley, 150; climate in, 153; farming in, 155; resources in, 155

Calle Ocho celebration, *p139*

camelids, 305, *p304*

Canada, 175–98, *m177*; acid rain and, 195; agriculture in, 188; Alberta, 191; Blackfoot girl at First Nations Pow Wow, *p175*; bodies of water in, 182–83; British (1600s and 1700s), 186; British Columbia Parliament Buildings, *p176*; challenges for, 194–95; city and country life, 190–91; economic and political relationships, 192; economic growth, 188; environmental challenges for, 195; Europeans in early, 185; exploration and settlement, 185–87; First Nations people of, 184–85; immigration, 188; independence for (1931), 189; industry in, 188, 189; Montreal in Quebec Province, 191; national parks in, *p180*; National Tower in Toronto, *p191*; Newfoundland, 374; Northwest Territories, weather in, *p181*; Ottawa, 190; outdoor winter sports in, *p192*; physical landscape of, 178–83; polar bears in Arctic, *p176*; police on horseback in, 187; population doubling time in, 73; regions of, *m179*; relationship with the U.S., 192; Saskatchewan and Alberta Provinces, 187; size of, 178; timeline, *c176–77*; Toronto, 190–91; train system in, *p193*; unity and diversity for, 194; U.S. border with, 178; U.S. relations with, 167; westward expansion, 187

Canada, provinces of, 178–81; Atlantic, 179; British Columbia, 180–81; northern lands, 181; Prairie, 180, 189; Quebec and Ontario, 179

Canadian Rockies, 180

Canadian Shield, 179, 180, 181; lakes in, 183

canals, 57

canopy, 241

Cantabrian Mountains, 368

Cape Canaveral, 115

Cape of Good Hope, 375

capital: as factor of production, 97

Caracas, Venezuela, 281; urban center in, *p272*

Cara people, 280

Caribbean Islands, *m201*; climate of, 208; colonialism in, 214–15; Columbus reaches, 376; cultures of, 221; earthquakes and volcanoes in, 208; economies of, 220–21; European settlers in, 214; history of, 214–15; independence of countries, 215; indigenous people in, 214; major islands, 207–8; music in, 221; natural resources in, 209; physical geography of, 207–9; timeline, *c200–1*; tourism in, 221; turmoil in, 215

Caribbean Sea, 208; Colombia coastline along, 275; Gulf Stream and, 338–39; in Mexico and Central America, 204; Tropical North coastline, 276

Carib people, 249

Carnival, 260; in Tropical North, 289, *p289*

Carpathian Mountains, 397, 398

Cartagena, Colombia, 287

Cartier, Jacques, 185

cartographer, 29

cartography, *p30*

Cascade Mountains, 149–50, 153

cash crops, 213

Caspian Sea: in Eastern Europe and Western Russia, 396, 398, 411

Castilian language, 380

Castillo de San Marcos National monument, *p126*

Castro, Fidel, 215

Catalan, 380

Catalonia, 380

Catherine the Great (Russia), 404

Catholic Church: Andes and midlatitude countries, 314; conflict in Tropical North over, 283; missionaries in Tropical North, 281; Protestant Reformation and, 345; Southern Europe and, 381; in Tropical North, 289; Vatican City and, 376; in Western Europe, 352; Western Europe and, 343. *See also* Christianity; Roman Catholic Church

Cauca River, 276

Caucasian Americans, 138

Caucasus Mountains, 396, 397, 411

caudillos, 213

Celts, 351, 380

Central America, *m201*; bodies of water in, 204–5; climate of, 205–7; conflict in modern, 213; culture of, 219; early civilizations and conquest, 212; economies in, 218–19; history of, 212–13; independence from Mexico, 212–13; landforms in, 203–4; natural resources in, 207; physical geography of, 202–7; population growth in, 219; rain forests in, 51; Spanish empire in, 376; timeline, *c200–1*

Central America Free Trade Agreement (CAFTA-DR), 219

Central Asia: Mongols of, 403. *See also* entries for individual countries

central city, 257

Central Highlands, Mexico, *p97*

Central Plateau, Mexico, 203

Central Powers, 405

Charleston, South Carolina, 133

Chau, Stephen, *q35*

Chávez, Hugo, 273, 277, 291

Chechnya, 411, 417

Chekhov, 414

Cherokee people: forced migration of, 130–31; Sequoyah's language symbols, *p125*

Chibcha people, 280

Chicago, Illinois, 134

Chile: as Andean country, 298; desert in, 299; mining in, *p303*; political instability in, 311; road collapse after earthquake in, 54; Spanish colonies in, 308; Torres del Paine National Park, *p298–99*. *See also* Andes and midlatitude countries

China: population concentration along, 76

chinampas, 211

Chinese New Year, 79

chinook, 154

Chocó region, 277

Christianity, *c84*; in Europe, 352, 372; French settlers to convert native groups to, 185; Rome as seat of, 376; Scandinavia converts to, 374; spread in Brazil, 251; in the U.S., 138; Western Europe and, 343

Christian missionaries: in Tropical North, 281

"Christ the Redeemer" statue: in Rio di Janeiro, *p238*

Chunnel, 336; tunnel-boring machines used for, *p56*

CIA World Factbook: on Brazil, *q270*

city-states, 372

civil rights, 137

civil war: Greece, 377; in U.S., 131, 134

Clark, William, 158

climate, 23; changes to, 51; climate regions, 49–51; climate zones, 49–50, *p49*; defined, 48; Eastern United States, 122; elevation and, 47; factors that influence, 46–49; landforms and, 48–49; Mexico and Central America, *g206*; Northern vs. Southern Europe, 369–70; Tropical North, 276–77; weather vs., 48; Western Europe, 338–39; western United States, 153–54; wind and ocean currents and, 47–48. *See also entries for specific countries*

climate change: worldwide, 51. *See also* global warming

coastal area, 62

coastal plain, 56, 121

coffee: as Brazilian export, *p246*, 247, 254; as Venezuela cash crop, 279

Cold Temperate climate, 49–50

Cold War, 349, 357, 407–8

collectivization, 406

Colombia, *m273*, 274; Chocó region, 277; as coal producer, 278; main rivers, 276; Panama's independence from, 213; population groups in, 286; Spanish settlements in, 280–81; struggles in, 291; tensions with Ecuador, 285; U.S. relations, 291. *See also* Tropical North

colonialism: in Brazil, 250–52; *m252*; defined, 212; Spanish, in Mexico, 212

colonization. *See* **colonialism; European colonization**

Colorado: cowboys on horseback near Cimarron River, *p162*; Denver, early agriculture in, 163;

Index

information science companies in, 171; Treaty of Guadalupe Hidalgo and, 159

Colorado Plateau, 150

Colorado River, 150, 152; dams along, 152, 155; Hoover Dam on, 152, *p152*

Columbia Basin, 150

Columbian Exchange, 214–15, *m214*, 376

Columbia River, 152, 155

Columbus, Christopher, 208, 248, 365, 376, *p71*

command economy, 96

commercial fishing. *See* **fish and fishing**

communism, 406; civil war in Greece and, 377; food lines in Eastern Europe under, *p416*

Communist party, in Russia, 395, 405–7

compadre **relationship,** 315

compass rose, 28

compulsory, 255

condensation, 64

Congo Basin, 264

coniferous tree, 179, 341

Connecticut: in New England subregion, 117, *m117*. *See also* United States, east of the Mississippi River

conquistadors, 211

Constantine (Emperor), 343, 374

Constantinople, 374

contiguous, 149

continent, 52

Continental Divide, 152

continental shelf, 60

continents, 59, *m16–17*

Copacabana Beach: Rio de Janeiro, Brazil, *p257*

Copenhagen, Denmark, 379, *p379*

Copernicus, Nicolaus, 345, 375

cordillera, 149, 299

core: Earth's, 43

Corpus Christi, Texas, 165

Correa, Rafael (President of Ecuador), 291

Cortés, Hernán, 211

Costa Rica, *m201, p46*; farmer in, *p46*; peace in, 213; trade agreements, 219. *See also* Central America

Cotopaxi volcano, 275, *p275*

Cotswold Hills, 337

countries, boundaries of, 29

coup, 311

Crater Lake, *p151*

creole language, 289

Crete, 367

Croatia, *m395*; climate in, 399; as former Yugoslav republic, 409; religion in, 413; urban population of, 411. *See also* Eastern Europe

Crusades, 343–44

crust, Earth's, 43, *p43*

Cuba, *m201*; as Caribbean island, 207; command economy in, 96; economy of, 220; Fidel Castro and, 215; independence of, 215; salsa music in, 221; Spanish colonies in, 214. *See also* Caribbean Islands

Cuban Missile Crisis, 408

cultural blending, 79, 89

cultural region, 86

culture, 82–86; Caribbean islands, 221; Central America, 219; cultural changes, 88–89; cultural regions, 86; customs and, 84–85; defined, 82; economy and, 86; ethnic groups and, 83; global, 89; government and, 85–89; history and, 85; language and, 83; Northern and Southern Europe, 383; population growth and, 75; Renaissance and, 375. *See also* arts/culture

Cumberland Gap: Daniel Boone in, *p128*

Curacao, Willemstad, *p200*

currency, 101

customs, cultural, 84–85

Cuzco Valley, 306

Cyprus, 367. *See also* Southern Europe

czar, 404

Czechoslovakia: new government in, 409; reforms in (1968), 408; Soviet army occupation of, 407; in Warsaw Pact, 407

Czech Republic, *m395*; early Slavic groups in, 402; as EU member, 417; religion in, 413. *See also* Eastern Europe

daily life: Andes and midlatitude state, 315; Eastern Europe, 415–16; Northern Europe, 383; Southern Europe, 383; Tropical North, 289–90; Western Europe, 354; Western Russia, 415

Dallas, Texas, 165

dance. *See* arts/culture

Danube River, 337, 398; fishing industry and, 401; roadways along, *p338*

Darién, 275, 277

Darwin, Charles, 276

Davies, Norman, *q392*

da Vinci, Leonardo, 375

Dead Sea, 61

death rate, 72–73

de Blij, H.J, *q38, q420*

de Champlain, Samuel, 185

deciduous forest, 341

deciduous trees, 179

Declaration of the Rights of Man and of the Citizen, 346

Deepwater Horizon disaster, 168

deforestation, 264

de Gama, Vasco, 376

Delaware, 117, *m117*. *See also* United States, east of the Mississippi River

Delhi, India, 81

delta, 62

democracy, 86–87

Denali, 150

Denmark, 366–67; electricity generation in, 371; ethnic and language groups, 380; as EU member, 385; Lego toys, 382, *p382*; literacy rate in, 383. *See also* Northern Europe

Denver, Colorado, 164; as "mile High City", 22

dependence, 218

desalinization, 61

de San Martín, José, 282

desert: as biome, 50; formation of, 43; plant life in, 155. *See also* entries for individual desert regions

desert climate, 49–50

desert plants, 155

Desmond, Viola, *p187*

Destin Beach, *p118*

developed countries, 98

Devil's Island, 283, *p283*

devolution, 415

dialect, 83

diamonds: Tropical North, 279

Dias, Bartholomeu, 375

dictatorship, 87

dikes, 336

disease: Native Americans and European, 157, 211, 308–9; population growth and, 74–75

Disney World, *p22*

diverse. *See* **ethnic groups**

Dnieper River, 398

Dniester River, 398

Dolomites, *p367*

Dominican Republic, *m201*, 207; Duarte Peak, 208; independence of, 215; Spanish colonies in, 214; trade agreements, 219. *See also* Caribbean Islands

Dominican Republic-Central America Free Trade Agreement (CAFTA-DR), 101

Dominion of Canada, 186

dormant volcanoes, 208

Dostoyevsky, 414

doubling time, 73

drugs: illegal, in Mexico, 218; pharmaceutical, making new, *i95*

Duarte Peak, Dominican Republic, 208

Dubček, Alexander, 408

Dublin, Ireland, 350; O'Connell Street in, *p351*

Dust Bowl, 167–68

Dutch Guiana, 282, 284

dynamic, 20

Earth, 42–51, *p44*; atmosphere of, 51; climate on, 46–51 (*See also* climate); crust of, 43–44, *d43*, 53; deepest location on, 60; effects of human actions on, 57; erosion and, 55; forces of change, 52–57; inside, 43; layers of, 43–44, *d43*; orbit of, 42, 45–46; physical systems of, 44; plate movements on, 53–54; population growth on, 72–75; saltwater vs. freshwater on, *i60*, 61; seasons and tilt of, *d45*; seasons on, 45–46; sun and, 42–43; surface of, 52; tilt of, 45; water on, *i60*, 61–65; weathering of, 55

earthquake: Andes, 299; Caribbean Islands, 208; Central America, 219; defined, 54; Mexico and Central America, 204; plate movement and, 54; Southern Europe, 368

Eastern bloc, *m408*

Eastern Europe, 393–420, *m395*; arts in, 414–15; boundary disputes within, 417; changes under Gorbachev, 408–9; climates in, 399; Cold War and, 407–8; daily life in, 415–16; divisions and conflict, 409; early Slavic states in, 402–3; earning a living in, 416–17; economic changes in, 410; energy and minerals, 401; ethnic and language groups, 412–13; fishing industry, 401; forests and agriculture, 400–1; global financial crisis and, 417; history of, 402–9; landforms, 396–97; population centers in, 411–12; religion in, 413–14; rise of communism, 406–7; Slavic people in, *m413*; social and political changes, 411; timeline, *c394–95*; unrest in Soviet satellites, 408; war and revolutions, 405–6; waterways, 398

Eastern Hemisphere, 26, *c27*

Eastern Orthodox Church, 376, 413

East Germany, 407

Ebro River, 368

EC. *See* **European Commission (EC)**

economic systems, 96

economy: basic economic question, 94–96; Brazil, 261; Canada, 192, 195; culture and, 86; Eastern Europe, 410, 416–17; eastern United States, 138; economic activities, 97; economic organizations, 100–1; economic performance, 98; economic systems, 96–99; factors of production, 97, *d98*; Italy, 384; Mexico, 216–17; Northern Europe, 384–85; population centers and, 134; Southern Europe, 384–385; three economic questions, *i95*; Tropical North, 316–17; types of national economies, 98–99; United States, 138–39, 169–71; Western Europe, 356–57; Western Russia, 410–11, 416–17; world, 99–101

Ecuador, *m273*, 274; border dispute with Peru, 285; Cotopaxi volcano in, 275; farmer Juan Lucas in, *p271*; Galápagos Islands, 273, 276; government of, 284; Guayas River, 276; population groups in, 286; regions in, 288; struggles in, 291; tensions with Colombia, 285; territorial disputes, 317. *See also* Tropical North

Edmonton, Alberta, Canada, 191

EEC. *See* **European Economic Community (EEC)**

Egypt: population concentration in, 76.

Eire, 336

Elbe River, 337

elevation, 58, 274; climate and, 47; on maps, 29

Elk Mountains, *p150*

Ellis Island, immigrants arrive at, *p71*

El Niño, 304

El Salvador: in Central America, 213; earthquake in (2001), 204; trade agreements, 219. *See also* Central America

emerald mining, *i278*

emigrate, 78

encomienda, 281

energy resources, 74; Eastern United States, 123; western United States, 155

England, 336, *m336*; Holy Island in, 374; Hundred Years' War between France and, 344; Parliament, 345. *See also* Britain; Great Britain; United Kingdom

English Channel, 336, 343; tunnel beneath, 354 *See also* Chunnel

Enlightenment, 345

environment, the: Andes and midlatitude countries, 317; Brazil, 263; Canada's challenges, 195; climate zone and, 50; defined, 23; *Exxon Valdez* disaster, 168; fracking and, 140–41; human actions and, 23, 168; Mexico, 217; population growth's effects on, 74; society and, *t24*; United States, west of the Mississippi River, 168–69; U.S. Superfund program, 137

environmental hazards, *p23*

Environmental Protection Agency, "Smart Growth and Schools," *q104*

Equator, 21, *m21*, 26; climate around, 46; and equinoxes, 45–46; in northern Brazil, 243; seasons and, 245; warm air masses near, 47; water currents ad, 48

equinoxes, 46

Eratosthenes, 374

Eriksson, Leif, 374, 376

erosion, 55; United States west of the Mississippi River, 168–69

escarpment, 242

Estaiada Bridge, São Paulo, Brazil, *p258*

Estonia, *m395*; boundary disputes and, 417; religion in, 413. *See also* Eastern Europe

estuary, 301, 337

ethanol, 155, 247

ethnic groups: defined, 83; in eastern United States, 138

EU. *See* **European Union**

Europe: aging population in, 386–89; as continent, 52; emigration from, 78; as EU member, 385; feudalism in, *i404*; urbanization in, 81. *See also* Eastern Europe; Northern Europe; Southern Europe; Western Europe

Europe: A History (Davies), *q392*

European colonization, 125–27; in Tropical North, 282; west of the Mississippi River, 157–58

European Commission (EC), 350, *q359*

European Economic Community (EEC), 350

European Union (EU), 101, *m353*; "Basic Information," *q362*; creation of, 333; Eastern European countries in, 417; effectiveness of, 358–59; formation of, 351; Northern and Southern European countries and, 377; Northern European members, 385; Southern European members, 385

Euskera language, 380

evaporation, 63

experience, perspective of, 20

exports, 99; top 10 countries, *c225*

extinct, 160

extinct volcanoes, 208

Exxon Valdez disaster, 168

Facebook, 92, *g93*

factors of production, 97, *d98*

Falkland Islands, 311

fall line, 122

family. *See* **daily life**

FARC. *See* **Revolutionary Armed Forces of Colombia**

farming, *p76*; Eastern United States, 123; in rural areas, *p76*; technology changes, 129–30. *See also* agriculture

fault, 54

favelas, 257–58

Federal Emergency Management Agency (FEMA), 137

Felixtowe: cargo containers at, in U.K., *p359*

FEMA. *See* **Federal Emergency Management Agency (FEMA)**

Ferdinand, Francis (Archduke), 405, *p405*

Fernández de Kirchner, Cristina (President of Argentina), 311

fertile triangle: in Russia, 400

feudalism, 343; in Russia and Europe, *i404*

financial crisis, global, 356, 358

Finland: climate of, 369; ethnic and language groups, 380; as EU member, 385; lakes in, 368; red fox in wilderness of, *p364*; religion in, 380. *See also* Northern Europe

First Nations people, 184–85; autonomy for, 194; Blackfoot girl at Pow Wow, *p175*; in Canada, 189, 191; trade with French settlers, 185

fish and fishing: Canada, 191; overfishing, 191; Peru, 316

fjords, 368, *p368*

Florida: Destin Beach in Panhandle, *p118*; as peninsula, 366; in Southeast subregion, 117, *m117*; St. Augustine, 126. *See also* United States, east of the Mississippi River

Food and Water Watch, *q140*

food lines, in Eastern Europe during Communist rule, *p416*

Ford Company: first Model T cars, 115

forests: cutting and burning, *p74*

Forever Open Roads project, 355

Fort Worth, Texas, 165

fossil fuels: effects on water supply, 65; Tropical North, 278

fracking, 140–41

France: as agricultural producer, 340; as Ally in WWI, 405; ceding Canada to Britain, 177; Chunnel connecting U.K. and, *p56*; climate in southern, 339; Cold War and, 349; colonial empires and, 357; high-speed rail lines in, 354; Hundred Years' War with England, 344; Louisiana territory of, 158; on Northern European Plain, 335; reform in, 346; sending convicted criminals to French Guiana, 283; as wheat producer, *p341. See also* Western Europe

Franks, 343

free trade, 100

free-trade zone, 219

French explorers: in Canada, 185

French Guiana, 274, *m273*; European colonization of, 282; population groups in, 287; relationship with France, 284–85; as sparsely populated, 288. *See also* Tropical North

freshwater, 44, 61; saltwater vs., *i60*

frontier, 158

Galápagos Islands, 272–73, 276

Galician, 380

Galileo, 365, 375

Gaudí, Antoni, 381

GDP. *See* **gross domestic product (GDP)**

gems. *See* **mineral resources**

genocide, 405

geographer's skills, interpreting visuals, 25

geographer's tools, 26–33; geospatial technologies, 30–33; globes, 26; maps, 27–30

geographic information system (GIS), 31–32

geography: defined, 18; "father of", 374; five themes of, 20–24; human, 70–104; physical, 39–68; six essential elements of, 24–25, *c24*; uses of, *c24*

Geography (de Blij and Muller), *q38, q420*

Geography Activity: Andes and midlatitude states, *m320*; Brazil, *m268*; Canada, *m196*; continents, *m66*; Eastern Europe and Western Russia, *m418*; Mexico, Central America, and Caribbean Islands, *m226*; Northern and Southern Europe, *m390*; Spain, *m36*; Tropical North, *m292*; United States east of the Mississippi, *m142*; United States west of the Mississippi, *m172*; Western Europe, *m360*

Georgia (U.S. state): in Southeast subregion, 117, *m117. See also* United States, east of the Mississippi River

geospatial technologies, 30–33; geographic information system (GIS), 31–32; Global Positioning System (GPS), 31; limits of, 33; satellites and sensors, 32–33

Germany, 398; autobahns in, 355; in Central Powers (WWI), 405; Cold War and, 349; colonial empires and, 357; Danube in southwestern, 398; East vs. West, 407; on Northern European Plain, 335; as oil producer, 340; plant life in, 341; reuniting of East and West, 333; World War I and, 347–48. *See also* Western Europe

Gibraltar, 366. *See also* Southern Europe

Gilbert, Jonathan, *q322*

GIS. *See* **geographic information system (GIS)**

glacier, 56, 61; Finish lakes and, 368; on Olympic Mountains, 149

glasnost, 408–9

global economy: trade, 99–101. *See also* globalization

global financial crisis, 356, 358, 365; Eastern Europe and, 417; Europe and, 385

globalization: defined, 89; indigenous cultures and, 318–19. *See also* global economy

Global Positioning System (GPS), 31; satellites, *p32*

global warming: defined, 51

globes, 26, 27

Gogol, 414

gold and gold trade: Brazil, 252; South Dakota's Black Hills, 160; Tropical North, 278–79; U.S. west of the Mississippi, 155, 160–61; Yukon Territory, 187

Golden Gate Bridge, *p146*

Goldstein, Steve, *q294*

Google Maps, 35

Google Street View, 34–35

Gorbachev, Mikhail, 395, 408–9, *p409*

government: Canada, 190; Colombia, 291; Cuba, 220; European Union, 351, 385; forms of, 86–87; Russia, 411; Spain, 380; United States, 136–37

GPS. *See* **Global Positioning System (GPS)**

granary, 188

Gran Colombia, 282, 283

Grand Banks, Atlantic Provinces, Canada, 182

Grand Canyon, 55, 150

graph skills, GDP comparison, 99

grassland, 50

Great Bear Lake, 183

Great Britain: as Ally (WWI), 405; colonial empires and, 357; Falkland Islands and, 311; France cedes Canada to, 177; World War II and, 407. *See also* Britain; United Kingdom

Greater Antilles, 207, 208

Great Lakes: in Canada, 179, 183; in United States, 118

Great Migration, 131

Great Moravia, 402

Great Plains, 148; climate in, 154; dust storms in, 147; modern agriculture in, 169–70; Oregon

Trail over, 159; southern, energy resources in, 155; transformation from grasslands to farms, 160

Great Salt Lake, 61, 152, 153

Great Slave Lake, 183

Great Smoky Mountains, *p114*, 121

Greece: ancient, 372–73, *m373*; civil war in modern, 377; as democracy, 377; ethnic and language groups, 380; as EU member, 385; financial crisis in, 385; population in, 379; Roman conquest of, 374; ruins of ancient temple in, *p364*; soccer teams, 383. *See also* Southern Europe

Greek Empire, 372, *m373*

Greek Orthodox Church, 376

Greenland, 52; ice sheets covering, 56; Viking settlements in, 374

Greenwich, England, 21

Gregory VII (Pope), 355

Grieg, Edvard, 382

gross domestic product (GDP), 98; comparing, *g99*

Grosso Plateau, Brazil, *p243*

groundwater, 61

Guadeloupe, tourists on boat near, *p209*

Guanajuato, Mexico, *p77*

Guarani people, 314

Guatemala, *m201*; girl from highlands of, *p199*; independence for, 212–13; Maya people in, 210; natural resources in, 207; trade agreements, 219. *See also* Central America

Guayas River, 276

guerrilla warfare, 309

Guiana Highlands, 275, 277

guilds, 344

gulf, 62

Gulf Coast, 118; Hurricane Katrina in, 122

Gulf of California, 152, 204

Gulf of Mexico: currents through, 118; Deepwater Horizon disaster, 168; hurricanes in, 208; importance of ports in, 165; Louisiana and Texas border on, 151; Mexican peninsula in, 203; Mississippi River empties into, 119; oil and gas along coast of, 206; surrounded by landmasses, 62; western U.S. rivers flow toward, 152; Yucatán Peninsula in, 203

Gulf of St. Lawrence, 183

Gulf Stream, 208, 339; Norwegian current, 369

Gutenberg, Johannes, *p332*, 345, 365

Guyana, 274, *m273*; border dispute with Venezuela, 285; European colonization of, 282; independence for, 284; Makushi people, *p272*; Native Americans in, 287; as sparsely populated, 288. *See also* Tropical North

gypsies, 380

haciendas, 281

Hagia Sophia, *p375*

Haiti, *m201*; barren landscape of, *p220*; earthquake in, 208; economy of, 220; on Hispaniola, 208; independence of, 215. *See also* Caribbean Islands

Han Chinese, 83

Hawaiian Islands: climate in, 154; landforms of, 150; U.S. annexation of, 161

helmet, ancient iron and bronze, *p343*

Helsinki, Finland, 379

hemisphere, 26

Henry (Prince of Portugal), 375

Herzegovina, 409; agriculture jobs in, 416

Hidalgo, Miguel, 212

High Mountain climate, 49–50

Hillary, Edmund, *p17*

hills, as landforms, 23

Hinduism, *c84*; in the U.S., 137

hinterland, 256

Hispanic Americans, 138, *g138*

Hispaniola, 207–8

history: Andes and midlatitude countries, 306–11; Brazil, 248–55; Canada, 184–89; culture and, 85; Eastern Europe and Western Russia, 402–9; Mexico, Central America, and the Caribbean, 210–15; Northern and Southern Europe, 372–77; Tropical North, 280–85; United States, east of the Mississippi, 124–31; United States, west of the Mississippi, 156–63; Western Europe, 342–49

Hitler, Adolf, 348–49

Hobbes, Thomas, 345

Holocaust, 348, 352

Holy Island, 374

Homestead Act, 162

homogeneous, 380

Honduras, *m201*; hurricane in, 206; independence of, 213; trade agreements, 219. *See also* Central America

Honolulu, Hawaii, 151

Hoover Dam, 152, *p152*

Houston, Texas, 151, 165

How People Live, *q228*

Hudson Bay: Canadian Shield and, 179; in east central Canada, 183

Hudson River, 133

Hudson's Bay Company: in Canada, 186; in Canadian territory held by, 187

human actions: damage to water supply, 65; effects on Earth's surface, 57; environment and, *Exxon Valdez* disaster, 168

human-environment interaction, 23

human geography, 70–104; culture, 82–89; population changes, 76–77; population growth, 72–75; population movement, 77–81; timeline, *c70–71*

human rights, 87

human systems, *c24*

Humid Temperate climate, 49–50

humus, 340

Hungary, 397, *m395*; as EU member, 417; new government in, 409; religion in, 413; Soviet army occupation of, 407; in Warsaw Pact, 407. *See also* Eastern Europe

Huns, 343

Huron people, 184, 185

Hurricane Katrina, *p41*, 122; effects of, *p23*; New Orleans levee failures and, 135

hurricanes: in Caribbean Islands, 208; in Central America, 219; defined, 122; in Mexico and Central America, 206

hydroelectric power: United States west of the Mississippi, 155

hydrosphere, 44

Iberian Peninsula, 367

Ibiza Island, Spain, vacationers in, *p369*

Ice Age, 56; Canada and, 184; Northern European Plain and, 335

ice caps, 56, 61

Iceland, 367; Arctic Circle and, 369; bank failure in, 385; ethnic and language groups, 380; International Whaling Commission and, 385; musicians from, 382; population of, 379; terrain, 367; Viking settlements in, 374. *See also* Northern Europe

ice sheet, 56, 61

***Icon and the Axe, The* (Billington),** *q420*

Idaho, *m147*; modern agriculture in, 170. *See also* United States, west of the Mississippi River

IFG. *See* **International Forum on Globalization (IFG)**

Iglesia de San Francisco, Bogotá, Colombia, *p287*

Illimani mountains, Bolivia, *p300*

Illinois: in Midwest subregion, 117, *m117*; Ohio River and, 121. *See also* United States, east of the Mississippi River

IMF. *See* **International Monetary Fund (IMF)**

immigrate, 78

immigration: to eastern United States, 130, *p130*; EU and, 385; illegal, to U.S., 167; to Northern Europe, 380; Western Europe and, 352, 357

immunity, 281

import, 99

Inca Empire, 306–8; Spanish conquest of, 376

Incahuasi Island, gigantic cacti in, *p296*

Indiana: in Midwest subregion, 117, *m117*; Ohio River and, 121. *See also* United States, east of the Mississippi River

Indiana Connections: Democracy, 373; England's Political Heritage, 345; Location Capital Cities, 21; Nation-States, 344; Saving and Investing, 97; Tools of Discovery, 25

Indian Ocean, 62; relative size of, 62

indigenous, 125, 248

indigenous cultures, globalization and, 318–19

indigenous people. *See* entries for individual indigenous groups

Indo-European languages, 352

Indonesian Archipelago, 264

industrialized, 346

industrialized countries, 98

Industrial Revolution, 347, 356

industry: in Canada, 188, 189; defined, 97, 130; in Eastern United States, 123; in Mexico, 216. *See also entries for individual industries*

inflation, 410

intense, 54

International Date Line, 21, 26

International Forum on Globalization (IFG), *q319*

International Monetary Fund (IMF), 101

international trade, 99–101, *p100*

International Whaling Commission, 385

Internet, social media and, 90–93

Inuit people, *p185*

Ionian Sea, 398

Iowa, *m147*. *See also* United States, west of the Mississippi River

Iran: command economy in, 96

Ireland, 336, *m336*. *See also* Western Europe

Iroquois, 184

irrigation, 153

Islam, *c84*; Africa, 86; rise of, 376; Southwest asia, 86. *See also* Muslims

Island: formation of, 53; as landmass, 59; Mediterranean, 367

Istanbul, Turkey: Hagia Sophia in, *p375*

isthmus, 59, 202

Italian peninsula, 367, 373

Italy: as democracy, 377; economic reforms protest in, *p358*; economy of, 384; ethnic and language groups, 380; as EU member, 385; largest cities in, 378; Mount Etna volcano in, *p53*; soccer teams, 383; WWII and, 377. *See also* Southern Europe

Ivan IV (Czar of Russia), 404

Jamaica, *m201*; Blue Mountains, 208; as Caribbean island, 207; reggae music in, 221; tourism in, 221. *See also* Caribbean Islands

Jamestown, Virginia, 126

Japan: bombing of Hiroshima and Nagasaki, 349; International Whaling Commission and, 385; tsunami in (2011), 90

Jefferson, Thomas (President), 158

Jerusalem: pilgrimage to, 343

Jesuit priests, in Brazil, 251

Jewish population: in Western Europe, 352. *See also* Judaism

João, Dom, 253

John III (King of Portugal), 250–51

Judaism, *c84*; in the U.S., 137

Index

Jutland, 366
Kallawaya healers, 314, *p314*
Kansas, *m147. See also* United States, west of the Mississippi River
Kansas River, 152
Kazakhstan: boundary disputes and, 417. *See also* Central Asia
Kentucky: Ohio River and, 121; in Southeast subregion, 117, *m117. See also* United States, east of the Mississippi River
Kerepakupai Meru, 277
key, map, 28
Kiev, Ukraine, 411
Kievan Rus, 402
Kilauea volcano, 150
Klaus, Václav (President of Czech Republic), *q358*
Kodály, Zoltán, 414
Komi region, Russia, *p397*
Köppen, Wladimir, 49
Korea. *See* **North Korea**
Kosovo, 409, *m395. See also* Eastern Europe
labor, as factor of production, 97
Labrador Canada, 179, 188
La Coruña, Spain, 371
Lake Erie, 118
Lake Huron, 118
Lake Maracaibo, 278
Lake Mead, 152
Lake Michigan, 118
Lake Nicaragua, 204
Lake Ontario, 118
lakes, 61; freshwater, 61, 62; pollution of, 65, *p65*
Lake Superior, 118
Lake Tahoe, 152
Lake Titicaca, 301
Lake Winnipeg, 183
Lamb, Gregory M, *q294*
land: on Earth's surface, 44; as factor of production, 97; surface features on, 58–59
landforms, 23; bodies of water and, 59–60; changing, 52; climate and, 48–49; creation of, 56; ocean floor, 59–60; types of, 44, 58–60; weathering of, 55
landmasses, shifting, 52
Land Ordinance of 1785, 129
landscapes, 19; formation of, 43
languages and language groups: culture and, 83; English words from other, *c354*; Euskera, 380; Indo-European, 352; Latin, 343; Northern European, 380; romance, 352; Slavic, 412; Southern European, 380; Western Europe, 352. *See also entries for individual languages*
La Niña, 304
La Paz, Bolivia, 309
Lapps, 380
Larcomar Shopping Mall, *p315*

Las Vegas, Nevada, 165
Latin America, 202. *See also* Caribbean Islands; Central America; Mexico; South America
latitude, 21, *c21, c27*
Latvia, *m395;* boundary disputes and, 417; as EU member, 417; religion in, 413. *See also* Eastern Europe
Lavoie, Andrew, *q34*
leap year, 42–43
Lee, Alonso Silva, *q228*
Lego toys, 382
Lenin, Vladimir, 406
Lesser Antilles, *p207,* 208
levees, 120
Lewis, Meriwether, 158
Lewis and Clark expedition, 158
Libyan refugees, *p79*
Liechtenstein, *m333. See also* Western Europe
Lima, Peru: Larcomar Shopping Mall, *p315*
lithosphere, 44, 54, 149
Lithuania, *m395;* boundary disputes and, 417; as EU member, 417; religion in, 413. *See also* Eastern Europe
llamas, *p304,* 305
Llanos, 275, 277, 281–82; coal and oil deposits in, 278
location, types of, 20–21
Locke, John, 345
locks, 119
Lofoten Island fishing village, *p363*
Loire River, 338
London, England, 336, 350; Great Smog, 337
Long Beach, California, 151
longitude, 21, *c21*
longship, 374, *p374*
Los Angeles, California: as diverse city, 165; as Pacific port, 151; San Andreas Fault near, 169; U.S. citizenship ceremony in, *p166*
Louisiana, *m147;* energy resources in, 155; purchase of, 158; in Southeast subregion, 117, *m117. See also* United States, east of the Mississippi River
Louisiana Territory, exploration of, 158
Louverture, Toussaint-, 215
Louvre, *p355*
Lucas, Juan (farmer), *p271*
Luther, Martin, 345, 376
Lutheran Church, 380
Luxembourg: on Northern European Plain, 335. *See also* Western Europe
Macedon, 373
Macedonia, *m395;* climate in, 399; as former Yugoslav republic, 409; Roman conquest of, 374; urban population of, 411. *See also* Eastern Europe
Machu Picchu, 307, *p307*
Mackenzie River, 183

Madrid, Spain, 379
Magdalena River, 276
Magdalena River valley, 278
magma, 43, 53
Main-Danube Canal, 337, 398
Maine: Arcadia National Park in, *p137;* in New England subregion, 117, *m117. See also* United States, east of the Mississippi River
Main (MINE) River, 337
Makushi people, *p272*
Malta, 367. *See also* Southern Europe
Manhattan, 133
Manifest Destiny, 158
Manitoba, Canada, 180, 187
mantle, Earth's, 43
manufacturing. *See* **industry**
Maori tribe, 83
mapping technology, 30–33
map projections, 28–29
maps, 27–30; cartography as science of making, *p30;* distortion of, 27; elements of, 28; information on, 27; large- vs. small-scale, 29; map projections, 28–29; map scale, 29; physical, 28, 29; political, 29; types of, 29–30
Mapuche people, 306
maquiladoras, 216
Mariana Trench, 60
market economy, 96
Maroon Bells, *p150*
Maroons, 285
Marx, Karl, 406
Maryland, 136; in Mid-Atlantic subregion, 117, *m117. See also* United States, east of the Mississippi River
Massachusetts: in New England subregion, 117, *m117;* Plymouth, 126. *See also* United States, east of the Mississippi River
mature, 72–73
Mauna Loa volcano, 150
Maya civilization: in Central America, 212; in Mexico, 210; ruins in southern Mexico, *p200*
Maya people: organized team sports and, 85
Medieval Age, 343
Mediterranean Sea: climate of Western Europe and, 339; commercial fisheries in, 371; scenic islands in, *p369;* Southern Europe, 368; in Western Europe, 335
Medvedev, Dmitry, 411
megalopolis, 80–81, 132
Mercado Camon del Sur, 101
MERCOSUR, 261, 317. *See also* Mercado Camon del Sur
meridians, 21
mesa, *p275*
Meseta Central plateau, 368, 379
mestizos, 286
Métis, 187

metropolitan area, 132, 190, 257

Mexican food, 217

Mexico, *m201*; bodies of water in, 204–5; challenges for, 217–18; climate of, 205–7; culture of, 217; early civilizations in, 210–11; economy of, 216–17; food of, 217; history of, 210–12; independence and conflict in, 212; land features, 202–4; natural resources in, 206; outdoor market in Central Highlands of, *p97*; physical geography of, 202–7; population concentration along, 76; revolution and stability in, 212; Spanish empire in, 210, 376; timeline, *c200–1*; U.S. relations with, 167; U.S. war with, 159; village market in, *p217*

Mexico City, *p203*; on Central Plateau, 203; earthquake in (1985), 204; as megalopolis, 81; on Tenochtitlán site, 211

Michigan: in Midwest subregion, 117, *m117*; worker in auto assembly plant in, *p113. See also* United States, east of the Mississippi River

Mid-Atlantic Ridge, 60, 367

Mid-Atlantic subregion (United States), 117

Middle Ages, 343, 372

Middle Rhine Highlands, 335, 337

midlatitude countries. *See* **Andes and midlatitude countries**

Midwest subregion (United States), 117, *p117*

migration, *p79*; causes of, 78; to cities, 80; effects of, 79; forced, of Cherokees, 130–31. *See also* refugees

Milan, Italy, 378

Minas Gerais, Brazil, 254

mineral resources: Andean and midlatitude countries, 305; Brazil, 246; Canada, 189; Caribbean islands, 209; eastern United States, 123; Northern Europe, 370–71; Poland, 401; Romania, 401; Siberian Russia, 401; Tropical North, 278–79; western United States, 155

mining: Guanajuato, Mexico, *p77*; United States, west of the Mississippi River, 170. *See also* mineral resources

Minnesota, *m147. See also* United States, west of the Mississippi River

Minsk, Belarus, 411; teenagers in, *p415*

missions, 157

Mississippi: in Southeast subregion, 117, *m117. See also* United States, east of the Mississippi River

Mississippi River, 22, 119, *p120*; Amazon River vs., *p241*; areas east and west of, 148; as boundary, 120, 129; continental divide and, 152; early settlements along, 129, *m129*; flooding (2011), 115

Missouri, *m147*; early agriculture in Kansas City, 163. *See also* United States, west of the

Mississippi River

Missouri River, 119, 152; Lewis and Clark expedition along, 158

Mistral, Gabriela, 314

mixed economy, 96

Moche people, 306

Moldova, *m395*; religion in, 413. *See also* Eastern Europe

Monaco, *m333. See also* Western Europe

monarchy, 87

Mongols, 403–4

Monongahela River, 121

monsoon, 244

Montana, *m147.*; ranching in, 170 *See also* United States, west of the Mississippi River

Mont Blanc, 335

Montcalm, French General, in Battle of Quebec, *p186*

Montenegro, 409, *m395*; religion in, 413; urban population of, 411. *See also* Eastern Europe

Montreal, Quebec: rally against independence for, *p194*

Monument Valley Navajo Tribal Park: buttes in, *p146*

Moors, 376

Morales, Evo, 311

Mormons, settlement in Utah, 167

Moscow, Russia, 411

mountains: climate in western U.S., *p154*; elevation of, 58; formation of, 43, 53; as landforms, 23; tunnels through, 57; underwater, 60. *See also entries for specific mountains*

Mount Etna, *p53*

Mount Everest, 31

Mount Logan, 181

Mount McKinley, 150

Mount Olympus, 368

Mount Whitney, 149

Moyano, María Elena, 313

Mozart, Wolfgang Amadeus, 353

Muller, Peter O, *q38, q420*

multinational firms, 310

murals, 217

Muscovy, princes of, 403

music: Andean pan flute, *p290*; Caribbean islands, 221; Cuban salsa, 221; Jamaican reggae, 221. *See also* arts/culture

Muslims, 86; Albania, 414; Jerusalem and, 343; Ottoman Turks, 376; Southern Europe, 381; United States, 137; Western Europe, 352, *p352*, 357. *See also* Islam

Mussolini, Benito, 377

Mussorgsky, 414

NAFTA. *See* **North American Free Trade Agreement (NAFTA)**

Nambicuara people, 249

Naples, Italy, 378

Napoleon. *See* **Bonaparte, Napoleon**

NASA: Earth Observatory Web Site, *q362*

National Atlas of the United States, *q174*

National Map—Hazards and Disasters, The (U.S. Geological Service), *q38*

national park system, U.S., 137, *p137*, 155

Native Americans, 125; Amazon Basin, 241; Bolivia and Peru, 312; Brazil, 257; chief and boys, *p161*; civilizations, *m308*; as ethnic group, 83; European diseases and, 157, 211, 308–9; forced migration of, 130–31; Guyana, 287; indigenous religion of, 289; Lewis and Clark contact with, 158; Quito, Ecuador, 288; reservations, 161; Tropical North, 281; west of the Mississippi, history of, 156–57, 161. *See also* First Nations people *and entries for individual groups*

NATO. *See* **North Atlantic Treaty Organization (NATO)**

Natural Cuba (Lee), *q228*

natural disaster: emigration after, 78. *See also entries for individual natural disasters*

natural resources. *See* **entries for specific resources, countries, or areas**

Navajo people, 157

Nazi Germany, 407

Nazi Party, 348, *p349*

Nebraska, *m147*; elevation of, 148; Omaha, early agriculture in, 163. *See also* United States, west of the Mississippi River

Neruda, Pablo, 314

Netherlands, the: dairy farming in, 340; elevation of, 336; on Northern European Plain, 335, 397; oil production in, 340; windmills in, *p335. See also* Western Europe

Nevada, *m147*; Las Vegas, as resort city, 163; silver reserve in, 160; Treaty of Guadalupe Hidalgo and, 159; Virginia City, stagecoaches in, *p160. See also* United States, west of the Mississippi River

New Brunswick, Canada, 179, 186

New Delhi, India, 81

New England subregion, 116–17

Newfoundland, Canada, 179, 185, 188, 374

New France, 185

New Hampshire: in New England subregion, 117, *m117. See also* United States, east of the Mississippi River

New Haven, Connecticut, 133

New Jersey: in Mid-Atlantic subregion, 117, *m117. See also* United States, east of the Mississippi River

newly industrialized countries (NICs), 98–99

New Mexico: ranching in, 170; Santa Fe, early Spanish settlement in, 157; Treaty of Guadalupe Hidalgo and, 159

Index

New Orleans, Louisiana, 134–35; as French settlement, 158; as Gulf port, 151; Hurricane Katrina and, 135; Mardi Gras festivities, *p135*; as "the Crescent City", 22

New World, A. **(Quinn),** *q198*

New York (state): fossils of marine organisms found in, *p121*; in Mid-Atlantic subregion, 117, *m117. See also* United States, east of the Mississippi River

New York City, 133–34, *p133*; boroughs in, 133; Freedom Tower Memorial, 115; as large city, *p80*; Spanish Harlem in, 86; Times Square in, *p19*

New York Harbor, immigrants arrive in, *p130*

New Zealand: Maori of, 83

Nicaragua, *m201*; clothing factory in, 218; Hurricane Mitch in, 219; independence for, 213; Lake Nicaragua in, 204; natural resources in, 207; trade agreements, 219. *See also* Central America

Nicholas (Czar), 405–6

NICs. *See* **newly industrialized countries**

Nielsen, Carl, 382

Nile River: population concentration along, 76

nomadic people: early Great Plains, 156

nomads. *See* **nomadic people**

nonrenewable resources, 95

Nordic countries. *See* **Northern Europe**

Norgay, Tenzing, *p17*

Normandy, Viking conquest of, 374

North America: as continent, 52. *See also* Canada; Caribbean Islands; Central America; Mexico; United States, east of the Mississippi; United States, west of the Mississippi

North American Free Trade Agreement (NAFTA), 101, 167, 222–25; advantages and disadvantages, 222; Canada in, 177, 192; Mexico and, 216; North American Free Trade Association (NAFTA), statistics on effects of, 224

North Atlantic Treaty Organization (NATO), 193, 407; Kosovo intervention, 409

North Carolina: in Southeast subregion, 117, *m117*; waterfall in Pisgah National Forest, *p122. See also* United States, east of the Mississippi River

North Cyprus, 367

North Dakota, *m147. See also* United States, west of the Mississippi River

Northern Europe, 363–92, *m365*; climate in, 369; economy of, 384; ethnic and language groups, 380; European Union and, 385; history of, 372–77; landforms, 366–68; literacy rate in, 383; minerals and energy, 370–71; modern conflict and war, 377; natural resources in, 370–72; population of, 378–79; Protestant Reformation and, 380; religion and the arts, 380–82; sea resources in, 371; timeline, *c364–*

65; vegetation, 370; Vikings of, 372; waterways, 368; welfare capitalism in, 384; winter sports in, 383

Northern European Plain, 335, 340, 397

Northern Hemisphere, 26, *c27*; seasons in, 45

Northern Ireland, 336, *m336*

North Korea: command economy in, 96

North Pole, 21; climate in, 46; Earth's axis and, 42; region around, 52; seasons and, 45, *d45*

North Sea, 335–36; fishing in, 371; natural resources of, 384; petroleum deposits in, 370; worker on oil rig in, *p331*

Northwest Territories (Canada), 181, 183

Norway, 366–67; employment in, 384; ethnic and language groups, 380; fjords, 368; hydroelectric power plants in, 371; International Whaling Commission and, 385; literacy rate in, 383; oil production in, 340; Sautso Dam, 371, *p370. See also* Northern Europe

Norwegian Current, 369

Norwegian fisherman, *p363*

Nova Scotia, Canada: as Atlantic Province, 179; Cape Breton Highlands National Park, *p180*; in Dominion of Canada, 186

Novaya Zemlya archipelago, 400

Nunavut, Canada, 181

Nunavut territory, 194

Occupational Outlook Handbook, *q144*

ocean currents: climate and, 48; effect of, 56

ocean floor: landforms on, 59–60; mining of, 63; plate movements and, 53

oceans, 61–62

Octavian, 374

Ohio: in Midwest subregion, 117, *m117*; Ohio River and, 121. *See also* United States, east of the Mississippi River

Ohio River, 119, 121

Oklahoma, *m147*; early oil industry in, 163. *See also* United States, west of the Mississippi River

Old Faithful geyser, *p53*

oligarchs, 410

Olympic Mountains, 149

Ontario, Canada, 179, 190

opportunity cost, 96

orbit, 42

Oregon, *m147*; Cascade Mountains in, 149–50; claimed by U.S. and Britain, 159; Crater Lake in, *p151*; farming in, 155; modern agriculture in, 170. *See also* United States, west of the Mississippi River

Oregon Trail, 159

Orinoco River, 275, 278

Orlando, Florida, Disney World in, *p22*

Oslo, Norway, 379; Parliament building, *p384*

Ottoman Empire: Balkan under rule of, 402; World War I and, 405

Ottoman Turks, 374, 376

outer space, 44

overfishing, 191

Oxford University, 355

pachamanca, 315

Pacific Coast Ranges, 153

Pacific Islands, Aboriginal people of, 85

Pacific Ocean, 62; British Columbia, 181; Canada, 182; Central America, 204; Colombia, 275; First Nations people and, 184; Mariana Trench, 60; Mexico, 204; relative size of, 62; Ring of Fire in (see Ring of Fire); Tropical North, 276; western U.S. coast, 151

pagan religion, 374

pampas, 242, 300

Panama, *m201*; independence from Colombia, 213, 290–91; as isthmus, 59. *See also* Central America

Panama Canal, 59, 205, 218, *p213*

Pan-American Highway, 317, *m316*

pan flute, *p290*

Pantanal wetlands, 300

Paraguay: and urbanization, 81; climate in, 302; Guarani people in, 314; hydroelectric power plants in, 305; territorial disputes, 317. *See also* Andes and midlatitude countries

Paraguay River, 300–1

Paramaribo, Suriname, 288

páramo, 277

Paraná River, 300–1

Paris, France, 350; Louvre in, *p355*; as primate city, 81

Parliament, 345

Patagonia desert, 300

peacekeeping, 193

Pedro I (Emperor of Brazil), 253

Pedro II (Emperor of Brazil), 254

Pemón people, 277

peninsula(s), 59; defined, 366; European, 366

Pennsylvania: in Mid-Atlantic subregion, 117, *m117*; nuclear accident at power plant in (1979), 115. *See also* United States, east of the Mississippi River

perestroika, 409

Perón, Eva, 310, *p310*

Perón, Juan (President of Argentina), 310–11

Persia. *See* **Iran**

Persian Wars, 372

Peru: agriculture in, 316; as Andean country, 298; Atacama Desert, 300; desert in, 299; Native American populations in, 312; *pachamanca,* 315; Spain conquers, 307–8; woman carrying water in mining town, *p319. See also* Andes and midlatitude countries

Peru Current, 300

Peter the Great (Russian Czar), 404, *p394*

petroleum: in United States west of the Mississippi, 155

Phoenix, Arizona, 165

physical geography, 39–68; changing Earth, 52–57; land, 58–60; planet Earth, 42–51; water, 60–65. *See also entries for individual regions*

physical map, 29

physical systems, *c24;* Earth's, 44

Pico de Aneto, 335

pilgrimage, 343

Pinochet, Augusto (Chilean dictator), 311

Pisgah National Forest: waterfall in, *p122*

Pizarro, Francisco, 307, *p308*

place, *c24;* characteristics of, 19–20; as geography theme, 22

plains, 58; creation of, 56

Plains people, 157

plantation agriculture, 213, 252

plant life: in salt water, 61

plateau: defined, 58, 241; in western United States, 150

plate movements, 53–54

plate tectonics, Pacific coast mountains and, 149

Plato, 373

Platte River, 152

Plovdiv, Bulgaria: boulevard in, *p414*

Plymouth, Massachusetts, 126

Poland, 397, *m395;* as EU member, 417; mineral resources in, 401; miner in southern, *p400;* religion in, 413; Solidarity union in, 395, 408; Soviet army occupation of, 407; urban population of, 411; in Warsaw Pact, 407. *See also* Eastern Europe

polar climate, 49–50

polar ice caps, 61

polders, 336

political map, 29

pollution: air, in Mexico, 217; effects on Earth, 57; of lakes and rivers, *p65;* population growth and, 74–75

Popocateptl Volcano, *p204*

population. *See* **entries for specific regions**

population centers, 134

population changes, 77–81

population density, 76

population distribution, 76

population growth, 72–75; causes of, 72–73; challenges of, 74; effects on environment, 74; rates of, 73, 75

population movement, 77–81

population pyramid, *g73*

Po River, 368

Portland, Oregon, as Pacific port, 151

Portugal: conquest of Brazil, 250–51; as democracy, 377; as EU member, 385; soccer teams, 383; Tagus River in, 368; Treaty of Tordesillas and, 248. *See also* Southern Europe

Portuguese Empire, Rio de Janeiro as capital of, 253

Portuguese language, 259

postindustrial, 357

poverty: in Mexico, 218; population growth and, 74–75

Prague, Czech Republic, 411

Prairie Provinces, Canada, 180, 189

precipitation: defined, 48; evaporation and, 64

primate city, 81

Prime Meridian, 21, *c21,* 26, *c27*

Prince Edward Island, Canada, 179

Princip, Gavrilo, 405

productivity, 98

Protestant Reformation, 345, 352, 376, 380

provinces, 178

Provo, Utah, 165

Ptolemy, *p16*

Puccini, 382

pueblo, 156; ancient, near Taos, new Mexico, *p157*

Pueblo people, 156

Pueblos jóvenes, 313

Puerto Rico, 207, 208, *m201;* independence of, 215; Spanish colonies in, 214. *See also* Caribbean Islands

Puig, Manuel, 315

Push-pull factors for migration, 78–79

Putin, Vladimir, 411

Pyrenees mountains, 335, 367; skiing in Spain's, *p383*

Quebec, Canada, 179; in Dominion of Canada, 186; French settlement in, 185; Montreal and Quebec in province of, 191; rally against independence for, *p194*

Quebec Act, 186

Quechua language, 314

Quechuan family, *p318*

Queens (New York borough), 133

Quinn, Arthur, *q198*

quipu, 307, *p307*

Quito, Ecuador, 281, 288

Quitu people, 280

railroad: from the Mississippi to the Pacific, 160; Transcontinental, 187; Western Europe, 354–55

rainfall. *See* **monsoon**

rain forest, 264–67, *m267;* as biome, 50; climate and, 51; defined, 241. *See also* Amazon rain forest; tropical rain forest

rain shadow, 49, *c48*

rainwater, 64

recession, 385

refugees: defined, 78; Libyan, *p79*

reggae music, 221

region, 22, *c24*

relative location, 20–21

relief, on physical maps, 29

religion: Brazil, 259–60; Christian missions in colonial U.S., 157; culture and, 84; Eastern Europe and Western Russia, 413–14; eastern United States, 138; major world, *c84;* pagan, 374; population growth and, 75; Tropical North, 289; Western Europe, 352; west of the Mississippi River, 167. *See also entries for individual religions*

remittance, 221

remote sensing, 32

Renaissance: defined, 375; Italy as birthplace of, 381

renewable resources, 95

representative democracy, 87

reservations, Native American, 161

reserves, energy and mineral, 401

resources: defined, 23; nonrenewable, 95; renewable, 95; wants and, 94–95. *See also* mineral resources

revolution, 212; of Earth around the sun, 42–43

Revolutionary Armed Forces of Colombia (FARC), 291

Revolutionary War (U.S.), 127, 186

Reykjavík, Iceland, 371, 379

Rhine River, 337

Rhode Island: in New England subregion, 117, *m117. See also* United States, east of the Mississippi River

Rimsky-Korsakov, 414

Ring of Fire, *m40–41;* Andes mountains in, 299; defined, 54; Mexico and Central America in, 203, 208

Rio Bravo, 204, *p205*

Rio de Janeiro, Brazil, 243, 245, 253; Carnival in, 260; "Christ the Redeemer" statue, *p238;* favelas in, 257–58; Sugarloaf Mountain and Copacabana Beach in, *p257*

Río de la Plata, 301

Rio Grande River, 152, 204; Spanish settlements along, 157. *See also* Rio Bravo

rivers: freshwater, 62; liquid water in, 61; mouth of, 62; as natural boundaries, 120; in western United States, 152. *See also entries for individual rivers*

Riverside, California, 165

Robinho (Brazilian soccer player), *p237*

Rocky Mountains, 149, 152; Oregon Trail over, 159; resources in, 155; as vacation spot, 163. *See also* Canadian Rockies

Roman Catholic Church: Black Death and, 345; in Brazil, 259; in Eastern Europe, 413–14. *See also* Catholic Church

Roman Catholicism, 22

Romance languages, 352

Index

Roman Empire, 342–43, 373, *m373*

Romania, 397, *m395*; agriculture jobs in, 416; as EU member, 417; mineral resources in, 401; new government in, 409; population of, 412; religion in, 413; Soviet army occupation of, 407; in Warsaw Pact, 407. *See also* Eastern Europe

Romanov, Michael, 404

Romans, 351

Rome, Italy, 378; Vatican City within, 376

Rom people, 380

Rós, Sigur, 382

Rossini, 382

Rotterdam, 337

Roughing It (Twain), *q68*

Rousseff, Dilma (President of Brazil), 255, *p239*

Royal Canadian Mounted Police, 187, *p188*

rural areas, 77; farming village, *p76*; U.S., life in, 166

Russia: agriculture in, 400; as Ally (WWI), 405; Communist party in, 395; Czarist, overthrow of, 395; feudalism in, *i404*; fishing industry, 401; Imperial, 403–4; Komi region in, *p397*; population of, *g412*; Putin as president of, 411; Slavic ethnic groups in, 412, *m413*; U.S. purchase of Alaska from, 161; Volga River in west central, *p394*; Western, 393–420. *See also* Western Russia

Russian Orthodox Church, 411

Russian Plain, 396, 397

Russian Siberia. *See* **Siberia**

Russian Slavs, 403, 412, *m413*

"Rust Belt," 139

Sagrada Familia Church, *p381*

saguaro cactus, 155

Saint Domingue, 215

salt water, 61; freshwater vs., *i60*

samba, 260

Sami people, 371, 380

San Andreas Fault, 169

San Antonio, Texas, 165

Sánchez, Óscar Arias, 213

sand dunes. *See* **desert**

San Diego, California, 151

San Francisco, California: Golden Gate Bridge, *p146*

San Marino, *m365*. *See also* Southern Europe

San Martín, José de, 309

San Xavier del Bac, *p157*

São Miguel das Missöes: church in southern Brazil, *p250*

São Paulo, Brazil, 242, 245, 254, *p242*; diverse population in, 254; Estaiada Bridge in, *p258*; favelas in, 258

Sardinia, 367

Saskatchewan, Canada, 180, 187

Satellites: remote sensing, 32

Saudi Arabia: command economy in, 96

Sautso Dam, 371, *p370*

Saxons, 351

scale, map, 29

scale bar, on map, 28

Scandinavia: standard of living and freedoms in, 377; Viking age, 374

Scandinavian Peninsula, 366

Schabacker, Sam, *q140*

Schubert, 353

Scotland, 336, *m336*

scrubland, 370

seasons, 245; Earth's tilt and orbit and, 45–46, *d45*; Equator and, 245

Seattle, Washington, 151

Second Continental Congress, *p86*

Seine River, 338

Selva, 241

separatists, 194

Sequoyah, *p125*

Serbia, 409, *m395*; agriculture jobs in, 416; climate in, 399; religion in, 413. *See also* Eastern Europe

serfs, 404

service industries, 97, 139, 171

Seward, William, 161

Shakespeare, William, 353

shield, 179

Shows as He Goes (Native American chief), *p161*

Sibaja, Marco, *q270*

Sibelius, Jean, 382

Siberia: energy reserves in, 401; oil from Western, 416

Siberian Plain, west, 396

Sicily, 367; Roman conquest of, 374

Sierra Madre Occidental Mountains, 203

Sierra Madre Oriental Mountains, 203

Sierra Maestra mountains, 208

Sierra Nevada de Santa Marta mountains, 275

Sierra Nevada Mountains, 149, 153

Sikhism, *c84*

silver: in Virginia City, Nevada, 160

Sirionó people, 314

slash-and-burn agriculture, 249

slave trade, *g253*. *See also* African enslaved people

Slavic language, 412

Slavic states, early, 402–3

Slovakia, *m395*; as EU member, 417; religion in, 413; urban population of, 411. *See also* Eastern Europe

Slovenia, *m395*; as EU member, 417; as former Yugoslav republic, 409; religion in, 413; urban population of, 411. *See also* Eastern Europe

slums, 75

smallpox, 308–9

smelting, 342

Smoky Mountains. *See* **Great Smoky Mountains**

soccer, *p85*; Andean and midlatitude countries, 315; Brazil, 260, *p261*; Southern Europe, 383; Western Europe, 354

social media, 90–94

Socrates, 373

solar panels, *p75*

solar power, 155

solstice, 45

South America: birthrates in, 75; as continent, 52; rain forests in, 51; Spanish empire in, 376. *See also* Andes and midlatitude countries; Brazil; Tropical North *and entries for individual countries*

South Carolina: in Southeast subregion, 117, *m117*. *See also* United States, east of the Mississippi River

South Dakota, *m147*; gold in Black Hills of, 160; wind power in, 155. *See also* United States, west of the Mississippi River

Southeast subregion (United States, east of the Mississippi River), 117

Southern Europe, 363–92, *m365*; ancient Greece, 372–73, *m373*; ancient Rome, 372–74, *m373*; climate in, 369–70; economy of, 384; ethnic and language groups, 380; European Union and, 385; history of, 372–77; landforms, 366–68; modern conflict and war, 377; natural resources in, 370–71; population of, 378–79; Protestant Reformation and, 380–81; religion and the arts, 380–82; Roman Empires, 373–74, *m373*; sea resources in, 371; soccer in, 383; timeline, *c364–65*; vegetation, 370; waterways, 368

Southern Hemisphere, 26, *c27*; seasons in, 45, *d45*

Southern Ocean, 62

South Pole, 21; climate in, 46; Earth's axis and, 42

Soviet Union: breakup of, 395; Cold War and, 349; Lenin as premier of, 406; World War II and, 407. *See also* Union of Soviet Socialist Republics (USSR)

space, perspective of, 18–20

Space, Time, Infinity: The Smithsonian Views the Universe (Trefil), *q68*

Spain: conquest of Peru, 307–8; democracy, 377; empire in the Americas, 376; ethnic and language groups, 380; EU member, 385; financial crisis in, 385; map of, *m36*; Meseta Central plateau, 368; population in, 379; Roman conquest of, 374; soccer teams, 383; WWII and, 377. *See also* Southern Europe

Spanish conquistadors, 211

Spanish Harlem, 86
Spanish settlements: in Tropical North, 280–81; west of the Mississippi River, 157
Sparta, 372
spatial, 18
spheres, 26, 27
sports: culture and, 85; first organized team, 85; in Northern and Southern Europe, 383; Western Europe, 354
St. Augustine, Florida, 126; early fort in, *p126*
St. Lawrence River: French explorer on, 185; Gulf of St. Lawrence as mouth of, 183; land along, 179; locks in, 119
St. Lawrence Seaway, 119, *p119*; Canadian economy and, 189; Montreal on, 191; Toronto on, 190
St. Petersburg, Russia, 411
St. Sophia Cathedral, *p394*
Stagecoaches: in Virginia City, Nevada, *p160*
Stalin, Joseph, 406, 407
standard of living, 98
staples, 210
Staten Island, 133
steppe, 397
Stockholm, Sweden, 379
Strait of Gibraltar, 368
streams, freshwater, 62
street-mapping technologies, 34–35
subregion: defined, 116; in United States east of the Mississippi River, *m116–17*
Sucre, Bolivia, 309
sugar industry, in Brazil, 247, 251–52, *p251*
Sugarloaf Mountain: Rio de Janeiro, Brazil, *p257*
sun, Earth and, 42 and seasons, 45–46, *p45*
Suriname, 274, *m273*; European colonization of, 282; independence for, 284; internal unrest in, 285; population groups in, 286–87; as sparsely populated, 288. *See also* Tropical North
surpluses, 210
sustainability, 101
Sweden: Baltic ports of, 368; ethnic and language groups, 380; as EU member, 385; hydroelectric power plants in, 371; literacy rate in, 383; in Northern Europe, 366–67. *See also* Northern Europe
Swiss Alps, 337; train in, *p332*
Switzerland, 354; hydroelectricity in, 340; plant life in, 341. *See also* Western Europe
Tacoma, Washington, 151
Tagus River, 368
TAPS. *See* **Trans-Alaska Pipeline System**
Taquile, handwoven cloth from, *p296*
Tarabuco, Bolivia: woman from, *p295*
tariff, 290

Tawantinsuyu, 307
Tchaikovsky, 414
technology, 30; computer industries, 171; farming (1700s), 129; globalization and, 89; for watering farm land, *p153*; widespread use of, *p89*
tectonic plates, 53; boundaries, *m40–41*
temperate zone, 245
Tennessee: in Southeast subregion, 117, *m117. See also* United States, east of the Mississippi River
Tennessee River, 120
Tenochtitlán, 211, *p211*
tepui, 275
territories, 178
Teusaquillo, *p281*
Texas, *m147*; early oil industry in, 163; early Spanish settlements in, 157; energy resources in, 155; information science companies in, 171; port cities of, 165; U.S. annexation of, 159. *See also* United States, west of the Mississippi River
Thames River, 337
thematic maps, 30
thermal image, *p51*
Thor's Hammer, *p55*
Tiber River, 373
tierra caliente, 205, *g206*, 303
tierra fría, 205, *g206*, 277, 303
tierra helada, 206, *g206*, 303
tierra templada, 205, *g206*, 277, 279, 303
timberline, 149
timelines: Brazil, *c238*–39; Canada, *c176–77*; Caribbean Islands, *c200*–1; Central America, *c200*–1; Eastern Europe, *c394–95*; geographer's world, *c16–17*; human geography, *c70–71*; Mexico, *c200*–1; Northern Europe, *c364–65*; Southern Europe, *c364–65*; Tropical North, *c272–73*; United States, east of the Mississippi River, *c114–15*; United States, west of the Mississippi River, *c146–47*; Western Europe, *c332–33*; Western Russia, *c394–95*
Times Square, *p19*
Tito, Josip Broz, *p395*
Tobago. *See* **Trinidad and Tobago**
Tolstoy, 414
topsoil, 168
Toronto, Ontario, 190–191; National Tower, *p191*
Torres del Paine National Park, Chile, *p298–99*
Tosca **(Puccini),** 382
tourism: Caribbean Islands, 221; Central America, 218, 221; defined, 133; Mexico, 217
traditional economy, 96
Trail of Tears, 131
Trans-Alaska Pipeline system (TAPS), 163

Transamazonica Highway, 262, *p263*
trans-Andean highway, 317
transcontinental railroad, 187
transform, 64
Transoceanic Highway, 317
trawlers, 371
Treaty of Guadalupe Hidalgo, 159
Treaty of Paris, 350
Treaty of Tordesillas, 248
Trefil, James S, *q68*
trench, 60
tributaries, 119, 240, *p241*
Trinidad. *See* **Trinidad and Tobago**
Trinidad and Tobago, economy of, 220–21
tropical climate, 49–50; Brazil, 243–45; Tropical North, 276–77
Tropical North, 271–94, *m273*; agriculture and fishing in, 279; climates in, 276–77; coastlines of, 276; early people in, 280; European colonization of, 280–82; fossil fuels in, 278; history of countries in, 280–85; independence of, 284–85, *m284*; independent countries in, 283; labor and immigration, 283; landforms, 274–75; language groups, 288–89; minerals and gems in, 278–79; natural resources in, 277–79; overthrow of colonial rule, 282; physical geography of, 274–79; population groups, 286–87; revolutions and borders, 285; river travel in, *p272*; timeline, *c271–72*; trade relations, 290; waterways, 275–76
tropical rain forest: in Hawaii, 154. *See also* Amazon rain forest; rain forest
Tropic of Cancer, 45, *d45*, 46
Tropic of Capricorn, 45, *d45*, 46
tsunami, 54
tundra: in biome, 50; Canadian, 181; defined, 369
Tupi people, 249
Turkey. *See* **Southern Europe**
Twain, Mark, *q68*
Ukraine, 396, *m395*, 397; boundary disputes and, 417; early Slavic states in, 402; protests against USSR, 406; religion in, 413; Slavic ethnic groups in, 412; St. Sophia Cathedral in, *p394*; waterways in, 398. *See also* Eastern Europe; Western Russia
UN. *See* **United Nations**
UNASUR. *See* **Union of South American Nations**
UNDP. *See* **United Nations Development Programme (UNDP)**
UNESCO, 273
Union of South American Nations (UNASUR), 261, 290, 317
Union of Soviet Socialist Republics (USSR): Lenin and, 406; satellites of, 407; in Warsaw Pact, 407. *See also* Soviet Union
United Kingdom, 336, *m336*; cargo containers at Felixtowe in, *p359*; Chunnel connecting France

Index

and, *p56*, 336; industry in, 356; monarchy in, 87, *p87*; oil producer, 340; popular sports in, 354. *See also* Britain; England; Great Britain; Western Europe

United Nations (UN): Department of Economic and Social Affairs, on population growth, 73; Development Programme (UNDP), *q318*; World Heritage program, 133

United States: border with Canada, 178; Capitol Building, *p114*; Constitution, 136, *q144*; Declaration of Independence approval, *p86*; early immigrants to, 78; East Coast of, 118; ethnic origins in (1980 to 2010), *g138*; government of, 136–37; Gulf Coast of, 118; relationship with Canada, 192; representative democracy, 87; Tropical North countries' relationship with, 290–91; World War II and, 407. *See also* entries for individual American territories

United States, east of the Mississippi River, 113–44; Appalachian Mountains, 121–22; Atlantic coastal plain, 121–22; bodies of water, 117–21; cities and states in, *m115*; climate in, 122; early America, 124–27; economy, 138–139; European colonization, 125–27; everyday life, 138–39; farming and industry, 123; Great Migration, 131; history of, 124–31; major metropolitan areas, 132; minerals and energy resources, 123; Native Americans, 125; physical features, 116–23; physical landscape, 121–23; religion and ethnicity, 138; settling the land, 128–31; subregions, 116–17, *m117*; timeline, *c114–15*; U.S. government, 136–37

United States, west of the Mississippi River, 145–74; bodies of water, 151–52; challenges facing, 166–69; cities and states, *m147*; climates, 153–54; diversity in schools, *p165*; early farming and ranching, 162–63; early industry, 163; early recreation and entertainment, 163; economy of, 169–71; history of, 156–63; human actions and the environment, 168; life in cities and rural areas, 164–66; modern-day cowhand, *p145*; nature and the environment, 168–69; physical features, 148–55; population changes in, 166–67; relations with neighbors, 167; resources of, 155; timeline, *c146–47*; water problem, 167–68; westward expansion of, 158–61, *m159*

University of Paris, 355

upland, 396

Ural Mountains: location of, 396, 397; oil and natural gas in, 416

urban areas, 77

Urban II (Pope), 343

urbanization, 80–81

Uruguay: plateaus and valleys in, 299. *See also* Andes and midlatitude countries

Uruguay River, 300–1

U.S. Geological Service, *The National Map-Hazards and Disasters, q38*

U.S. Supreme Court, 136

USSR. *See* **Union of Soviet Socialist Republics (USSR)**

Utah, *m147*; Great Salt Lake in, 152, 153; information science companies in, 171; Mormon settlement in, 167; Treaty of Guadalupe Hidalgo and, 159. *See also* United States, west of the Mississippi River

Valdez, Alaska, 151, 154

Valencia, 368, 380

valleys, 56, 58–59

Vancouver, British Columbia, 191

Vargas, Getúlio, 254–55

Vatican City, 376. *See also* Southern Europe

Venezuela, 274, *m273*; Angel Falls in, 275; challenges in, 291; coffee as cash crop for, 279; oil producer, 278; people along coast of, 288; political problems, 284; population groups in, 286; Spanish settlements in, 280–81. *See also* Tropical North

Vermont: in New England subregion, 117, *m117*. *See also* United States, east of the Mississippi River

Vienna, Austria, 398

Vienna basin, 397

viewing Earth spatially, 18–19

Vigo, Spain, 371

Vikings, 185, 351, 372, 374

Vinland, 374

Virginia, 136; Jamestown, 126; in Southeast subregion, 117, *m117*. *See also* United States, east of the Mississippi River

Visigoths, 343, 351

visuals, interpreting, 25

volcanoes: active, in Central America, 203–4; Andes, 299; Apennines, 368; Caribbean Islands, 208; Cascade mountains, 149; eruption of, 43; extinct and dormant, 208; Hawaii, 150; plate movement and, 54; Tropical North, 274; underwater, 60

Volga River, 398, 416, *p394*

Wales, 336, *m336*

Walesa, Lech, 395

wants, resources and, 94–95

Ware, John, *p187*

Warsaw, Poland, 411

Warsaw Pact, 407

Washington (state), *m147*; Cascade Mountains in, 149–50; claimed by U.S. and Britain, 159; farming in, 155; information science companies in, 171; modern agriculture in, 170; Olympic Mountains, 149; port cities in, 164. *See also* United States, west of the Mississippi River

Washington, D.C.: absolute location of, 21; Mid-Atlantic subregion, 117, *m117*; planned city, 136; U.S. Capitol Building in, *p114*. *See also* United States, east of the Mississippi River

water: bodies of, 61–63; earning a living on, *p62*; Earth's surface, 44; freshwater vs. salt, *i60*, 61; problem of, in U.S. west of the Mississippi, 167–68; recreational activities on, *p62*; supply of, 63–65; three states of matter, 61; use to humans, 63, *p62*. *See also* entries for specific bodies of water

Water Availability in the Western United States, q174

water cycle, defined, 63, *d64*

waterfall: Angel Falls, 275, *p276*; in Appalachian Mountains, 122; in Pisgah National Forest, *p122*

water supply, 63–65

waves, ocean, 56

weather, 23; climate vs., 48

weathering, 55

welfare capitalism, 384

well, pumping water from, *p61*

Welland Canal, *p119*

Westerlies, 338–39

Western Europe, 331–62, *m333*; beginnings of, 342; Catholic Church's power in, 345; challenges for, 357; change and conflict in, 346–49; Christianity and, 343; climate in, 338–39; Cold War and, 349; culture in, 353; daily life in, 354; as diverse, 357; economy of, 356–57; education in, 355; empires in (1914), *m356*; energy sources, 340; Enlightenment and, 345; ethnic and languages groups in, 351–52; European Union, 351; history of, 342–49; Hundred Years' War, 344–45; immigration and, 357; Industrial Revolution and, 347; low-lying plains in, 334–35; Middle Ages, 343–44; mountains and highlands in, 335; Muslims working in fast-food restaurant in, *p352*; natural resources, 340–41; physical geography of, 334–41; plants and wildlife in, 341; post-WWII, 350; railways and highways in, 354–55; religion in, 352; Renaissance period, 375–76; rich soils in, 340; Roman Empire in, 342–43; soccer in, 354; timeline, *c332–33*; waterways in, 335–36; World War I and, 347–48; World War II and, 348–49

Western Hemisphere, 26, *c27*

Western Russia, 393–420, *m395*; arts in, 414–15; boundary disputes within, 417; changes under Gorbachev, 408–9; climates in, 399–400; Cold War and, 407–8; connections with Europe and Asia, 417; daily life in, 415–16; divisions and conflict, 409; early Slavic states in, 402–3; earning a living in, 416–17; economic changes in, 410; energy and minerals, 401; ethnic and language groups, 412–13; fishing industry,

401; forests and agriculture, 400–1; history of, 402–9; landforms, 396–97; people of, 403; population centers in, 411; religion in, 413–14; rise of communism, 406–7; Slavic people in, 412, m413; social and political changes, 411; timeline, c394–95; unrest in Soviet satellites, 408; USSR in, 406–7; war and revolutions, 405–6; waterways, 398. *See also* Russia

West Germany, 407; in NATO, 407

West Virginia: Ohio River and, 121; Southeast subregion, 117, m117. *See also* United States, east of the Mississippi River

wetland ecosystems, 56

Whitney, Eli, 129

wildfires: United States west of the Mississippi River, 168–69

wildlife: Alaska, p50

wind: climate and, 47–48

windmills, in the Netherlands, p335

wind patterns, m47

wind power, 155, 335

Wisconsin: in Midwest subregion, 117, m117. *See also* United States, east of the Mississippi River

Wolfe, General: in Battle of Quebec, p186

Woosley, Lloyd H., Jr, q174

world: changing, 20; interconnected, 24; in spatial terms, 18–20, c24

World Bank, 100–1

world economy, 99–101

World Heritage List, Galápagos Islands to, 273

World Heritage program, UN, 133

World Trade Organization, 100

World War I, 333; assassination triggers, 405, p405; Western Europe and, 347–48

World War II, 333; axis control during, m348; beginning of, 395; end of, 365; Soviet Union and, 407; western European countries post-, 350; Western Europe and, 348–49

Wyoming, m147; energy resources in, 155; Treaty of Guadalupe Hidalgo and, 159. *See also* United States, west of the Mississippi River

Yanomami people, 264

Yellowstone National Park: Old Faithful geyser in, p53

Yeltsin, Boris, 411

Yucatán Peninsula, 203, 208; Mayans on, 210

Yugoslavia, ethnic tensions in, 409

Yukon Territory: Canada, 181; gold rush in, 187

Yupanqui, Capac and Pachacuti Inca, 306

zones, climate, 49–50, p49

Index

McGraw-Hill Networks™ meets you anywhere—takes you everywhere. Go online at MHEonline.com.

Circle the globe, travel across time. How do you access networks?

1. Log on to the internet and go to MHEonline.com
2. Get your User Name and Password from your teacher and enter them.
3. Click on your networks book.
4. Select your chapter and lesson. Start networking.